THE INFINITY OF GOD

The Infinity of God

New Perspectives in Theology and Philosophy

Edited by

BENEDIKT PAUL GÖCKE

and

CHRISTIAN TAPP

University of Notre Dame Press

Notre Dame, Indiana

University of Notre Dame Press
Notre Dame, Indiana 46556
www.undpress.nd.edu

Published in the United States of America

Publication supported by DFG Deutsche Forschungsgemeinschaft German Research Foundation

Library of Congress Cataloging-in-Publication Data

Names: Göcke, Benedikt Paul, 1981– editor.
Title: The infinity of God : new perspectives in theology and philosophy / edited by Benedikt Paul Göcke and Christian Tapp.
Description: Notre Dame : University of Notre Dame Press, 2018. | Includes bibliographical references and index. |
Identifiers: LCCN 2018049301 (print) | LCCN 2018051447 (ebook) | ISBN 9780268104153 (pdf) | ISBN 9780268104160 (epub) | ISBN 9780268104139 (hardback : alk. paper) | ISBN 0268104131 (hardback : alk. paper)
Subjects: LCSH: God (Christianity)—Immutability. | God (Christianity)—Eternity. | God. | Infinite.
Classification: LCC BT153.I47 (ebook) | LCC BT153.I47 I54 2018 (print) | DDC 212/.7—dc23
LC record available at https://lccn.loc.gov/2018049301

∞*This paper meets the requirements of ANSI/NISO Z39.48-1992 (Permanence of Paper).*

CONTENTS

	Preface	vii
CHAPTER 1	Introduction *Benedikt Paul Göcke and Christian Tapp*	1
PART I	**Historical Approaches to the Infinity of God**	
CHAPTER 2	The Concept of the Infinity of God in Ancient Greek Thought *Franz Krainer*	21
CHAPTER 3	Infinity in Augustine's Theology *Adam Drozdek*	37
CHAPTER 4	Aquinas on Creation and the Analogy of Infinity *William E. Carroll*	54
CHAPTER 5	Spinoza and Leibniz on the Absolute and Its Infinity: A Case Study *Christina Schneider*	78
CHAPTER 6	Kant and the Infinity of Reason *Ruben Schneider*	97
CHAPTER 7	Infinity and Spirit: How Hegel Integrates Science and Religion, and Nature and the Supernatural *Robert M. Wallace*	122
CHAPTER 8	Bolzano's Concept of Divine Infinity *Christian Tapp*	150
CHAPTER 9	Cantor and the Infinity of God *Bruce A. Hedman*	167

PART II Systematic Approaches to the Infinity of God

CHAPTER 10 God Almighty: Divine Power and Authority in the Biblical and Patristic Periods 187
Bernhard Lang

CHAPTER 11 God's Omnipotence 212
Richard Swinburne

CHAPTER 12 Infinite Power and Finite Powers 233
Kenneth L. Pearce

CHAPTER 13 Infinite God, Open Future 258
William Hasker

CHAPTER 14 Infinity and God's Atemporality 276
Paul Helm

CHAPTER 15 Infinite Goodness 296
Brian Leftow

CHAPTER 16 Divine Infinity and Personhood 317
Ken Perszyk

CHAPTER 17 Divine Infinity and the Trinity 341
Thomas Schärtl

CHAPTER 18 (A)symmetries between God and World: Process Philosophy, Postmodern Theology, and the Two Families of Infinity Argument 364
Philip Clayton

CHAPTER 19 The Quantitative and the Qualitative Infinity of God 385
Benedikt Paul Göcke

Contributors List 410

Index 415

PREFACE

Most of the studies contained in this volume were first presented at the Infinity of God conference held at Ruhr-University Bochum (Germany), August 8–11, 2013. The conference was part of the Emmy-Noether research group's project "Infinitas Dei" and was generously supported by the *Deutsche Forschungsgemeinschaft* (DFG). We are grateful to all the participants at the conference for valuable and thought-provoking discussion, and also to Max Brunner, Elisabeth Petersen, Magdalena Ruschkowski, Alfonso Savarino, Annegret Sock, Andrea Strickmann, and Daniel Tibi for their meticulous work on the manuscript.

Benedikt Paul Göcke and Christian Tapp

CHAPTER 1

Introduction

The Infinity of God

BENEDIKT PAUL GÖCKE AND CHRISTIAN TAPP

In analytic philosophy of religion, the existence of God and the classical divine attributes of omniscience, omnipotence, and moral perfection have received extensive treatment over the last few decades. The infinity of God, in contrast, has received comparatively little scholarly attention.[1] There is no single edited volume dealing exclusively with the infinity of God.[2] To rectify, we have brought together philosophers and theologians to grapple exclusively with the infinity of God from historical and systematic points of view. Since our authors come from different philosophical and theological backgrounds, we hope to provide a fruitful stimulus to discussion of the infinity of God. In this introduction, we briefly clarify the question(s) at stake in the volume.

THE INFINITY OF GOD?

Whoever asserts that "God is infinite" brings together two of the most complicated terms of the humanities and the natural sciences. The statement needs clarification in at least three ways.

The "God" Problem. It is not clear which concept the term "God" expresses. "God" is used in different senses in different philosophical and theological contexts. And so it is unclear whether these uses are intended to refer to the same being or not, or whether they refer to a single entity.[3] In order to understand the question of whether God is infinite we have to clarify, at least roughly, which concept of God we have in mind throughout our volume.

The "Infinity" Problem. It is not clear which concept the term "infinity" expresses.[4] In different sciences, it is deployed to articulate concepts that are not always obviously related to each other. But in mathematics the infinite is dealt with almost everywhere, such as in the conception of real numbers as infinite sets of natural numbers. In the philosophy of mathematics, there is an inquiry into the nature of transfinite sets and if there is an actually infinite set of numbers.[5] Answers to these questions cover positions as diverse as mathematical platonism, according to which there is an infinite realm of mind-independent mathematical entities, and intuitionism, which entails that mathematical objects are mind-dependent entities and therefore unlikely to be infinite. On the other hand, in physics and the philosophy of physics, based on the various mathematical concepts of infinity, questions concerning the infinitely large and the infinitesimally small are discussed. It is asked whether space is of infinite extension and whether there is a limit to spatial divisibility. Singularities that appear in the mathematical description of physical processes are called "infinities."

In contrast, most theologians and philosophers do not think of infinity primarily in terms of quantity but as a unique quality of the Divine or the Absolute, a quality not directly connected with number and measurement.[6]

The Relation Problem. There are at least three ways of understanding the statement that God is infinite. First, that God is infinite is an abbreviated way of referring to features of God, and *nothing in addition to* those features: we can sum up whatever is true of God by saying that *God is infinite*. Second, saying that God is infinite means that infinity is an independent feature of God *in addition to* other features He might have, that is, whatever else is true of God, *that He is infinite* is a further qualification of God. Third, that God is infinite means that certain features of God are themselves infinite: there is at least one divine *attribute* that is itself infinite.[7]

Since, in the first way of understanding, the statement *that God is infinite* is just a *façon de parler* that adds no further content to the analysis of the divine attributes, we bracket this way of speaking of the infinity of God. Two options remain: the statement that *God is infinite* refers to a feature additional to other attributes of God, or it entails that at least one of the divine attributes is itself infinite.

Since, in the first case the infinity of God is considered to stand on its own, but, in the second case it needs to be qualified by some divine attribute that is putatively infinite, we call the first account *the categorematic approach* to the infinity of God and the second *the syncategorematic approach.*[8] Infinity is treated in a different way in each case, but the categorematic and the syncategorematic approaches are not mutually exclusive. It is not inconsistent, prima facie, to argue that the infinity of God is *both* an extra feature of God *and* a feature of at least one of His attributes.

CONCEPTS OF GOD

Depending on one's philosophical and theological commitments, the term "God" is deployed to articulate different concepts. We would probably obtain as many concepts of God as the number of philosophers and theologians we asked. For the sake of clarity, we provisionally use "God" with a minimal determination that allows us to unify positions as diverse as classical theism, panentheism, process theology, and open theism.

To do so, we distinguish a theological from a philosophical use of the term "God." From a theological point of view, "God" denotes the deity mentioned in the holy scriptures of a particular monotheistic religion.[9] Because the focus in this volume is on the Christian understanding of God, in its theological use, "God" refers to the deity mentioned in the Bible. Even though throughout the books of the Bible the concept of God is subject to development and thus to change, we assume theologians agree that there is a minimal set of necessary features any concept of God based on the Bible must entail to be adequate. Such a minimal theological consensus at least includes that God is the loving creator of all that is, and He wants us to be saved.[10]

What philosophers of religion primarily have in mind when they deploy the term "God" is neither scripture nor revelation, but either *the ultimate source of everything* or *the ultimate goal of everything.*

Although these philosophers often criticize theological concepts of God for inconsistency, they are mainly interested in the question whether any theological concept of God corresponds to a philosophical concept of the ultimate source or goal that is based not on revelation and faith, but on reason alone. They ask: Can reason confirm that the theological concept of God is an adequate concept of the ultimate source or goal of everything?

Based on a combination of the theological and the philosophical uses of "God," we suggest the following minimal account of "God":

> God (*def.*): deity that has the essential features of the God of the Bible and is the ultimate source and goal of existence, insofar as such a source or goal is available to purely philosophical argument.

Based on this understanding, whether God is infinite turns out to be the following question:

> Q1: Is the deity mentioned in the Bible, specified by the minimal theological consensus, and rationally accessed as the ultimate source or goal of everything, *infinite*, and, if so, *in what sense*?

THE CATEGOREMATIC APPROACH TO THE INFINITY OF GOD

We will briefly analyze categorematic notions of infinity here before we turn to syncategorematic notions of infinity in the next section.

The problem with the categorematic understanding of infinity is that even when intended to express an independent and additional quality, "infinite" is used in ways that express mutually exclusive concepts. Historically and systematically, popular interpretations of infinity used to refer to a genuine property of an entity that include the notions of *to be boundless*, *to be unlimited*, and *to lack finitude*.[11]

We can exclude boundlessness and unlimitedness because the predicates "being without bound" and "being without limit" need to be qualified by stating the respect in which something is without bound or limit. For instance, something can be "without *spatial* limit" or "without *tem-*

poral bound." However, since the intelligibility of spatial and temporal infinity depends on a possible measure by spatial or temporal units, spatial and temporal infinity belong to the class of syncategorematic notions of infinity and do not specify anything categorematic. This leaves us with the negation of finitude as a possible criterion of categorematic infinity.

Infinity as the denial of finitude can either be understood in a negative or in a positive way. In Plato's thought, paradigmatically representing the mind of classical Greece, infinity as the denial of finitude is expressed by *apeiron*, "absolute formlessness" and "absolute ontological underdetermination." Being finite consists in possessing a certain form of ontological determination. Since anything that lacks form *eo ipso* is inaccessible to the mind and is of the lowest ontological status, being infinite for Plato is a very bad thing indeed—whatever is infinite is without form and therefore lacks intelligibility.[12]

Gregory of Nyssa as influenced by Neoplatonism saw things in a different light. Paradigmatically representing early Christian orthodoxy, he thought of infinity as expressing *total superabundance* and *the fullness of being*.[13] Since in our volume here God is understood philosophically as the ultimate source or goal of everything, we can exclude the Platonic interpretation of categorematic infinity and adopt that of Gregory of Nyssa. On the categorematic concept of infinity, whether God is infinite therefore turns out to be expressed in the following question:

> Q2: Does the deity mentioned in the Bible, specified by the minimal theological consensus, and rationally accessed as the ultimate source or goal of everything, *possess total superabundance and the fullness of being*?

SYNCATEGOREMATIC APPROACHES TO THE INFINITY OF GOD

On the syncategorematic approach, the infinity of God is not a quality in addition to other divine attributes but consists in the infinity of at least one divine attribute. The difficulty here is that there is more than one way of understanding the syncategorematic infinity of a divine attribute. There are two relevant directions we can take. First, "syncategorematic infinity" can be used to refer to *a quantity*. Second, it can be used to express *the mode of givenness* of a particular attribute.

The first way of understanding syncategorematic infinity is, again, open to two interpretations, depending on the concept of quantity we have in mind. "Quantity" can refer to an extension or to an intension, that is, to the size of the class of objects having the respective property or to the degree to which an object has that property.[14] (This is the sense of "intension" that the Scholastics had in mind when they considered the *intensio* of a property. It is not to be confused with the Fregean concept of intension, which is the sense a certain predicate expresses.)

If we understand quantity as extensional quantity, then we obtain a notion of quantitative infinity on which the extension of a property is putatively infinite. Since an infinite extension can either be understood to be an *infinite continuum* or to be an *infinite extension of discrete units*, there are two further ways of spelling out quantitative extensional infinity: (1) that a property *F* is infinite according to its extensional quantity means either that there are infinitely many *F*s (an infinite multitude) or that there is an infinite continuum of *F* (an infinite magnitude).[15] For example, one might take *divine omniscience* to entail infinite knowledge, in the sense that God knows infinitely many true propositions. Or, (2) one might take *divine omnipotence* to entail infinite power, in the sense that what God can do has no limits in space or time.

If we understand quantity not in an extensional way but according to its *intensio*, then we obtain a different notion of syncategorematic quantitative infinity. Whereas the extensional account is concerned with infinitely many *F*s or infinitely much of *F*, this account of syncategorematic infinity is concerned with *the degree to which a property is realized* in an object. The classical example is "infinite whiteness" (if there were such), which, in the extensional sense, means infinitely many white things, or an infinitely extended white surface, and, in the intensional sense, means unlimited or infinite degree of whiteness (the brightness or, in terms of physics, the capacity to reflect all colors of the visible spectrum). "Mary is infinitely wise" cannot mean that Mary's wisdom extends to infinitely many or infinitely large entities, but means that the degree of her wisdom is unlimited. Some philosophers take omnipotence to entail infinite power in the sense of God's power not being limited to any degree.

Finally, we can understand syncategorematic infinity as referring to the *mode of givenness* of a certain property. What does it mean to say that the mode of givenness of a property is infinite? Could the mode of givenness of some property *F* be infinite? This is the most puzzling of the notions discussed in this introduction, but there are some uses in which it seems clearly to make sense. For instance, some philosophers take God's omnipresence to presuppose presence in a mode quite different from the presence of physical objects in space, which can be said to be present at a place by being contained by it. "Presence" applies to God—if it does at all—in a sense that requires dropping the element of limitation by containment.

Since to exemplify a property *F* in an infinite mode means to possess this property irrespective of any limitations of the exemplifying entity (as in the case of presence), the infinity of the mode of givenness of a property *F*, exemplified by an entity, is the archetype of what it means to possess *F* tout court. Any other mode of givenness of this property is consequently a restriction of the archetype in question.

It follows that we obtain different questions concerning the syncategorematic infinity of the divine attributes depending on which interpretation of syncategorematic infinity we have in mind.

The question of whether God is infinite, according to the *extensional quantitative approach that refers to discrete units* (multitude), is this:

> Q3: Does the deity mentioned in the Bible, specified by the minimal theological consensus, and rationally accessed as the ultimate source or goal of everything, *exemplify a property the extension of which consists of infinitely many discrete units*?

According to the extensional quantitative approach that deploys a *continuous notion of extensional quantitative infinity*, the question is as follows:

> Q4: Does the deity mentioned in the Bible, specified by the minimal theological consensus, and rationally accessed as the ultimate source or goal of everything, *exemplify a property the extension of which is an infinite continuum*?

According to the *intensional quantitative approach*, the question of whether God is infinite can be stated as follows:

> Q5: Does the deity mentioned in the Bible, specified by the minimal theological consensus, and rationally accessed as the ultimate source or goal of everything, *exemplify a property to an infinite degree*?

Finally, if we do not have quantitative infinity in mind but instead the *mode of givenness* of a particular divine attribute, the question is this:

> Q6: Does the deity mentioned in the Bible, specified by the minimal theological consensus, and rationally accessed as the ultimate source or goal of everything, *exemplify a property such that the mode of givenness of this property is infinite or archetypical*?

THE INFINITY OF GOD FROM A HISTORICAL AND SYSTEMATIC POINT OF VIEW

All of the chapters in this volume implicitly or explicitly grapple with the elaborated questions concerning the infinity of God. We proceed in two parts. The studies of the first part mainly deal with historical appreciations of the concept of infinity and the various assessments of its capacity to function as a categorematic or syncategorematic attribute of God.

The first part, titled "Historical Approaches to the Infinity of God," is opened by Franz Krainer in chapter 2, "The Concept of the Infinity of God in Ancient Greek Thought." Krainer provides a brief analysis of the variety of concepts of qualitative and quantitative infinity found in ancient Greek thought reaching from Plato to Gregory of Nyssa. Although in this period the term "infinity" often has more than one meaning, sometimes even within the work of a single philosopher, Krainer argues that at least two contrary conceptions of God's infinity can be identified: infinity as expressing divine indeterminacy or perfection. On both these accounts, however, there is a subtle agreement that God's qualitative infinity is strongly related to God's incomprehensibility. Therefore, according to Krainer, the concept of an infinite God in ancient Greek thought naturally led to the development and support of negative theology in which it is disputable whether the infinity of God allows us to formulate any intelligible statement about God at all.

In chapter 3, "Infinity in Augustine's Theology," Adam Drozdek provides an analysis of Augustine's stance on the infinity of God. Although Augustine at first thought about God as an infinite corporeal being, he later became convinced that God had to be understood in an incorporeal manner. Drozdek argues that although in this respect Augustine struggled with both the problems concerning quantitative notions of infinity related to infinite space and time and the possibility of infinite divine knowledge—for which every infinite quantity, according to Augustine, is finite and thus comprehensible—he never explicitly developed an account of divine infinity as such. Instead, Augustine stressed that God's essence lies in his immutability and eternity, both of which indicate that God is beyond infinity and finitude.

William Carroll's chapter 4, "Aquinas on Creation and the Analogy of Infinity," argues that for Thomas Aquinas there is a close connection between the concept of creation and the concept of the infinity of God—whereas creatures are identified as creatures by the reception of being, and thus are always in certain respects finite, unreceived being is the hallmark of God the Creator. To show that for Thomas unreceived being is absolutely infinite and fully determined, Carroll analyses what Thomas says about divine infinity in *Scriptum super IV libros Sententiarum*, *Summa contra Gentiles*, and *Summa theologiae*, all of which deal with the concept of God as subsistent being. Carroll concludes that for Thomas divine infinity is a natural consequence of his concept of creation and can be known, at least in an analogical way, by reason alone.

In her "Spinoza and Leibniz on the Absolute and Its Infinity: A Case Study," Christina Schneider in chapter 5 analyzes the entailments of different concepts of infinity for the concept of the Absolute and its relation to the world. To do so, she compares Spinoza's and Leibniz's conceptions of the infinity of God. Based on the assumption that divine infinity, formally, is used to express a divine perfection, Schneider distinguishes between two concepts of divine infinity as a perfection: on the first meaning, divine infinity is understood to exclude negation, whereas on the second meaning divine infinity refers to the highest degree of an attribute, as found, for instance, in God's omniscience and omnipotence. Spinoza, according to Schneider, operates with the first concept of divine infinity, Leibniz with the second. Once this is clarified, she spells out some problems for Leibniz's attempts to avoid Spinozism: to escape Spinozism, Leibniz both conceived God to be completely independent of His creatures and introduced the concept of monads. However, since monads are not intellectually accessible by God, Schneider identifies a crucial problem: either the concept of God's omniscience has to be modified to include a kind of Spinozistic omnisubjectivity or God cannot be omniscient—the latter of which is not consistent with Leibniz's concept of divine infinity.

Ruben Schneider in chapter 6, "Kant and the Infinity of Reason," deals with Kant's account of the existence and essence of God understood as the infinite being that is the ground of the world order. In contrast to traditional interpretations in which Kant rejects every attempt to construe a metaphysical theory of God, Schneider argues in a first step that Kant firmly presupposed the existence of God, but he argued against philosophical attempts to grasp divine attributes as they are in themselves. In a second step, Schneider investigates Kant's concept of infinite divine reason and spells out some possible consequences of Kantian philosophy. It seems that Kant's philosophy is at least open to a panentheistic interpretation, according to which the difference between God and the created world is within God and the finite mind of creatures is participating in the absolute spirit of God.

In chapter 7, "Infinity and Spirit: How Hegel Integrates Science and Religion, and Nature and the Supernatural," Robert M. Wallace shows how Hegel employs his conception of infinity in order to try to integrate science and religion and also both the natural and the supernatural realm of being. He first argues that Hegel's concept of true infinity should be understood as "the finite's own going beyond its finitude." He then spells out how based on this notion of the ascent of finitude, science, religion, ethics, art, and philosophy can all be understood as metaphysically necessary aspects of a single self-determining reality that is properly referred to as the divine being. Everything that constitutes this ultimate reality is part of an ascent above its initial opinions and appetites and frees the individual of determination by things that are not itself. Thus it enables a true self-determination of reality.

Christian Tapp in "Bolzano's Concept of Divine Infinity," chapter 8, argues that infinity is central to each of the three areas in which Bolzano had expertise: mathematics, philosophy, and theology. He concentrates on Bolzano's analysis of the infinity of God and shows that on Bolzano's view the concept of quantitative infinity is best understood as follows. A series is quantitatively infinite if and only if it has no last term and every finite series can be mapped one-to-one onto a part of it. According to Tapp, on Bolzano's view this concept of quantitative infinity is more basic than all other concepts of infinity. Therefore, even qualitative concepts of the infinity of God have to be related in one way or the other to the elaborated quantitative concept of infinity. Tapp ends by identifying further tasks that must be dealt with to fully specify an adequate conception of divine infinity.

In chapter 9, "Cantor and the Infinity of God," Bruce A. Hedman elaborates on Cantor's different concepts of infinity and briefly summarizes his theories of ordinal numbers, cardinal numbers, and his theory of sets and transfinite numbers that revolutionized mathematics. Furthermore, he shows that it was of the utmost importance to Cantor that his stance on the various concepts of infinity was in accordance with Christian faith: Cantor distinguished his concept of the transfinite firmly from the absolute infinity of God and was relieved that a leading papal theologian confirmed that his theory did not contradict faith. On Cantor's theory, according to Hedman, God is the absolutely infinite that is both the ontological ground of the transfinitum and its repository. Precisely because of this, however, it is mathematically indeterminable in itself.

The second part of the volume, titled "Systematic Approaches to the Infinity of God," has a systematic focus and provides different accounts of divine infinity, both in terms of quantity and quality. The first studies of the second part deal with the infinity of divine attributes, such as omnipotence, omniscience, everlastingness, and perfect benevolence, whereas the second group of studies deal with the infinity of God as such, divine simplicity, the Holy Trinity, and, finally, the use of infinity in current scientific theories.

In chapter 10, "God Almighty: Divine Power and Authority in the Biblical and Patristic Periods," Bernhard Lang first examines how the Israelites describe God's exercise of power, and second how Israel's God came to be called "almighty." Almightiness, according to Lang, in the biblical context is to be understood as having infinite power and could be defined as the unrestricted faculty to do anything that one wants to do.

There was no distinct concept of divine almightiness in the earlier biblical texts, but the notion of God changes in the late biblical period, which, according to Lang, is clearly shown in the Apocalypse of John: almightiness ever since has been perceived to be the essential attribute of God that encompasses all other attributes.

Whereas Lang concentrates on the historical development of the doctrine of divine almightiness in the Bible, the next two chapters spell out the biblical notion of almightiness in philosophical terms. First, in chapter 11, "God's Omnipotence," Richard Swinburne develops a definition of omnipotence in terms of God's ability to bring about events. After a careful analysis of simultaneous and backward causation, both of which Swinburne rejects, he argues for the adequacy of the following definition of omnipotence: *S* is omnipotent during some period of time if and only if *S* knows all metaphysically necessary propositions and all metaphysically contingent true propositions about every event at any time earlier than the beginning of his action and all propositions that those propositions entail; he is not moved by any nonrational influences, and is able to cause by an act beginning at any instant t and ending at any instant t_2, both during that period, any metaphysically contingent event M beginning at any instant t_1 later than t and ending at t_2, which does not require him to be influenced by nonrational influences in order to do that act. Based on this definition, Swinburne ends by arguing that apart from divine aseity and everlastingness, each other divine attribute is entailed by this concept of omnipotence.

In contrast to Swinburne's analysis of omnipotence, Kenneth L. Pearce in chapter 12, "Infinite Power and Finite Powers," provides an analysis of divine omnipotence in terms of infinite power. In a first step, he argues that being infinitely powerful means possessing power over the truth of a proposition tout court; it does not mean to possess an infinity of particular powers. In a second step, he shows that being absolutely powerful to perform an action *A* means to have perfect efficacy and to have perfect freedom with respect to *A*. According to Pearce, then, omnipotence is a relatively simple concept, whereas the concept of finite powers is the complicated one: it can only be established by adding a series of limitations to God's infinite power. It seems, therefore, that for Pearce, finite powers of creatures are participations in and restrictions of the infinite power of God.

In chapter 13, "Infinite God, Open Future," William Hasker turns away from the explicit analysis of omnipotence and concentrates on the analysis of God's relation to time. In the first part of his study, Hasker provides a systematic overview of some of the concepts of infinity that were deployed by Plato, Aristotle, Scotus, and Hegel. Hasker, though, is especially critical of Hegel's conception of the true infinite and argues that it looks defective. In his second part, he turns to the analysis of divine infinity as it is operative in the paradigm of open theism.

According to Hasker, the main concern of open theism is to maintain a robust realism concerning the character and activities of God as described in the Bible, where the most significant feature of open theism is God's dynamic omniscience summed up both by the assumption that God exists in time and by the supposition that God's knowledge changes as time develops into an open future. Hasker analyzes different concepts of the infinity of God's dynamic omniscience and argues that on each of them God turns out to be the infinite and everlasting creator of a universe with an open future.

In contrast to Hasker's analysis of God's infinity as suggesting a temporal existence of God, Paul Helm's chapter 14, "Infinity and God's Atemporality," deals with God's infinity understood in a classical way as entailing that for God to have immediate access to all places and all times of his creation, He has to be eternal, that is, atemporal. Helm defends this view against recent objections by analyzing divine infinity in terms of God's atemporality. There are two main strategies of argumentation for Helm. First, the infinity of God entails that God does not exist in time, nor is He subject to time, as his creatures are. Second, Helm argues that existing outside of time entails that there are things that God cannot do that his creatures can do. However, although it is often argued that this conclusion leads to a contradiction in the concept of God, Helm argues that these consequences do not constitute an objection to the infinity of God properly understood in terms of divine atemporality.

In chapter 15, "Infinite Goodness," Brian Leftow deals with another of the classical divine attributes: God's benevolence understood as infinite goodness. In the first part of his study, Leftow summarizes Aquinas's understanding of infinity, and based on this he defends the claim that God can have virtues. Leftow argues that even if one denies God's having emotions, which Leftow does not, one can think of God as virtuous. Once this is done, Leftow provides a definition of the perfect degree of a virtue and defines perfect benevolence in such a way that it is not contradicted by the intelligibility of a surpassable record of benevolent acts. His notion of perfect benevolence thus is defined qualitatively, not quantitatively. Finally, Leftow argues that in the case of God and His benevolence, even if He necessarily does something good, the necessity arises entirely out of his own internal states rather than being imposed from without, which means that we can still be grateful to God.

Leaving the analysis of particular divine attributes behind, Ken Perszyk in chapter 16, "Divine Infinity and Personhood," deals with the question whether it is possible at all that God is both an infinite entity and a person. To answer this question, he first analyzes different concepts of infinity and divine infinity before, in a second step, he turns to the analysis of the concept of personhood. Once he has clarified in what sense God could be said to be a person, Perszyk argues that there are several reasons that an infinite God cannot be a person. He concludes that being a person cannot literally apply to God if God is literally infinite.

In chapter 17, "Divine Infinity and the Trinity," Thomas Schärtl is concerned with the concept of the Holy Trinity and the difficulties of accounting for it in terms of divine infinity. He first discusses Gregory of Nyssa's reflections on infinity before he turns to Nicolaus Cusanus's theory of divine infinity. Schärtl then discusses divine simplicity and its relation to infinity: to derive simplicity from infinity, he argues, two tools are needed. The first tool is the concept of coextensionality: two properties are identical if and only if whatever fulfills the one concept in a possible world also fulfills the other in that possible world. For instance, the doctrine of coextensionality entails that the extension of "infinite goodness" is necessarily identical to the extension of "infinite wisdom," and vice versa. The second tool is paradigmatic predication. Paradigmatic subjects of predication instantiate the property they express but they do not have them in the usual sense of having a property. Based on these two tools, Schärtl first concludes that the divine attributes are paradigmatic subjects of predication and that God is the ultimate and most eminently paradigmatic subject of predication. Second, he argues that his concept of infinity safeguards divine simplicity and divine uniqueness.

In chapter 18, "(A)symmetries between God and World: Process Philosophy, Postmodern Theology, and the Two Families of Infinity Argument," Philip Clayton analyses six symmetries in the relation between God and the world suggested by Whitehead and how they have been dealt with in different postmodern theological traditions. According to Clayton, postmodern theologians share Whitehead's emphasis on the symmetry between God and world and the corresponding metaphysics of immanence, even though they rarely formulate their view in these terms. He argues that contrary to Whitehead's assertion, however, there are several asymmetries in the God–world relationship that in Western discourse have been spelled out by deploying two concepts of infinity: infinity as unlimited perfection that concentrates on infinite perfection, and infinity as absolute or perfect infinity. Keeping these differences in mind, Clayton argues that divine infinity offers a surprisingly effective bridge between classical metaphysics and the focus of contemporary thinkers on the Unnamable and Unspeakable.

In chapter 19, "The Quantitative and the Qualitative Infinity of God," Benedikt Paul Göcke analyses quantitative and qualitative notions of infinity. In a first step, he argues that quantitative and quasi-quantitative approaches to divine infinity are not sufficient to formulate a precise thesis of God's infinity. In a second step, he establishes a positive qualitative concept of divine infinity in which God exemplifies, in the unity of his being, contradictory properties. He shows that if one assumes that there is a single ultimate source of everything, one can draw the conclusion that there is a single qualitatively positive infinite entity, which, in contrast to finite entities, is not subject to the law of noncontradiction but instead is subject to paraconsistent logic. Based on this, Göcke argues that in the Christian theological tradition, the single ultimate source of everything is God and that as a positive infinite entity God is both distinguishable from the realm of finitude that is subject to the law of noncontradiction and at the same time the more indistinct insofar as He is distinct.

NOTES

1. The reason might be twofold. On the one hand, from a theological point of view, the Bible itself has very little to say about the infinity of God: the Bible speaks of the omnipresence and of the powerfulness of God, of his immense knowledge, but not of his substantial infinity. The infinity of God therefore might have

been considered to be of less interest than other, prima facie, more positive features of the divine being, such as omniscience or omnipotence. But because of the implicit quantifier "everything," even these terms are open to an interpretation in terms of infinity. On the other hand, from a philosophical point of view, the concepts of infinity and divine infinity are quite unclear in themselves, and substantial reflection is needed on these concepts and their different interpretations throughout the disciplines. Since our interest is not in the research history concerning the infinity of God, but rather in the infinity of God itself, we leave it open what might have been the reasons that the infinity of God was treated very little in the past few decades.

2. Michael Heller and W. Hugh Woodin (2011) edited a collection of studies on the concept of the infinite in which, for the most part, mathematicians, physicists, and philosophers deal with problems surrounding the notion of the infinite in relation to abstract entities, space, time, numbers, and aesthetics. Graham Oppy (2006) intends to clarify the notion of the infinite in order to prepare the ground for an analysis of cosmological arguments for the existence of God. However, the question of what it means to say that *God is infinite*, and *whether God is infinite or finite*, is barely addressed in these volumes. In Heller and Woodin's collection, there are only four chapters dealing straightforwardly with the infinity of God, whereas in Oppy's book the infinity of God itself is not dealt with.

3. Cf. Göcke 2013 for a clarification of the different contexts of use of the term "God" in philosophy and theology. Cf. Göcke 2012 for an analysis of the use of "God" in classical theism and panentheism.

4. As David Hilbert (1967) famously put it: "The infinite has always stirred the emotions of mankind more deeply than any other question; the infinite has stimulated and fertilized reason as few other ideas have; but also the infinite, more than any other notion, is in need of clarification" (371).

5. Cf. Oppy 2006 (7–19) for a brief overview of questions surrounding the notion of the infinite. Cf. also Tapp 2011b.

6. As Bombieri (2011) says: "What is infinity? Is it the inaccessible, the uncountable, the unmeasureable? Or should we consider infinity as the ultimate, complete, perfect entity?" (55).

7. Is the following a fourth option? *That God is infinite* means that we cannot understand what it is that God is. It seems not. Whoever asserts that the statement *that God is infinite* expresses the proposition that we cannot understand what it is that God is needs an account of why the infinity of God excludes our understanding of God. In order to justify this claim, though, he or she has to argue that our inability to understand God is because the infinity of God is either an extra feature or a qualifier of his features that entails our inability to understand what it is that God is. Cf. Tapp 2011b (95) for a further analysis of different interpretations of infinity, particularly for those that differentiate between a quantitative, eminent and metaphysically precategorical dimension of infinity.

8. Cf. Tapp 2011b (94), Tapp 2016 (96–97), and Bocheński 1970 (179–82).

9. From the point of view of religious studies, the term "god" can be used to refer to deities mentioned in many different religions that neither are monotheistic religions nor are based upon holy scriptures. However, since our focus is on monotheistic theology, we bracket this more relaxed way to use the term "god."

10. Since the biblical data do not entail a single unambiguous concept of the deity they deal with, much is left open with regard to the potential specification of this minimal theological consensus, which is why positions as diverse as process theology, classical theism, panentheism, and open theism are all prima facie consistent with scripture.

11. To these three conceptions of infinity and the following discussion, see Tapp 2015.

12. Cf. Hart 2011: "For Plato—and, really, for the entire classical philosophical tradition of Greece, including Stoicism—the infinite was solely a negative concept. Words like *apeiron* . . . were more or less entirely opprobrious in connotation; they were used to designate that which was 'indefinite' or 'indeterminate' and, hence, 'irrational' or 'unthinkable.' The infinite is that which lacks form, that which reflects no eidon and receives the impress of no morphe. As such, it is pure deficiency. Hence, Plato would never have called the Good beyond being 'infinite'" (258).

13. Cf. Hart 2011: "What Gregory understands 'infinity' to mean when predicated of God is very much (at least on the fact of it) what Plotinus understood it to mean in regard to the One: incomprehensibility, absolute power, simplicity, eternity. God is uncircumscribable . . . elusive of every finite concept or act, boundless, arriving at no terminus. . . . God is without opposition, as he is beyond nonbeing or negation, transcendent of all composition or antinomy; it is in this sense of utter fullness, principally, that God is called simple" (267).

14. Cf. Tapp 2016 (96–97).

15. In general, magnitudes are taken to be continuous, such as, for example, a 1-cm line, whereas multitudes are discrete, such as, for example, 25 points. This is *not* to say, however, that an infinitely large continuous object like a line extending infinitely through a Euclidean space cannot also be conceived of as an infinite sets of points.

REFERENCES

Bocheński, Joseph M. [3]1970. *Formale Logik*. Freiburg im Breisgau: Alber.

Bombieri, Enrico. 2011. "The Mathematical Infinity." In *Infinity: New Research Frontiers*, edited by Michael Heller and W. Hugh Woodin, 55–75. Cambridge: Cambridge University Press.

Göcke, Benedikt Paul. 2012. "Panentheism and Classical Theism." *Sophia: International Journal for Philosophy of Religion, Metaphysical Theology and Ethics* 52 (1): 61–75.

———. 2013. "An Analytic Theologian's Stance on the Existence of God." *European Journal for Philosophy of Religion* 5:129–46.

Hart, David Bentley. 2011. "Notes on the Concept of the Infinite in the History of Western Metaphysics." In *Infinity: New Research Frontiers*, edited by Michael Heller and W. Hugh Woodin, 255–74. Cambridge: Cambridge University Press.

Heller, Michael, and W. Hugh Woodin, eds. 2011. *Infinity: New Research Frontiers*. Cambridge: Cambridge University Press.

Hilbert, David. 1967. "On the Infinite" ("Über das Unendliche," 1926), translated by Stefan Bauer-Mengelberg. In *From Frege to Gödel*, edited by Jean van Heijenoort, 369–92. Cambridge, MA: Harvard University Press.

Oppy, Graham. 2006. *Philosophical Perspectives on Infinity*. Cambridge: Cambridge University Press.

———. 2011. "God and Infinity: Directions for Future Research." In *Infinity: New Research Frontiers*, edited by Michael Heller and W. Hugh Woodin, 233–54. Cambridge: Cambridge University Press.

Tapp, Christian. 2005. "On Some Philosophical Aspects of the Background to Georg Cantor's Theory of Sets." Special issue, *Philosophia Scientiae* 5:157–73.

———. 2011a. "Eternity and Infinity." In *God, Eternity, and Time*, edited by Christian Tapp and Edmund Runggaldier, 99–115. Aldershot: Ashgate.

———. 2011b. "Infinity in Mathematics and Theology." In *Theology and Science* 9 (1): 91–100.

———. 2014. "Absolute Infinity—A Bridge between Mathematics and Theology?" In *Foundational Adventures: Essays in Honor of Harvey M. Friedman*, edited by Neil Tennant, 77–90. London: College Publications.

———. 2015. "Unendlichkeit Gottes." In *Eigenschaften Gottes*, edited by Thomas Marschler and Thomas Schärtl, 129–51. Münster: Aschendorff.

———. 2016. "Infinity in Aquinas' Doctrine of God." In *Analytically Oriented Thomism*, compiled by M. Szatkowski, 93–115. Heusenstamm: Editiones Scholasticae, 93–115.

PART I

Historical Approaches to the Infinity of God

CHAPTER 2

The Concept of the Infinity of God in Ancient Greek Thought

FRANZ KRAINER

INTRODUCTION

Structure

In this chapter, I give a very brief sketch of the concept of infinity in ancient Greek thought, with an emphasis on Neoplatonism. Infinity as we conceive and discuss it today has, like so many other important philosophical ideas, its roots in ancient Greek thought. In contrast to other important philosophical concepts, the history of the concept of infinity shows no harmonious and continuous development within the history of ancient Greek ideas. The pre-Socratics sometimes attribute infinity to the origin of all things, but Plato and Aristotle do not consider infinity an important concept. They deny that infinity can be a property of God, since infinity is a lack of form and therefore a lack of perfection. Aristotle went even further and denied the existence of an actual infinity. In Neoplatonism, however, an increasing interest in infinity began, and being infinite is henceforth sometimes attributed to God. This culminates in the work of Gregory of Nyssa, who considers God to be, first and foremost, infinite. (Whether he considered God's being to be infinite, or if he merely meant to say that God's essence cannot be conceived, is open to discussion.) To give a brief overview of these major themes, I will first state the problem that Plato and Aristotle formulate. The next two sections are concerned with gradually developing positive concepts of divine infinity. I discuss Philo and Plotinus and then present Gregory's conception of the infinite in a little more detail than that of the other thinkers. Then in a systematic summary I try to give a wider picture of the themes common to the infinity of God in ancient Greek thought.

Scope

There are two methodological presuppositions. As this is an overview of a very long period of history, there are some philosophers and themes that have to be completely neglected. Apart from Aristotle, this is especially the case for all quantitative treatments of infinity, such as that of the Pythagoreans and Zeno's paradoxes. The terms of my chapter's title were chosen on purpose, and a short explanation can further clarify my scope here. Though the main focus is the concept of the infinity of God, this cannot be easily isolated from the concept of infinity itself, especially in ancient Greek thought. I will therefore also discuss some general thoughts regarding infinity, but this is always to obtain a better understanding of the concept of the infinity of God.

I have chosen not to restrict myself to the classical period, since this would have meant that I would end without any exposition of a proper concept of the infinity of God, which was not developed until Plotinus and Gregory of Nyssa. Since Gregory, especially, is more of a theologian than a philosopher, I treat the concept of infinity in ancient Greek thought widely construed.

Basic Terms and Concepts

A major problem for anyone who tries to understand what a certain philosopher thinks about infinity is that there is usually only one term—ἄπειρον, "infinitas," "infinity," "Unendlichkeit"—that is used to express at least two concepts. The first is the concept of mathematical or quantitative infinity. Under "mathematical infinity" I understand the infinity of mathematical entities, such as numbers and sets. Under "quantitative infinity" I understand either mathematical infinity or infinity that is not purely mathematical but for which there is a model in terms of mathematical infinity. The most common entities sometimes considered to be infinite in this quantitative sense, but not in the purely mathematical, are space and time, though there are numerous others, such as souls or the orbit of a planet.

The second major concept is that of qualitative infinity, which is much less precise than the quantitative concept. To explain it properly would take too much space, but a short overview is still necessary. There is infinity as negation of one property or the positive assertion that a certain entity or property is not limited in a certain sense. It can also be the negation of all the properties of an entity, which either amounts to being indeterminate, or being indefinite, or a concept of an absolute being.[1] In ancient Greek thought, ἄπειρον is usually used to express some understanding of infinity, while ἀόριστον is used most of the time for indeterminacy. However, this is not always the case. Sometimes they are used interchangeably, and what concept they express is often a question open to interpretation (cf. Sweeney 1992, 15–28, and my treatment of Gregory of Nyssa in this chapter).

PLATO, ARISTOTLE, AND THE PROBLEM OF AN INFINITE GOD

When the pre-Socratics tried to explain the origin of the world, they referred to various elements, but all of those were rather concrete entities that were not well suited to explain how or why there came to be a world. It was Anaximander who came up with a more sophisticated idea. He declared the infinite, or indeterminate, to be the origin of all things and used for it the term ἄπειρον. It is not exactly clear what he really thinks το ἄπειρον refers to. It could even be material. It is improbable, but not impossible, that he was already capable of thinking the infinite as an abstract infinity. What we can see clearly is the wish to affirm that the source of being lacks certain limits. It is, for example, not limited in time: it has neither beginning nor end. It is not limited in space. This does not necessarily mean that the source of being is outside space, but in Anaximander we find the motif that it could contain the world.[2] What we know is that το ἄπειρον has a function as origin of the world and a governing principle, and that it seems to be divine.[3] The most important philosophers after Anaximander disagreed not only about his account of the origin of the world but also about the infinity of God.

Plato and Aristotle did not contribute much to our understanding of divine infinity. Plato did not contribute much to the concept of infinity himself, but Aristotle made the most important contribution to the concept of infinity in Western thought, albeit a negative one, since he spoke of it in purely quantitative terms and, more importantly, he denied the existence of an actual infinite. This view of infinity haunted mathematics until Cantor's set theory, and, in the form of various finitist approaches to mathematics, is still present today. The denial of the existence of an actual infinity was to be a huge burden for the attribute of infinity within theological doctrines of God.

Plato agreed with Anaximander on the nature of the precosmic stage, but disagreed on the most important point. For him, form and determination did not come from the infinite, but from the demiurge. So, what there is in the beginning is not divine but is shaped into form by another divine entity (Plato, *Timaeus* 53a–b). For Plato and Aristotle, a lack of determination and form is a lack of perfection. Therefore God cannot be infinite. It has to be noted that neither Plato nor Aristotle seems directly interested in the infinity of God. That God cannot be infinite is more a corollary of more important themes. Still, their authority was enough to compromise later work on a concept of the infinity of God.

Though Aristotle was not interested in the infinity of God, he made the most important contribution of the understanding of infinity in Western thought. His position can be easily summarized. The infinite is quantitative and it does not actually exist. Both themes are already present in his definition of infinity as "that which it is impossible to traverse, because it is not the kind of thing to be traversed" (Aristotle 1983, 204a3). He does not offer us an argument why infinity has to be only quantitative, and so, not surprisingly, the qualitative concept was not eliminated by his treatment of infinity. Even if it is not my aim here, it might be interesting to ask the question what the major flaw in infinity is, such that it cannot exist actually, which contradicts Aristotle's principle of plenitude. The most important point seems to be the definition. There are only two options. A potentially infinite series is not traversed. Then it would not exist as an actual infinity. But, if it were to be traversed, it would be, by definition, not infinite! It seems as though Aristotle defined the infinite in a way that excludes its actual existence by definition. Though this is not directly related to the infinity of God, it underlines the point that a concept of the infinity of God did not receive much praise after Plato and Aristotle.

PHILO AND PLOTINUS

Philo is among those considered to be the first who attributed infinity to God after Plato's and Aristotle's negative assessment. Recently, Geljon has reinforced this view. (See Geljon 2005 with reference to the following statements.) The case for Philo as the first who reached a concept of the infinity of God can be made because of his emphasis on the incomprehensibility of God. A God who is not enclosed or limited might be viewed as an infinite God. Philo also believes God to be incomprehensible: the search for knowledge of God's nature does not end, and thus human beings are on the way to knowing God more and more, but this never ends. Even if this is not explicated in terms of infinity, it is interesting to note that it is the main topic of Gregory of Nyssa's assessment of the infinity of God. He explicitly links incomprehensibility to infinity.

Another topic for Philo is that God is not limited by anything, including space and time. Philo also thought that God's power, sovereignty, and goodness are unlimited. So there are some hints that God is unlimited and that some of his attributes are unlimited, but we have no explicit statement of God's infinity.

Philo believed certain attributes of God should be thought of as unlimited, and he emphasized God's incomprehensibility, which is a major shift away from Plato and Aristotle, but he still might not be the first who came upon a concept of the infinity of God. He might have implicitly thought of God as infinite, even if he did not come to the explicit conclusion that God's infinity is one of the attributes that need to be included in a doctrine of God. It is still remarkable how much he dared to depart from Plato—he obviously did not regard a lack of determination as a lack of perfection.

Half a millennium after Plato and Aristotle, Plotinus finally dares to call his One "infinite." This infinity was infinity as indetermination.

Indeed, every determination counts as a limitation and, therefore, a lack of perfection. This is the complete opposite of Plato and Aristotle. An accompanying important theme is the incomprehensibility of the One (Plotinus, *Enneads* 5.5.6). Since it has no determination, it cannot be comprehended. It is interesting to note that Plotinus also describes the One's power as infinite (6.9.6). This cannot be underestimated, because it is—apart from absolute indetermination—not clear what "infinite" means when it is not used as a predicate. The most important theme seems to be ontological. The infinite is not limited by anything, be it space or other entities. It usually determines everything else. This is, however, an important strand of thought according to which infinity is about power, and Plotinus could be counted as subscribing to it. However, he still makes more use of "infinity" in an ontological sense.

Additionally, Plotinus describes the One as being the source of everything else. And that which is the source, that which is prior, "contains" everything else. So there is a somewhat panentheistic understanding of the infinity of God.

One major drawback of Plotinus's concept of the infinity of God is that he also views other entities as infinite, for example matter and the soul. Plotinus also attributes infinity to evil (*Enneads* 1.8.3). One possible solution is that, in the case of evil, infinity is indetermination, a privation. In a way, this resembles Plato's view that a lack of determination is a lack of perfection. That means that it is a lack of determination where there should be determination. In the case of the One, there is no determination where there should be no determination since every determination of the One would be a limitation and therefore a lack of perfection.

In spite of this difference between the infinity of the One and the infinity of evil, it is obvious that Plotinus did not arrive at an understanding of infinity that reaches the lofty heights of attributes, such as perfection, omnipotence, or moral perfection, that are exclusively predicated of God. This will change with Gregory of Nyssa.

There is not much to summarize in Philo and Plotinus except that it was a negative theology that enabled them to attribute infinity to God and the One. This means the concept of infinity is—with the exception of his infinite power—rather close to the concept of indetermination. It is also interesting that though the concept of infinity in ancient Greek thought is rather difficult and diverse, there could be at least a coherent development in Neoplatonism. The attribution of infinity to God—accompanied by a negative theology—becomes gradually more important within Neoplatonism.

GREGORY OF NYSSA

Gregory of Nyssa is famous for being one of the three Cappadocians who made important contributions to the doctrine of the Trinity. Ever since the Church historian Ekkehard Mühlenberg claimed in 1965 that Gregory is the first to formulate a proper concept of the infinity of God, he has been referred to in philosophical discussions of infinity. Gregory spoke of God as being infinite in numerous passages and considered the ascription of infinity to be the most important thing that can be said about God. His treatments of infinity have a strong spiritual connotation:

> The barrier which separates uncreated nature from created being is great and impenetrable. One is finite, the other is infinite; the one is confined within its proper measure as the wisdom of its maker determined, the limit of the other is infinity. The one stretches out in measurable extension being bounded by time and space, the other transcends any notion of measure, eluding investigation however far one casts the mind. (Gregory of Nyssa 2007, 2.69–70)

There is, however, a major problem. We do know that he believed God first and foremost to be infinite. One could argue whether "be" is not already a shift toward one line of interpretation, but what exactly that means remains unclear. According to one interpretation, he considered God to be infinite in his essence. In the *Brill Dictionary of Gregory of Nyssa*, we read in the entry "infinity": "Therefore, God is limitless (ἀόριστον), i.e. infinite (ἄπειρον), by his very essence" (Karfíková 2010, 424; cf. Karfíková 2001). Another commentator writes: "In his very entity he is infinitely perfect" (Sweeney 1992, 487); "God is still Being and infinity continues to be an attribute of that very Being" (493); "God's very reality is infinite because He is being, goodness, beauty and other perfections" (501). According to this interpretation, God's nature is infinite, and this should be interpreted as a positive statement. Most proponents of this interpretation add additional positive statements about God regarding his goodness or his simplicity. Mühlenberg even goes so far as to say this: "There is a positive concept, that expresses temporal unlimitedness and which is synonymous with το ἄπειρον, respectively that of eternity" (Mühlenberg 1966, 111).

To shed some light on the problem of the meaning of ἀόριστον and ἄπειρον in ancient Greek thought, one can take a look at the interpretations of both terms in Gregory of Nyssa. Mühlenberg's interpretation leads to a view where ἀόριστον and ἄπειρον have different meanings.[4] Ascribing infinity to God would then not be a statement about God's indetermination and incomprehensibility but something else, though it is not clear exactly what this would be. But there is also another line of interpretation, according to which Gregory thought that "the Divine is . . . above all expressions of words, having but one name that can represent His proper nature, the single name of being 'above all name'" (Brightman 1973, 104–5; the quote is from *Contra Eumomium* [*CE*] 1.683). Brightman continues that Gregory would criticize the following questions: "Does the lack of diastema define the nature of God? Does the word 'infinite' define God? Does the endlessness of God along with his beginninglessness describe his essence?" (106).

Another proponent of this view is Geljon:

> But his interpretation stands in sharp contrast to Gregory's theology regarding the definition of God's essence. In *CE* 2.529 he clearly states that he does not define any negative attribute as God's essence. In his view, God's essence is totally unknown to the human mind and cannot be expressed by any term. The only thing the human intellect can know about God is that he exists. . . . The notion of infinity is certainly important for Gregory but must be seen as part of his apophatic theology. . . . He does not, however, argue that infinity expresses God's essence or is a unifying idea of God. (Geljon 2005, 168)

This view leads to an interpretation of ἀόριστον and ἄπειρον as synonymous: Regarding Gregory's terminology we observe that, where he deals with the subject of infinity, he mostly uses both ἄπειρος and ἀόριστος. ἄπειρος means "without limit" (πέρας = limit) and ἀόριστος means "without limit or determination" (ὅρος = limit or determination).

He remarks that both terms mean the same (*CE* 1.169) (Geljon 2005, 168). Accordingly, he translates ἀόριστος φύσις as "infinite nature" (160; quotation from *CE* 1.667–78).

I will now try to make a case for the latter interpretation. What definitely supports the first interpretation is that Gregory explicitly calls God "infinite" or "infinitely good." But he also often says that God's nature cannot be known. So we have a textual basis for both interpretations. But I believe there are three points that support an interpretation in the sense of a consequent negative theology.

The first is the context of the relevant texts. Most quotations that seem to suggest Gregory taught that God's nature is infinite stem from his discussion with Eunomius. But there he had the goal of refuting him, so he wasn't just stating his own position. He wanted to refute Eunomius at all costs. His other works, such as the *Life of Moses*, are rather mystical and focus on the complete incomprehensibility of God. It is possible that he takes on the convictions of Eunomius himself, or principles he thought to be generally true, when he starts an argument about the infinite goodness of God in order to prove his infinity.

The second point is that he argues against the belief that a single name defines God's nature. Eunomius thought that names define the nature of things. There is a certain term that defines the nature of an entity. In the case of God, this would be "unbegottenness." So, a begotten entity cannot be God. One way of tackling this argument is to note that Gregory denied that a single term defines Gods nature and assume that this also includes "infinity."

A third point is that Gregory argues for God's incomprehensibility on the basis of his infinity. An argument from the infinity of his nature for the incomprehensibility of his nature would hardly be valid. This tension is also present in the interpretation according to which God's essence really is infinite. Sweeney writes: "God's very reality is infinite because He is being, goodness, beauty and other perfections" (1992, 501). At the same time, he comments: "Gregory undertakes next to establish that although we know that God the Creator exists, we do not know what He is in His being (οὐσία): He is incomprehensible" (489). Sweeney has a good reason for writing these conflicting statements because both themes are present in Gregory.

One possible solution is that Gregory definitely lays out a negative theology but fails to consequently adopt it. And in every negative theology that does not end in complete silence, sooner or later contradictions will arise, because some attributes, such as God's infinite goodness, are too precious to be negated. His treatises that are not meant as arguments for the divinity of Christ, and include only his own meditations, also seem to support the latter interpretation: "He learns from what was said that the Divine is by its very nature infinite, enclosed by no boundary. If the Divine is perceived as though bounded by something, one must by all means consider along with that boundary what is beyond it" (Gregory of Nyssa 1978, 236).

What we can say for sure is that Gregory held God to be infinite, whatever that means exactly. And he links his incomprehensibility to his infinity. It is interesting to note that Plato's and Aristotle's position that infinity is a lack of form and determination, and therefore lack of perfection, is completely reversed in Gregory. The most natural thought about God is that he is infinite. If this were not the case, this would be a lack of perfection. It's hard to see whether there is an argument involved here, because this seems more like a rhetorical device, but it is by no means a bad one.

Thus Gregory did consider God to be infinite. He spoke more of divine infinity than anyone before him, and he considered infinity to be the most important thing that could be ascribed to God, which gives him a place among the few theologians and philosophers in the history of Western thought who give priority to the attribute of infinity in their concept of God. Among these thinkers, he is the first one who gave a detailed account of divine infinity. If it is blurry, it has to be granted that this might not be Gregory's fault, since the concept of infinity is itself ambiguous. Neither he nor his predecessors felt able to use different terms to distinguish between the various concepts of infinity and related concepts. They deployed one or, at most, two terms.

SYSTEMATIC SUMMARY

I now try to summarize what can be learned from ancient Greek thought about the concept of infinity of God. The most obvious observation is that the concept of infinity itself is not a precise one. The term ἄπειρον is used to express different concepts.

The Different Concepts of Infinity

Mathematical and Quantitative Infinity

The most precise concept of infinity is that of mathematical infinity. Though many paradoxes easily arise, it is at least clear what someone who speaks of mathematical infinity means. This is also the case for nonmathematical quantitative infinity, which is usually, but not always, described in terms of mathematical infinity, for example, the infinity of time or space.

Infinity as Indetermination

There are, however, other concepts expressed with the same term. Usually they are summarized as "qualitative infinity." The most important nonquantitative concept of infinity in ancient Greek thought is that of indetermination, which, unfortunately, is not exclusively expressed with the term ἀόριστον, but also often with ἄπειρον. There are various kinds of indetermination. There is an indetermination of one value or property and an indetermination of all properties. One can further distinguish between epistemic and ontological indetermination. The more interesting concepts for divine infinity are those that are committed to an indetermination of all properties. Ancient Greek philosophers preferred ontological indetermination for two reasons. One was a systematic function for the origin of the world, as it seemed obvious to some that all the diversity did not arise from another diversity but from complete indetermination. Another is that there was no need to view the divine as a person. Being a person would have been a determination and should thus be viewed as a restriction, a lack of perfection. Theologians and proponents of theism were committed to the personhood of the divine. Accordingly, there is at least one property that determines the divine being and would, in the case of absolute indetermination, limit it, which would be an imperfection. This makes the interpretation of theologians and theists rather complicated if they argue for absolute indetermination, so one has to assume implicitly that this cannot be what they really wish to argue for. An obvious hermeneutical device is the introduction of the incomprehensibility of God, so that arguments that seem to suggest an absolute indetermination can be interpreted as arguments for the mere incomprehensibility of God.

The Problem of Different Concepts of Infinity

So far I have identified three groups of concepts of infinity: (1) mathematical/quantitative infinity, (2) indetermination/negation of a certain property, and (3) indetermination/negation of all (limiting) properties. The second of those can be used for divine infinity (for example, being not limited by space or time), and the third is used exclusively for divine infinity (the Divine as the Absolute that is not limited or determined by anything but determines everything else). All of these concepts are expressed with the same term, which is confusing enough, but it becomes worse. There can be an ambiguous use of ἄπειρον within one author. When Aristotle states that "nature flees the infinite" (*De generatione animalium* 715b13–15), he means that it seeks determination.

There is also a problematic passage in Gregory of Nyssa. He says about Moses's spiritual ascent to God:

> The great Moses, as he was becoming even greater, at no time stopped in his ascent, nor did he set a limit for himself in his upward course. Once having set foot on the ladder which God set up . . . , he continually climbed to the step above and never ceased to rise higher, because he always found a step higher than the one he attained. (Gregory of Nyssa 1978, 227)

This ascent to God is moral and intellectual. It is an unending quest. But this infinite ascent that never ends has stages. This is a quantitative concept of infinity. Now, one might argue that there is no problem here because what is quantitative is the moral and intellectual ascent, not God himself. But one still has the impression that the infinite ascent can be mapped onto God's infinity.

This should not lead to a critique of those who initiated the inquiry into infinity at the beginning of Western thought, but it should be a reminder of the need to clarify what "infinity" means exactly and which different concepts really are involved.

Infinity as Ontological Category

Now, it might still be interesting to find a positive concept of infinity. Does the attribution of nonquantitative infinity say anything other than that an entity is indeterminate or incomprehensible? I believe that one can find some small hints that infinity, if not understood in a quantitative sense, has an ontological connotation. Let me underline this point with a long quotation from Gregory:

> For certainly that which is bounded leaves off at some point, as air provides the boundary for all that flies and water for all that live in it. Therefore, fish are surrounded on every side by water, and birds by air. The limits of the boundaries which circumscribe the birds or the fish are obvious: The water is the limit to what swims and the air to what flies. In the same way, God, if he is conceived as bounded, would necessarily be surrounded by something different in nature. It is only logical that what encompasses is much larger than what is contained. Now it is agreed that the Divine is good in nature. But what is different in nature from the Good is surely something other than the Good. What is outside the Good is perceived to be evil in nature. But it was shown that what encompasses is much larger than what is encompassed. It most certainly follows, then, that those who think God is bounded conclude that he is enclosed by evil. . . . Therefore, he who encloses the Divine by any boundary makes out that the Good is ruled over by its opposite. . . . Therefore, no consideration will be given to anything enclosing infinite nature. (Gregory of Nyssa 1978, 236–38)

In the light of these metaphors, one would believe that, in order to preserve God's majesty, we have to think of him as containing everything else. It's interesting that Hegel later stated that every concept of infinity that presupposes that the infinite has something outside of it is a bad infinity. Some seem to think that infinity when properly thought through leads to a version of panentheism. Of course, someone could try to defend classical theism, conjoined with the predication of infinity to God, by pointing out that God's infinity means that every entity other than God does not have the power to exist without him, that their existence depends on him. So those entities do not need to exist in God—whatever that might mean—but outside him, but not in a way that restricts his majesty, power, or nature.

Common to both lines of reasoning is the use of infinity as an ontological category. Nothing exists outside God. Nothing exists independently of God.

The Conceivability of Infinity

All ancient Greek philosophers agree that a philosophy of the infinite is not about the proper concept of an abstract but hardly conceivable infinity, but about the human mind's ability to grasp infinity. Aristotle is especially skeptical about the human mind's ability to conceive infinity. But even those who do believe that there is an infinity emphasize the human mind's inability to really understand it. Gregory writes: "It is not in the nature of what is unenclosed to be grasped" (Gregory of Nyssa 1978, 238). So, if an intuitionist mathematician is inclined to discuss whether the human mind can grasp infinity or infinite entities, rather than trying to find the proper abstract concepts, he is within the rich tradition of ancient Greek thought.

Negative Theology and the Infinity of God

So far we have two lines of thought regarding the infinity of God. According to Plato and Aristotle, God cannot be infinite since this would be a lack of form and determination and, therefore, a lack of perfection. According to some pre-Socratics and Neoplatonists, God is infinite, and this view is accompanied by a negative theology. It is interesting to note that this does not suggest a disagreement about the relation of negative theology to the infinity of God. Both lines of thought seem to agree that a God that is pure form, and for whose concept we have an exact definition, cannot be infinite. We have no clear statements from Plato and Aristotle, but one could suggest that they would not argue against statements such as, "If God is infinite, we cannot know him." This poses an interesting task for those who wish to maintain God's infinity and his comprehensibility or other attributes. It is by no means clear how a God who is infinite in the sense of "indeterminate" can be known, or how he can be a person.

SUMMARY

There is much to be said about the concept of divine infinity in ancient Greek thought, but I will limit my final summary to two points.

The use of the term ἄπειρον is extremely complicated and confusing. There is a whole variety of different concepts that is expressed with this term, and sometimes this ambiguity occurs even within the work of one philosopher or a single text. This, it should be noted, is not limited to ancient Greek thought, and it still happens today. Whether the fault lies in the forging of the different concepts summarized under the term ἄπειρον in ancient Greek thought or whether it lies in the nature of infinity itself is not easy to judge, though I tend to the latter opinion.

There is universal agreement among ancient Greek philosophers that God, should he be infinite, can hardly be known. He might even have no determinations at all. This means that in the eyes of ancient Greek philosophers the concept of the infinity of God is accompanied by a negative theology. One could argue that even the personhood of the divine becomes a problem. This means that theologians and theists who adopt a doctrine of the infinity of God have to explain why they claim to have rather specific knowledge of him and how this does not contradict their concept of infinity.

NOTES

1. For other concepts of infinity, see the introduction to this volume. One worth mentioning is Hegel's concept of true infinity: there is nothing outside, or independent of, the infinite, such as the finite. In contrast to a purely qualitative concept of infinity, Hegel's true infinite therefore includes being finite.

2. So far this is a qualitative sense of infinity, but in the Pythagoreans and in Anaximander we also find a quantitative concept of infinity. So, at the very beginning, both concepts are present, and (rather unfortunately) the same term, ἄπειρον, is used for them both. Although ἀόριστον can also be used for the qualitative concept of infinity, ἀόριστον has for most authors a wider meaning, including "being indeterminate."

3. Diels-Kranz 12B2, 12B3.

4. "Auf diese Weise wird der Ausdruck ἀόριστον, statt mit dem Unendlichen identifiziert zu werden, in die Richtung des Unbestimmten gedrängt. Zum mindesten wird die Ansicht nicht durchgehalten, daß το ἀόριστον = το ἄπειρον eine eindeutig durchdachte Prädizierung von Gottes Wesen ist" (Mühlenberg 1966, 146).

REFERENCES

Aristotle. 1983. *Physics, Books III and IV.* Translated with introduction and notes by Edward Hussey. Oxford: Oxford University Press.

Brightman, Robert S. 1973. "Apophatic Theology and Divine Infinity in Gregory of Nyssa." *Greek Orthodox Theological Review* 18 (2): 97–114.

Geljon, Albert-Kees. 2005. "Divine Infinity in Gregory of Nyssa and Philo of Alexandria." *Vigilae Christianae* 59 (2): 152–77.

Gregory of Nyssa. 1978. *The Life of Moses.* Translation, introduction, and notes by Abraham J. Malherbe and Everett Ferguson. New York: Paulist.

———. 2007. *Contra Eunomium II.* Translated by Stuart George Hall. Leiden: Brill.

Karfíková, Lenka. 2001. "Die Unendlichkeit Gottes und der unendliche Weg des Menschen nach Gregor von Nyssa." *Sacris Erudiri* 40:47–81.

———. 2010. "Infinity." In *The Brill Dictionary of Gregory of Nyssa*, edited by Lucas Francisco Mateo-Seco and Giulio Maspero, translated by Seth Cherney, 423–26. Leiden: Brill.

Mühlenberg, Ekkehard. 1966. *Die Unendlichkeit Gottes bei Gregor von Nyssa: Gregors Kritik am Gottesbegriff der klassischen Metaphysik.* Göttingen: Vandenhoeck u. Ruprecht.

Sweeney, Leo. 1992. *Divine Infinity in Greek and Medieval Thought.* New York: Peter Lang.

CHAPTER 3

Infinity in Augustine's Theology

ADAM DROZDEK

INFINITY OF THE MANICHEAN GOD

In his spiritual journey, after he was awakened from his theological slumber by Cicero's *Hortensius* (*Confessiones* 3.4.7), Augustine fell under the influence of the Manicheans (3.6.10). He accepted their vision of the corporeal God, whose divinity consisted primarily of God's eternity and spatial infinity. There were two kinds of eternally existing infinite matter: one good, one evil. God was "the most luminous mass," infinite in all respects, except where He was opposed and thus limited by evil mass (5.10.20). Augustine saw God as "something corporeal in space, either infused into the world, or infinitely diffused beyond the world, and also incorruptible, inviolable, and unchangeable" (7.1.1), and as "a large [entity] penetrating through infinite spaces from all sides the entire mass of the world [the way light penetrates air], and beyond it in all directions, through immensity without end" (7.1.2, 7.14.20). Augustine viewed the created universe as a large but finite mass that was surrounded and permeated by God, who was "in every way infinite, as if there were a sea everywhere and on every side through immensity only the infinite sea, and it contained within itself some sponge, huge, but finite, so that the sponge would in all its parts be filled from the immense sea." And Augustine, a finite creature, saw himself filled with "infinite You," infinite God, who was "the whole, true, highest, and infinite good" (7.5.7).

Even with this materialistic view of God, Augustine was troubled by the fact that a larger bulk of matter would contain more of God that a small bulk (*Conf.* 7.1.2), since "every mass is smaller in a part than in its totality, and if it were infinite, it must be smaller in the part limited by a certain space than in infinity, and is not wholly everywhere, like Spirit, like God" (3.7.12).

After he was influenced by "the books of the Platonists" (*Conf.* 7.9.13), he was encouraged to seek the incorporeal truth (7.20.26), and he freed himself from the idea that "the truth is nothing because it is not diffused through space, either finite or infinite" (7.10.16; cf. Plotinus, *Enneads* 5.5.10). He still saw God as infinite, but differently than before (7.14.20): "infinite, and yet not diffused in spaces finite or infinite" (7.20.26). However, Augustine did not specify how different his new view of divine infinity was. It appears that Augustine wanted to retain infinity as God's attribute, which was the preeminent attribute of the Manichean God, but because God was now for him incorporeal, this infinity should be characterized in an incorporeal manner. We now turn to what this characterization should be, and what Augustine apparently hoped to find in the books of the Platonists.

PLOTINUS

In the *Timaeus*, Plato describes the creation of the world: the eternally, timelessly existing divine Demiurge used eternally existing ideas/forms to mold the eternally existing prime matter in the eternally existing space. Although the world was finite, molding it required infinite knowledge of the Demiurge and an infinity of ideas (Drozdek 2008, chap. 10).[1]

Another and a more important source for Augustine was Plotinus, whose *Enneads* he read in Latin translation.

On the top of Plotinian ontological hierarchy there is the One, the source of the Intellect that becomes the source of the Soul. The Intellect is the seat of intelligibles/ideas, and from the Soul proceed individual souls. And there is also matter, a substrate of individuals in the intelligible and the material world.

Matter is unformed substrate, undetermined, thereby dark and obscure. Matter, by definition, is what has no form (*Enneads* 2.4.13, 1.8.9).

As such, it is infinite, and infinity is not just matter's attribute but its essence: matter is infinity itself (2.4.15). Also, because of its formlessness, matter becomes identified with metaphysical evil (1.8.4).

The Intellect is the source of all individual intelligible beings (ideas) (*Enneads* 5.8.7); it has infinity of infinite powers (6.2.21–22). Therefore, the nature of Intellect is infinite on account of its strong connection with the intelligibles, of which it is the source. A similar claim can be made about the Soul. The Soul is the source of individual souls and as such it is an infinity of power in respect to all these souls (6.4.14, 5.7.1); one Soul exists and it includes all souls that are separate without partition, so "infinity is its nature" (6.4.4, 6.4.14).

Because the Intellect emerged from the One and the Soul from the Intellect as unities, they are determined and thus finite, determined by the act of their sources. However, they are also infinite multiplicities and thus evil. Therefore, in the Intellect and the Soul evil is inextricably linked with the good.

Whereas the Intellect and the Soul are finite and infinite in their nature at the same time, the situation drastically changes in the case of the One.

The One "is not being. . . . The nature of the One is generating all things, yet it is none of them; neither a thing nor quality nor quantity nor intellect nor soul; not in motion, not at rest, not in place, not in time, but is itself, a singularity [μονοειδές], better yet, formless, existing before all form, before movement, before rest, since they concern being and make it the manifold" (*Enneads* 6.9.3). However, although the One is ineffable (5.3.13), Plotinus spoke about its causality and did mention that "its power is infinite" (5.5.10). However, he warned that calling the One a cause did not mean attributing causality to the One, but to ourselves. Strictly speaking, we should not apply any particular term to it: "We are circling around it on the outside, as it were, trying to explain it with the experience of our own; sometimes we come close, sometimes we step away from its difficulties" (6.9.3). Even calling it the One is inadequate; by calling it thus we indicate that it is not multiplicity, but calling it the One does not adequately covey its nature (5.5.6). Ascribing no attributes should also include ascribing no infinity and no finitude. Moreover, because of the association of infinity with evil and because the One is also the supreme Good, God, "the most perfect of all things" (5.4.1), the One cannot be infinite.

Plato believed, if only implicitly, that God is infinite in respect to existence and in respect to knowledge. Plotinus considered the power of the divine One to be infinite, while the One itself was above the infinite and the finite. These ideas were reflected in Augustine's theology.

UNHAPPINESS

In his 387 letter to Nebridius, written soon after his conversion and while penning his *Soliloquies* that he mentioned twice in this letter (*Epistola* 3.1, 3.4), Augustine expressed a feeling of unhappiness. The remarkable thing about it is that the reason for his unhappiness was the problem of infinity. He could not consider himself to be happy without knowing why the world had the size that it did, since nothing prevented it from being larger or smaller.[2] Since matter is infinitely divisible, the world can be divided into even smaller parts. Moreover, for each natural number it is possible to generate mentally a larger number, and thus the sequence of natural numbers is infinite: 1, 2, 3, and so on; as Augustine phrased it, "intelligible number increases infinitely."[3] However, there is the smallest natural number; therefore, the intelligible number "cannot be infinitely decreased." On the other hand, "the sensory number . . . infinitely decreases, but cannot infinitely increase" (3.2). In this, Augustine did not consider fractions as mental or intelligible numbers but empirical or sensory numbers only, which directly corresponded to infinite divisibility of matter or to a possibility of generating a smaller corporeal quantity for each corporeal quantity. In this way, the number of parts constituting the world can conceivably change from one moment to another. In sum, the size of the world can grow indefinitely, the intelligible numbers can grow into infinity, and the number of parts of the world can also grow without end since sensory numbers can be decreased into infinity. Infinity can affect the world in both directions: in increasing and in decreasing. In the letter to Nebridius, Augustine ended his discussion with the somewhat flippant "I have to [go to] sleep" (3.3). However, he investigated the problem of infinity of numbers in some of his early dialogues. Although he stated in his *Soliloquies* that he was interested in knowing God and his soul—and nothing else (1.2.7, 1.15.27), he was interested in other subjects as avenues leading to his ultimate goal of self-knowledge and knowing God. For instance, he explicitly stated that what was needed for theology was the science of good disputation, that is, dialectic, and the power of numbers, at least one of these two areas (*De ordine* 2.18.47).[4] The power of numbers is the power of infinity.

Surprisingly, infinity is commonplace or, rather, something accessible to every person. Even the most sinful soul "is affected by numbers and manages numbers" (*numeris agatur, et numeros agat*) (*De musica* 6.17.56), which can be taken to mean that the mind can comprehend infinity. And yet, intellect is finite for itself, it does not want to be infinite, but it could be since it wants to remain in its own knowledge, that is, it wants to know itself, since it loves itself (*De diversis quaestionibus* 15). Also, can intelligence of a thing be extended to infinity? Since there is a perfect way of comprehending something, intelligence of a thing cannot extend to infinity (32). It is an interesting situation: the mind can handle infinity, but the mind itself is finite, and for understanding to be understanding, that is, to be rational, understanding must be finite. Therefore, the finitude of the mind can harness the infinity of numbers. This ability of handling infinity by the finite mind is named in various ways by Augustine. It is simply an imagination: every corporeal body can be thought of in infinite number or it can be infinitely extended, like light (*De vera religione* 20.40). The mind has images of large objects, of heaven and earth, with no need of room for them. The mind thus is not contained in images of the largest spaces, but these images and the space are contained in the mind "by an ineffable power and faculty," by which it can shrink them and enlarge them immensely (*Contra epistolam Manichaei* 17.20).[5] And, again, it is possible to see that in the sequence of numbers, $1, 2, \ldots, n, \ldots, 2n, \ldots$, the distance between a number n and the beginning of the sequence is the same as the distance between the number n and the double of that number $2n$, that is, $2n - n = n$ (or rather, to account for *distance*, $2n - (n + 1) = n - 1$). The unfailing truth of this relation between an innumerable number of cases can only be seen by "the inner light which the corporeal sense does not know" (*De libero arbitrio* 2.8.23). This inner light can pronounce the existence of arithmetic laws that hold for infinity of numbers. It is this "light of the mind" that properly assesses the result of an arithmetic operation on numbers (2.8.21).

This ineffable mental power is, like many mental endowments, good by itself since it was given by God, but it can be abused or misused, as can any bodily and mental faculty (*De libero arbitrio* 2.18.48). And thus, through its abuse, Manicheans extended immense light into an infinite space (*De vera religione* 49.96), and the Epicureans spoke about infinite worlds (*Contra epistolam Manichaei* 18.20).

INFINITE KNOWLEDGE

If humans can deal with infinity of numbers through their ineffable faculty, then surely can God—and Augustine was rather impatient with those who thought otherwise.

Some claim, said Augustine, that "the infinite cannot be comprehended by any knowledge (Aristotle, *Metaphysics* 14); therefore, they argue that God has in his mind finite conceptions of all finite things which He makes." In other words, the claim states that "no knowledge can comprehend any infinity." Such a claim results from the fact that "they measure the divine mind—which is absolutely immutable, infinitely capacious, and numbers all innumerable [sequences] without succession—with their own human, mutable, and narrow [mind]" (*De civitate Dei* 12.17).

Seemingly, this appears to undermine Augustine's early admission that the human mind through an ineffable faculty can deal with infinity. However, this faculty would still reside in a finite human mind, and thus, although it would allow humans to acknowledge the existence of infinity, at least, the infinity of numbers, the human mind would not be able to see all the numbers and all numerical relations at the same time.

However, the denial of the ability of the divine mind to understand infinity would be impious, because God would not know numbers, at least not all of them, since there is an infinity of numbers; it is obvious that there is an infinity of numbers, since for each number n there is a number $n + 1$ and, consequently, if there are n numbers then there can also be $2n$ numbers and even $m \cdot n$ numbers for any m. The assertion that God does not know all numbers would belie biblical statements: "You have ordered all things by number, and measure, and weight" (Wisd. 11:20); "He brings out their host by number" (Isa. 40:26 [Vulg.]); "The very hairs of your head are all numbered" (Matt. 10:30). The psalmist also stated of God that "there is no number of His understanding" (Ps. 146:5

[Vulg.]), which can be taken to mean that God's understanding is infinite. However, Augustine struggled here with terminology: "The infinity of number[s], although there is no number for infinities of numbers, is yet not incomprehensible by Him of whose understanding there is no number. And thus, if what is comprehended in knowledge is made finite by the comprehension of this knowledge, then all infinity is in some ineffable way finite to God, for it is not incomprehensible to His knowledge" (*De civitate Dei* 12.18). Thus, since only what is bounded can be comprehended, Augustine resorted to the use of paradoxical terminology that infinity becomes finite by the act of comprehension. A bit more carefully, he also stated that "His understanding exceeds all arithmeticians [*numerarii*]. We cannot number it and who can number the number itself? Whatever things are numbered, they are numbered with a number. If something is numbered, it is numbered with a number; there is no number for the number [itself]; in no way can the number be numbered. . . . Could anyone measure, or number, or weigh number itself, and measure itself, or weight itself with which God ordered all things?" (*Enarrationes in Psalmos* 146.11). At one point, Augustine decided that God makes the infinite finite in His divine comprehension. At another point he left open the problem of assigning a number to the set of natural numbers, which appears to indicate that making the infinite finite is not quite a satisfactory explanation and that there is another quality beyond the infinite and the finite that characterizes only the divine mind. This would point to the possibility that the divine mind can comprehend the infinite because it itself is above the infinite and the finite. It is beyond human ability to number the number itself with which God organized the universe. Because all natural numbers are finite, this number itself would be something of a different category than regular numbers and even the set of all numbers. This number itself would belong to the divine realm that surpasses even the infinity and is used to deal with infinity. In this way, what in our times is called a "transfinite" number would for Augustine be also a "transinfinite" number. In this sense also we can understand the paradoxical phrase that there is measure without measure, number without number, weight without weight God used in creating the universe (*De Genesi ad litteram* 4.3.8, 4.4.8). Hence, because of God's ability to comprehend perfectly infinity, He is the source of such a number that can be used to measure the cardinality of infinite sets. The number without number is the supreme number (*summus numerus*), that is, God Himself (*De Genesi contra Manichaeos* 1.16.26), which is a number of unique divine quality whereby God transcends all infinity and finitude.[6]

ETERNITY

When Augustine posed the question, "Is the truth nothing because it is not diffused through space, either finite or infinite?," he heard God's answer, "I am who I am" (*Conf.* 7.10.16). Also, because he was certain at that time that God was infinite (7.20.26), the divine infinity was for him no longer spatial infinity, which would be meaningless in the case of God who was not a spatial being; it was not temporal infinity, either, but what lay in the everlasting existence. "I am who I am" indicates God's eternal present, the ground of existence, the existence that has no beginning or end, which is an eternal, unending, unlimited "now." As eternal, God alone has immortality (12.11.11), and true immortality, true incorruptibility, and true immutability is eternity itself (*De Trinitate* 4.18.24). God "is not encompassed by finite or infinite space nor changed by finite or infinite volume of time" (*De Genesi ad litteram* 8.19.38). Time, of course, was created by God, but Augustine indicated here that should time be truly infinite, infinitely extending to the past and to the future, this would not affect God, who would surpass even the infinity of time. God, however, precedes all times past, and lives on after all future times (*Conf.* 11.13.16); no times are coeternal with Him (11.14.17). "Eternity and time are rightly distinguished by this, that time does not exist without some movement and change, while in eternity there is no change (cf. Plotinus, *Enneads* 3.7.5); . . . God, in whose eternity there is no change at all, is the creator and regulator of time" (*De civitate Dei* 11.6). Thus, Augustine identified immutability with eternity: "Eternity itself always stays put" (*Epistola* 7.1.2); "nothing passes in eternity, the whole is present" (*Conf.* 11.11.13). Also, "true immortality is immortality of Him in whose nature there is no change. That is also true eternity which is immutable God, without beginning, without end and, consequently, incorruptible. It is thus one and the same thing to call God eternal, or immortal, or incorruptible, or immutable; and it is also one and the same thing to say that He is living and intelligent, that is, to say wise" (*De Trinitate* 15.5.7). All of this is summarized by the statement that "eternity is God's very substance" (*Enarrationes in Psalmos* 101.2.10) and His nature (9.11). Immutability means incorruptibility and immortality: what does not change exists forever, exists eternally; that is, in eternity there is no before and no after, and thus even speaking about limits or boundaries in eternity is meaningless. The everlasting presence thus surpasses the infinity of temporal existence, or, at best, infinity of temporal existence can only be a faint reflection of the reality of eternity.

It is interesting that at one point Augustine said he quoted Hilary: "Eternity in the Father, form in the Image, the use in the Gift" [*aeternitas in Patre, species in Imagine, usus in Munere*] (*De Trinitate* 6.10.11), whereas Hilary actually had said that "infinity in Eternity, form in the Image, the use in the Gift" [*infinitas in aeterno . . .*] (Hilary, *De Trinitate* 2.1), which is a deliberate modification since Augustine hoped that he properly rendered Hilary's understanding of eternity. Since for Hilary, "I am who I am," that is, the fact that God is, sufficiently defines infinity (Hilary, *De Trinitate* 1.6) and because of the fact that God is His eternity (1.5), God's infinity signified for Hilary God's eternity, and thus the phrase "infinity is eternity" is rather pleonastic, and thus Augustine's modification is justified. Moreover, Augustine thereby stressed that God's essence lies in His eternity, not in infinity. Also, by saying "eternity in the Father" and not "infinity in eternity" or maybe even "infinity in the Father," Augustine wanted the meaning of eternity to embrace the meaning of infinity or maybe even eliminate the latter from the former: it makes sense to say about something that it is infinite if it can conceivably be also finite—infinite power, infinite extension, infinite temporary duration, and so on. Can eternity be finite? If not, then it would appear incongruous to speak about infinite eternity.

BEYOND INFINITY

Augustine believed that God's knowledge did not have any bounds, that God could comprehend infinity. Did it mean that God Himself in His essence was infinite, that infinity itself was God's attribute? A possibility of seeing God's eternity, which is His essence, in the supra-infinitistic way indicates that infinity did not have to characterize God's nature.

First, we have to notice that Augustine did not always use infinity as a technical term, as signifying something without limit, without bounds, without end. For example, he spoke about infinite genealogies (*Contra adversarium legis et prophetarum* 2.1.2); about extending an oration to an infinite volume (*Contra Iulianum opus imperfectum* 1.34); about an infinite variety of Latin translations of the Bible (*De doctrina Christiana* 2.11.16); about pagans seeking an infinite variety of pleasure (*De civitate Dei* 1.30); about an infinite task to refute anti-Christian objections (2.1); about infinite zeal of discussion of old philosophers (*Epistola* 135.1); he said that a person should watch not to become a grain in the midst of infinity of chaff (*Sermo* 5.3); that there is an infinite number of words (32.6.6) and the infinite pleasures of the body (361.5.5); that the soul captures images of infinite things (*Enarrationes in Psalmos* 145.4); after saying that God knows the number of grains of sand on earth and the number of hairs on human heads, Augustine said that what in this world is infinite for men is not infinite for God (146.11). In all these cases, infinity refers to something finite, but very large.

On the other hand, Augustine also mentioned Christ's or God's light of infinite greatness (*Epistola* 119.2; *Sermo* 4.5). However, when commenting on the verse "of Your greatness there is no limit" (Ps. 144:4 [Vulg.]), Augustine stated that "we can praise Him without understanding Him, since if we understood Him, His greatness would be limited and if His greatness is not limited, we can understand Him in something, but we cannot understand God in His totality. Being deficient in the face of His greatness, to be invigorated by His goodness, let us look at [His] works and praise the Maker in [His] work" (*Enarrationes in Psalmos* 144.6). Incidentally, the saved will know God's greatness, even though it is without end, by praising it and praising it without end (144.8). However, there are very few mentions that could be considered as statements, not about the infinity of God's attributes, but about the infinity of God.

In a 410 letter to Dioscorus, after mentioning Anaxagoras and his divine Mind, Augustine said that "it is manifest that the order and mode of all things stem from it [the Mind understood as "the truth and wisdom itself"], and that it may not incongruously be called infinite not with respect to its extension in space, but to its power which human thinking cannot comprehend" (*Epistola* 118.4.24). God, the truth and wisdom itself, can be called infinite on account of His power, which is an echo of

Plotinian understanding of divine infinity. In the same letter, Augustine also said that what is incorporeal is called a whole "since it is understood to be without limits in space: it is whole because of its completeness, it is infinite because it is not surrounded by limits in space" (118.4.24). In this, Augustine once again rejected the Manichean spatial God, but because he spoke indiscriminately about any incorporeal being, this infinity would also refer to souls and angels. In this understanding, incorporeal beings would be infinite because they are incorporeal, but this is because in the statement in this letter, limits that make something finite are understood only in a corporeal sense. Augustine's interest in this letter lies in the demarcation between the corporeal and the incorporeal—to stress the incorporeality of God—rather than between the finite and the infinite in the general sense.

Around the year 406, Augustine gave sermons on the Gospel according to John. In the opening sermon, he asked his listeners rhetorically, "What, then, is in your heart when you think of a certain living, eternal, omnipotent, infinite, everywhere present, everywhere whole, nowhere confined substance?" The answer: "When you think about these [attributes], this is the word about God in your heart" (*In Ioannis Evangelium* 1.1.8). Since the sermon was given only a few short years before the letter to Dioscorus, infinity could mean here infinity of power. It could also mean the freedom from *corporeal* limits, that is, as an equivalent of incorporeality.[7] In the same section of the sermon, Augustine wondered how anyone can say that God is immutable, because God is "above all creatures"; therefore, the infinity of God in this sermon may refer to the incomprehensibility of God. These possible interpretations are all the more likely because in his rhetorical question Augustine also mentioned freedom from confinement, which normally would be included in the meaning of infinity; that is, in this question "infinite" and "nowhere confined" would be two different characteristics.

In his major theological work, Augustine at one point remarked about the three persons of the Trinity: "Those three both seem to be mutually determined by each other and in themselves are infinite [*illa tria et a se invicem determinari videntur et in se infinita sunt*]. But here, in corporeal things, one thing is not as much as three together, and two are something more than one thing; but in the supreme Trinity, one is as much as the three together, nor are two anything more than one, and in themselves they are infinite [*in se infinita sunt*]. So, both each are in each, and all in each, and each in all, and all in all, and all are one" (*De Trinitate* 6.10.12).[8] Augustine speaks here about unity in diversity and diversity in unity, and it is from this perspective that the use of infinity in this passage should be interpreted.

When the Father thinks about the Son, He determines Him as the Son, thereby distinguishing Himself from the Son and the Son from Himself. The act of contemplating the Son is at the same time the act of determination of another member of the Trinity. The same can be said about any other Person of the Trinity contemplating any other Person of the Trinity. On the other hand, when the Father thinks about Himself, He sees the one divine substance in which there are no distinctions; He thinks about Himself as one God, as the unity, as infinity in the sense of indistinguishability between the members of the Trinity, as the unity with no *fines*, no *termini*, that is, no boundaries, no limits between the Persons of the Trinity. There are three of them if they are, as it were, viewed from the outside, each one viewing another. However, when viewed from the inside, when each Person views Himself, the distinctions between them disappear; they are one. This viewing from the inside of one Person and from the outside takes place at the same eternal moment; it is one and the same act. Moreover, the word *videntur* may be important here: the three Persons only *seem* to determine one another, thereby *seemingly* separating one from another. Finally, the statement about determination tout court requires a contrasting determination with infinity, and if infinity is understood as infinity of existence or of power, then the meaning of the statement becomes incomprehensible: What would mutual determination have to do with infinite existence or with infinite power?

Can we say that, for Augustine, God is infinite? The isolated and unelaborated mentions of infinity made by Augustine appear to point in the direction that God was not for him infinite, nor was He finite, but that He was beyond and above infinity and the finitude.[9] It is telling that when the problem of God's infinity was directly addressed, Augustine made at best a noncommittal statement.

Faustus, Augustine's Manichean opponent, asked him straight on, "Is God finite or infinite?" God, according to Faustus, is called the God of Abraham and Isaac and Jacob, and for Faustus the mark of circumcision, which separated these men from other people, signified also the limit of God's power that extended only to them. Thus, God, whose power is finite, cannot be infinite. Moreover, evil exists in the world; therefore, God is not infinite; otherwise, there would not be any evil (*Contra Faustum* 25.1).

In his response, Augustine blames Faustus for asking these questions by pointing to the underlying Manichean assumption that God is a spatial being.

> For when someone begins to ask, how God, whom no space can contain, can be finite; how [He can be] infinite, whom the Son knows perfectly; how [He can be] finite, [and yet] immense; how [He can be] infinite, [and yet] perfect; how [He can be] finite, who is without measure; how [He can be] infinite, who is the measure of all things, then all carnal thinking vanishes; and if [something] wants to become what it not yet is, it must first become ashamed of what it is. Therefore, as to the propositions concerning God as finite or infinite, it is better to end this by remaining silent [about it] until you cease going so far astray from the end of the law which is Christ. (*Contra Faustum* 25.2)

In this Augustine focused on the proper starting point, namely, the incorporeality of God. When this assumption is made, a person can be in the right mind and then questions can be asked as to how can God be infinite when the Son knows Him perfectly. To remain silent is a rather unsatisfactory, even evasive, answer. In any event, by the two parts of each question Augustine indicated a tension or even contradiction between two things that can be stated about God: spirituality and finitude; infinity and perfect knowledge; finitude and immensity; infinity and perfection, finitude and measurelessness; infinity and measurability. One possibility is to allow for both opposites to hold and try to theologically explain it. Another possibility is to remove one opposite from each pair and retain the other. In this way, spirituality, perfect knowledge, immensity, perfection, measurelessness, and measurability would be retained, and the other opposites discarded: finitude and infinity. If so, God would be neither finite nor infinite; He would be beyond finitude and infinite; He would be immense.[10] The fact that Augustine did not give a simple answer to Faustus's simple question would indicate that he was considering such a solution. Just as "God is above every measure of the creature, above every form, above every order, not by spatial location, but by ineffable and singular power" (*De natura boni* 3), where "measure" can also mean "limit" (22), so God would also be above the infinite and the finite.

Augustine thought the problem of the infinity of the world disturbed his happiness. It is thus quite certain that he also thought about the infinity of God. However, although he frequently discussed God's attributes, he never included infinity as such an attribute.

In an early work, *De doctrina Christiana*, he presented these attributes of God: eternity, immutability, majesty, power (1.5.5), ineffability, excellence, immortality (1.6.6), life itself, wisdom, truth (1.8.8), goodness, justice, being the highest being (1.32.35). Infinity of God is mentioned only once, and derisively at that, when speaking about those for whom God is something luminous and infinite beyond the world (1.7.7).[11]

In the *Confessions*, Augustine provided this list of attributes: "most high, most excellent/good, most capable, most omnipotent, most merciful and most just, most hidden and most present, most beautiful and strongest, stable and incomprehensible, immutable, changing all things; never new, never old; renewing all things . . . , always working, always at rest; gathering and lacking nothing; supporting and pervading and protecting, creating and nourishing and perfecting, seeking, yet lacking nothing" (1.4.4). Infinity is not mentioned here.

In the voluminous *De Trinitate*, where Augustine constantly speaks about God's attributes, there are only two marginal references to infinity while speaking about God, or just one if we consider that the same phrase is used twice in the same paragraph. At one point, he provided a list of twelve to which all God's attributes can be reduced: "eternal, immortal, incorruptible, unchangeable, living, wise, powerful, beautiful, righteous, good, blessed, spirit" (15.5.8; cf. another list at 5.11.12). No infinity.

From the very beginning of his theological reflection, Augustine saw God as infinite. In the Manichean period, the corporeal God was infinitely extended in space, and infinity was more associated with the spatial infinity than with God. This, in fact, was for him a stumbling block in his transition from the Manichean phase to his Platonist theology: If God should be considered an infinite being, how can divine infinity be retained without seeing God as a being extended in space? A breakthrough inspired by Plotinus came when the spatial aspect of God was abandoned and God was viewed as a purely spiritual being. What about God's infinity?

When studying Plotinus, Augustine discovered a very strong statement that identified matter, infinity, and evil. Augustine rejected identification of matter and evil: matter (*hyle*) as God's creation, was something inherently good (*De natura boni* 18). Also, evil for Augustine was an absence of good, but apparently he did not quite know what to do about infinity itself. He comfortably spoke about God's infinite understanding, since God can hold in His mind an infinity of entities at the same time, in particular, an infinity of numbers, knowledge of which God used in creating the world. Thereby, infinity became finite for God. He sometimes mentions the infinity of other divine attributes, but apparently cannot bring himself anywhere to say that God is infinity itself. The lingering reason appears to be the very negative Plotinian characterization of infinity that identified it with evil. It thus would appear that making God's nature infinite, identifying God with infinity, could, very uncomfortably, sound like fusing evil into the divine nature. So, Augustine, in the Plotinian vein, saw God as surpassing the finite and the infinite. Although he did not state it explicitly, he surely was leaning toward this position, considering the fact that although God's infinity was so very important to him in the Manichean phase, associating infinity with God is conspicuously absent in his later writings, except for isolated, marginal mentions. It is either that it faded away from his theological consciousness or he did not see infinity as part of the divine nature, and thus infinity was not even worth mentioning.[12] Only the Greek Church Fathers saw God's nature as infinite and infinity as something positive, which also became much later an accepted position among the Latin Fathers, most prominently, by Bonaventura and Aquinas.

NOTES

1. Augustine knew only two of Plato's dialogues and only in Latin translation, and three dialogues only from excerpts from Cicero (Combès 1927, 10). Thus, it appears that he did not know Plato's subtle discussion of infinity given in the *Philebus*.

2. An interesting parallel can be found in the twentieth century when Henri Poincaré asked himself how he would know that the world did not become overnight 1,000 times larger (Poincaré 1953, 99).

3. More poetically, "the forest of numbers grows infinitely" (*Sermo* 270.3).

4. In a later work, Augustine moderated this statement by saying that it is useful for a Christian to study some sciences found among the pagans, which include "the science of disputation and of number," but with moderation (*De doctrina Christiana* 2.39.59).

5. Thus, "the soul discovers in its imagination and in its memory the infinity which it projects onto the outside world" (Hadot 1990, 65).

6. A similar sentiment can be detected in the statement that "the perfection of creation . . . by the number six and by measure, number, and weight does not mean that God is identical with them as they are understood within the creation, but rather that God is the source of these perfections in himself, and that he is above them as they are manifest in his creation" (Dunham 2008, 94).

7. This seems to be the case when Augustine rather colorfully said that believers were taught that the nature of God is "not corporeal, not enclosed in any place, not extended across an infinite space like some mass, but entire everywhere and perfect and infinite, without the gleaming of colors, without outlines of figures, without markings of letters, without series of syllables" (*In Ioannis Evangelium* 96.4; cf. *De Trinitate* 5.1.2). No important attributes are mentioned here, such as eternity, omnipotence, or love. Most of them are about incorporeality of God and so infinity appears to be also.

8. In the *Confessions*, Augustine asks the question as to "whether these three [to be, to know, and to will] are there in the Trinity, or whether the three are in each so that the three belong to each, or whether, in some amazing way, both in simplicity and in multiplicity, [God] is himself infinite but finite/bounded to himself [*infinito in se sibi fine*]." And he answers only with another question, "Who would dare to answer it?" (13.11.12).

9. It has been suggested that, according to Augustine, God unites the finite and the infinite into "a superior unity," which would include both transcendental finitude and transcendental infinity (Hadot 1990, 72). This superior unity can be considered something that is beyond the infinite and the finite.

10. As a Manichean, Augustine apparently considered immensity and being synonymous with infinity (cf. *Conf.* 7.5.7). Also, when in *De Trinitate* he spoke about the immensity of God's "republic of creations" (3.4.9), he meant infinity. However, here, in *Contra Faustum*, when contrasting finite with immense instead of with infinite, Augustine may have hinted at immensity to be something else than infinity. Possibly in the same direction points a remark that "God is of infinite and immense light" (*Sermo* 4.4.5), not just infinite but even beyond infinity. Interestingly, for Plotinus, the One was of "immense nature" (ἄπλετος φύσις) (5.5.6). In fact, Augustine did talk about God as being immense (*Contra Maximum haereticum* 2.23.7, 2.26.13; *In Ioannis Evangelium* 63.1). Later, in the fourteenth century, Jean de Ripa explicitly contrasted immensity and infinity by saying that "immensity of the real presence of God exceeds immensely infinity of all the possible void" (Combes, Ruello, and

Vignaux 1967, 200, 235). He was not an isolated case: in the Middle Ages, the possibility of the essence of God being neither finite nor infinite was seriously investigated (Côté 2002, 142, 143, 152, 154, 157, 199, 215).

11. Infinity is not considered to be an attribute of God, since Augustine "was not yet certain as to what divine infinity might mean in the light of his discovery, set forth in *Confessions*, book 7, of God as completely incorporeal" (Sweeney 1995, 201).

12. Étienne Gilson did not want to "risk adventurous conclusions" (1954, 574) concerning the issue, but the tone of his article, including its title, quite clearly indicates that in his view Augustine considered God to be infinite.

REFERENCES

Combes, André, Francis Ruello, and Paul Vignaux. 1967. "Jean de Ripa I sent. dist. XXXVII: de modo inexistendi divine essentie in omnibus creaturis." *Traditio* 23:191–267.

Combès, Gustave. 1927. *Saint Augustin et la culture classique*. Paris: Plon.

Côté, Antoine. 2002. *L'infinité divine dans la théologie médiévale (1220–1255)*. Paris: Vrin.

Drozdek, Adam. 2008. *In the Beginning Was the* Apeiron*: Infinity in Greek Philosophy*. Stuttgart: Steiner.

Dunham, Scott A. 2008. *Trinity and Creation in Augustine: An Ecological Analysis*. Albany: State University of New York Press.

Gilson, Étienne. 1954. "L'infinité divine chez saint Augustin." In *Études augustiniennes*, 1:569–74. Paris: Vrin.

Hadot, Pierre. 1990. "La notion d'infini chez saint Augustin." *Philosophie* 26:58–72.

Poincaré, Henri. 1953. *Science and Method.* Mineola, NY: Dover.

Sweeney, Leo. 1995. "Divine Attributes in *De doctrina Christiana*: Why Does Augustine Not List 'Infinity'?" In *De doctrina Christiana: A Classic of Western Culture*, edited by D. W. H. Arnold and P. Bright, 195–204. Notre Dame, IN: University of Notre Dame Press.

CHAPTER 4

Aquinas on Creation and the Analogy of Infinity

WILLIAM E. CARROLL

There is for Thomas Aquinas a close connection between his understanding of creation and what it means to speak of the infinity of God. Even though his discussions of divine infinity usually occur well before he addresses in detail the doctrine of creation, an understanding of Thomas's analysis of what it means for God to be Creator allows us to see more clearly how he argues that God is infinite. I am reminded of Josef Pieper's claim that "the notion of creation determines and characterizes nearly all the basic concepts of Thomas's philosophy of Being" (Pieper 1999, 48). For Thomas, to create is to cause existence, and the Creator is *ipsum esse subsistens*: in which "to exist" is what it means to be the God who creates. God's existence, so understood, does not involve the participation in or reception of being. And it is the reception of being (*esse*) that identifies creatures as creatures. Unreceived being, which characterizes the Creator, is truly unlimited, and properly called "infinite," so long as one recognizes that the term "infinite" is being used in a special analogical sense. Indeed, Thomas says that God is absolutely infinite and fully determined.

Thomas in his first magisterial discussion of creation, in his *Writings on the "Sentences" of Peter Lombard* (*Scriptum super IV libros Sententiarum*; hereafter cited as *In Sent.*), raised the question of whether the power to create could be communicated to creatures (*In II Sent.* d. 1, q. 1, a. 3). Although Thomas, relying on the authority of faith, denied that God had communicated the power to create to any creature, he did admit that, from a philosophical point of view, it was possible. As Thomas reflected more on this topic, he changed his mind. We see in the position he finally adopts, namely, that it is impossible for God to communicate the power to create to creatures, the close connection between a proper understanding of creation and divine infinity, starting from a discussion of the need for an infinite power for creation to occur.[1] No creature, inasmuch as being a creature, can possess such infinite power.

Illustrative of this connection is what Thomas writes in the *Summa theologiae* (*ST*, I, q. 45, a. 5) when he takes up the topic: Whether it belongs to God alone to create. Here he notes the objection that, if we agree that the power of the maker is viewed according to the measure of what is made, and then recognize that every created being is finite, it would seem to follow that only a finite power is needed to create. Thus, according to this objection, it would not be impossible for a creature to create (obj. 3). But Thomas observes that to produce something from nothing whatsoever requires an infinite power, and since "no creature has simply an infinite power, any more than it has an infinite being . . . it follows that no creature can create."

In speaking of the power to create as being an infinite power, Thomas is well aware of the various senses of infinity that need to be distinguished from one another. For example, as a participant in the wide-ranging debates in the thirteenth century concerning creation and the eternity of the world, Thomas thought that reason alone could neither affirm nor deny the world's being eternal or its having a beginning of its duration. With respect to those arguments affirming a temporal beginning based on the impossibility of an actual infinity of the past, Thomas, following Aristotle, notes the distinction between infinity in terms of succession, which they both allow, and an actual infinity, *totam simul*, which they both deny. In his famous arguments for an unmoved mover and uncaused cause, Thomas employs as a key element the impossibility of an infinite regress of essentially ordered causes. Thomas also accepts Aristotle's famous distinction from the *Physics*, between actual and potential infinity: the latter referring to infinity by addition or to infinity by division of a magnitude.

When Thomas argued for the philosophical intelligibility of a universe, created and eternal, he was, of course, careful to distinguish between predicating eternity of the world and any identification of that sense of eternal with what it meant to speak of God's being eternal. He will make similar distinctions between different ways in which infinity can be predicated with respect to creatures and what it means to speak of the Creator as infinite. Throughout his discussion of creation, Thomas always maintains what he said in his *Writings on the "Sentences" of Peter Lombard*: "Not only does faith affirm that there is creation, reason also demonstrates it" (*In II Sent.* d. 1, q. 1, a. 2). That there is a philosophical sense of creation, accessible to human reason, corresponds to his claim that divine infinity is also knowable by reason alone. In fact, the fundamental philosophical arguments for the reality of a Creator are part of the arguments for God's being infinite.

What comes to be an increased emphasis on divine infinity in discussions in the medieval Latin West occurs in the context of the availability in the twelfth century of translations of texts from several Greek Church Fathers, which contained passages emphasizing the unknowability of God based, in part, on the incommensurable distance between Creator and creatures: the radical distinction between the uncreated and the created.[2] In particular, works of Pseudo-Dionysius, Gregory Nazianzus, and John Damascene were influential in this regard, and they played an important role in the development of negative or apophatic modes of discourse in theology.

From the patristic era onward, Christians were concerned with how to understand the diverse names attributed to God in both the Hebrew scriptures and the Christian New Testament. Pseudo-Dionysius in his *Divine Names* asks how it is that we can know God "when He cannot be perceived by the mind or the senses and is not a particular being" (7.3). He suggests that we do not know God according to His nature, for this is "unknowable and beyond the reach of all reason and intuition." Yet "by means of that ordering of all things which (being as it were projected out of Him) [we] possess certain images and semblances of His Divine Exemplars, [and thus we are able to] mount upwards (so far as our feet can tread that ordered path), advancing through the Negation and Transcendence of all things and through a conception of an Universal Cause, towards That Which is beyond all things" (Rolt 1977, 110).[3]

John Damascene, who was frequently cited by Scholastics in the thirteenth century, wrote that finally what we might say of God "is by no means indicative of His essence—no more than is the fact of His being unbegotten, without beginning, immutable, and incorruptible, or any of those other things which are affirmed of God or about Him. These do not show what He is, but, rather, what He is not."[4] And, "The Divinity, then, is limitless and incomprehensible, and this, His limitlessness and incomprehensibility, is all that can be understood about Him. . . . [As] some limitless and boundless sea of substance. He contains all being in Himself."[5]

What should be noted here is the linking in Damascene of "limitless" and "incomprehensible." For some in the Latin West, such a linkage meant that the beatific vision could not mean a vision of the divine essence. Since Damascene observed that the divine essence was "uncircumscribed," it would be "invisible" to the intellect of creatures. Some theologians in the twelfth century were attracted to Eriugena's use of Pseudo-Dionysius in arguing that even in heaven the blessed only see God indirectly, by way of "theophanies" (Tugwell 1988, 50). Even if we were to reject what many considered an excessive negative theology, it still left unresolved how it is that a created intellect can come to see the Creator. What sense is one to make of various biblical passages that promise that the blessed will see God face-to-face? How can the creature come to see the Creator?

As the specific debate about the beatific vision and God's unknowability entered the thirteenth century, what seemed crucial was the clear affirmation that indeed the blessed in heaven see God and that too excessive a negative theology about the unknowability of God threatened the possibility of the beatific vision. In January 1241, the masters of theology at Paris, under the authority of William of Auvergne, bishop of Paris, condemned the proposition that "the Divine essence in itself will not be seen by any man or angel."[6] Antoine Côté, who has written extensively on the emergence of the doctrine of divine infinity in the thirteenth century, argues that three principal factors help to explain this emergence: (1) the controversy about the beatific vision, given the great gulf between Creator and creature; (2) the availability of the texts of authors like Pseudo-Dionysius and John Damascene; and (3) the availability of translations of the works of Aristotle, especially the *Physics* (Côté 2002, 217).

Theologians in the thirteenth century worked within the 1241 Parisian affirmation about the beatific vision and increasingly came to identify the distinction between creature and Creator with that between the finite and the infinite. Here they are using the distinction set forth in Damascene, Chrysostom, and Pseudo-Dionysius. Commentary about the beatific vision is occurring at the same time that the works of Aristotle, translated into Latin, have begun to have increasing influence on philosophical and theological discourse. Important here is Aristotle's oft-repeated point that there is no proportion between the finite and the infinite, and hence no commensurability between the two. Aristotle's discussion of infinity will be important for Aquinas as he comes to understand what it means for God to be infinite. Although Thomas develops his understanding of divine infinity with no specific reference to the discussion of the beatific vision, he will use his analysis when he does offer an account of what it means for the blessed in heaven to see God face-to-face.[7]

For some scholars in the thirteenth century, one consequence of affirming that the blessed in heaven really do see God was to deny divine infinity. Guerric of Saint-Quentin, a Dominican who taught at Paris from 1233 to 1242, argued that the infinity that authors such as John Damascene attributed to God concerned only God's power, not His essence (which is not infinite). The divine essence is simple and finite (Sweeney 1992, 417–19). Others, like Richard Fishacre, who taught at Paris circa 1236–48, address divine infinity when speaking of the beatific vision and move from arguments about divine power to divine infinity. Fishacre connects infinity with simplicity: "A substance which is completely separated from all else, a being who is simple in himself and who enters into composition with nothing outside, he is virtually infinite in his infinite separation from impediments and matter" (Sweeney 1992, 422). The predominant feature of this argument was to proceed from God's infinite power, manifested, for example, in creation *ex nihilo*, which *then* implied that God is infinite in his essence, since for God, who is absolutely simple, being and power are one.

Before turning to Thomas, I should mention the thesis of Anne Ashley Davenport, who thinks that the stimulus in the thirteenth century for discussions of divine infinity was not so much the debate about the beatific vision or the availability of Aristotle's *Physics*, but rather it was the threat of two heresies, pantheism and Manicheanism. Both heresies "attacked at its root the Catholic doctrine of divine immensity" (Davenport 1999, 5). There were some, like Amaury of Bène (ca. 1200), who thought that if God is immense, He is everywhere, and since everywhere, God is in every place—in every stone, in every piece of bread, just as much as He is in the consecrated host (5). The Cathars (medieval Manichees), as Davenport describes them, were "unable and unwilling to reconcile the suffering of this life with a just and benevolent God, [and thus] they split the deity in two and postulated two eternal and opposite principles: a benign God 'of light' from whom the spiritual realm radiates eternally, and an evil principle 'of darkness' responsible for material creation and time" (7). Already in 1215, the Fourth Lateran Council, in part in response to medieval Manicheanism, offered the first doctrinal definition of creation *ex nihilo*. Indeed, it was the challenge of Manicheanism that Thomas thought was a major threat to a correct understanding of creation. His treatment of creation in his *Writings on the "Sentences" of Peter Lombard* begins with the question whether there is only one first principle.

Discussions of the beatific vision, the biblical references to God's greatness and immensity,[8] the role of Aristotelian philosophy for Christian theology, the ways in which Greek Church Fathers spoke of infinity with respect to the Godhead, and the dogmatic definition of creation were all features of the intellectual landscape in which Thomas worked. The specific question of divine infinity is part of the much larger set of questions concerning predications of God and, also, of the even broader topic of what if anything one can know about God.

Although at the beginning I commented on Thomas's reference to the power to create as being an infinite power, I think it is important to recognize that Thomas argues for divine infinity directly and not based on manifestations of God's power.[9] Davenport, commenting on this, observes that Thomas, having first established "by philosophical means that God's very substance is unrestricted . . . concludes further that God's 'participatory' attributes—power, goodness, and wisdom—are unrestricted as well" (Davenport 1999, 52). However, she distinguishes Thomas's argument for divine infinity—which she identifies as "a naturally demonstrable truth"—from an argument based on creation, which she says "rests on Revelation" (53). But Thomas, as I have already indicated, thinks that there is a philosophical sense of creation, the arguments for which are those that do indeed underpin his understanding of divine infinity.

References to divine infinity can be found throughout Thomas's works, but there are three major texts in which Thomas addresses *directly* the topic of divine infinity that I wish to consider: (1) *In I Sent.* d. 43, q. 1; (2) *Summa contra Gentiles* I, c. 43; and (3) *Summa theologiae* I, q. 7, a. 1. The presentations in all these texts rest on the affirmation that God is *ipsum esse subsistens* (Balas 1981, 94–95). It is this view of God as "subsistent being" that is the hallmark of Thomas's understanding of God as Creator. It is already evident in an early work, *De ente et essentia* (c. 4), in which one can note the strong influence of Avicennian metaphysics. This connection between the metaphysical claim that God is *ipsum esse subsistens* and His being infinite is also clear in the *Writings on the "Sentences"* where the explicit question is whether God's power is infinite (*utrum potentia Dei sit infinita*; *In I Sent.* d. 43, q. 1, a. 1, sol.). From the fact the divine *esse* is *absolutum et nullo receptum in aliquot*, and, hence, *infinitum simpliciter*, it follows that the divine power is also infinite.[10] In his analysis, Thomas notes that the "status of power as infinite or finite is decided by whether the essence from which that power issues is infinite or finite, and the infiniteness or finiteness of essence is understood explicitly in terms of the absence or presence in it of matter and potency."[11] Such a recognition that in creatures essence and existence are distinct and that in God they are identical is at the heart of Thomas's understanding of who the Creator is and what it means to create.

Everything that is finite *in its very nature* is finite by way of that which determines and, as it were, restricts its essence (Sweeney 1992, 435). The "determination," to which Thomas refers, is what distinguishes the finite and the infinite; furthermore "determination" here has two senses: "form (or, more generally, act) determines matter (or potency) by perfecting it, whereas matter (or potency) determines form (or act) by limiting it" (434).

Thomas recognizes the need to distinguish among different senses of "infinity." His discussion of infinity begins with his recognition that infinity, by its nature (*ratio*), is proper to the category of quantity. It is here, with respect to quantity, that we first employ the notion of *finis*, of bounds and limits—of being bounded (*finitum*) or of being unbounded (*infinitum*). Initially, thus, infinity so understood suggests a kind of imperfection (LaMountain 1956, 327). Early on, Thomas notes the analogical character of the predicate "infinite" even when speaking of quantity: "The same kind of infinity cannot be predicated of an infinite body and of an infinite series" (327). Additional analogical analysis will be necessary to distinguish between infinity referring to entitative perfection or imperfection—and then again when referring to divine infinity: to that infinity properly said of creatures and that properly said of the Creator.

At times in his *Writings on the "Sentences,"* Thomas distinguishes between "extensive infinity," which pertains to number and quantity, and "intensive infinity," which pertains to quality and perfection. Later, in *De veritate* (q. 2, a. 9), he distinguishes between the ordinary use of the term "quantity" to refer to extension and a metaphysical use of it to refer to degree of perfection.[12] Also, he argues that "nothing prevents something from being infinite in one way and finite in another way." Using the heuristic example of an infinite white body, Thomas

> says that the whiteness of the white body might be called "infinite" because of the infinite extension of the body, "but its quantity *per se*, that is, its intensive quantity would nonetheless be finite, and this would likewise be the case for any form of an infinite body, for every form received in any matter is limited according to the mode of existence of the receiver, and so does not have infinite intension" (*De veritate* q. 2, a. 9). In short, the [hypothetical] infinite white body would be infinitely extended, but not infinitely white. (Tomarchio 2002, 167)

Being received into something else, the way, for example, whiteness "is received" in a body, is a limiting of what is received.

I want to provide a brief summary of what Thomas says about divine infinity in the *Summa contra Gentiles* (*S.c.G.*) before moving to a text in the *Summa theologiae* to which most people refer. In book 1, chapter 43, of the *Summa contra Gentiles*, Thomas notes that the sense of infinite applied to number or extension cannot be applied to God, but we can speak of infinite in the sense of "spiritual magnitude," that is, with reference to "power and the goodness or completeness [*completio*] of his nature." He says that "spiritual beings are called great according to the mode of their completion." It is this sort of magnitude—spiritual magnitude—that one uses in beginning to refer to God's infinity. To speak of

God's infinite power *follows from* God's nature. This is a movement—from divine essence to divine power—that we have already seen in the earlier text on the *Writings on the "Sentences" of Peter Lombard.* Thomas also notes that "infinite" in discussions about God must be taken in a negative sense rather than a privative sense. "Infinite," when applied to God, does not designate some imperfection, some incompleteness (the privative sense), but rather it is to be taken in a negative way: "There is no terminus or limit to His perfection. He is supremely perfect" (*S.c.G.* I, c. 43, n. 3). For Thomas, the most fundamental reason connecting perfection with infinity in God follows from his recognition, first, that any form (or what he terms "act") is limited insofar as it is received in matter, and, second, "God is act in no way existing in another, for neither is He a form in matter . . . nor does His *esse* inhere in some form or nature, since He is His own *esse*. . . . It remains, then, that God is infinite" (n. 5).

Whereas the *Summa contra Gentiles*, as is often the case, sets forth argument after argument, the treatment of divine infinity in the *Summa theologiae* is characterized by brevity. Following his claims to have demonstrated the existence of God in question 2 of the Prima Pars, Thomas, in the prologue to question 3, announces his intention to proceed by way of negation, since "we cannot know what God is, but rather what He is not." Rudi te Velde offers a good summary of the process Thomas describes:

> The *via negationis* results in a series of four negative attributes which point out, each in its own way, what the cause of all things, separated from all those things, is not: un-composed, in-finite, immutable, un-divided (one). These negative attributes point to the divine being as separated from the composed, finite, multiple and multiplied reality which is the proper domain of human knowledge. They determine God as existing outside the categorical structures of the object of human knowledge. . . . God does not fall under a genus; his essence cannot, therefore, be logically identified by means of genus and species. The only way of identifying God is negatively, by distinguishing him from all genera of things. (Te Velde 2006, 80)

After the discussion of divine simplicity in question 3, Thomas turns in question 4 to divine perfection lest we might be caught too much in a view of God as only a kind of negative transcendence. Perfection "accounts for the concreteness of God as a substantial and fully determined reality, which comprehends in itself the whole positive substance (or perfection) of created reality in the simple unity of being" (Te Velde 2006, 81). This discussion of perfection is linked to an understanding of God as Creator, as cause of the many and diverse perfections in the created order. Te Velde observes that "God is said to be perfect in the sense that the perfection of creatures, diversified over the many genera of things, are originally and unifiedly present in God as identical with his simple being. . . . God is not the being inherent in all things, considered abstractly, but the *ipsum esse* of the cause of all beings" (81). God is *ipsum esse subsistens*, "not abstract being, but being that is fully determinant in itself and subsistent, and from which all other beings derive their being" (81).

The emphasis upon God's simplicity and perfection, set forth in questions 3 and 4 of the Prima Pars, is the context for the rest of the discussion of divine attributes and names. "God's mode of being must be understood—according to our indirect and negative understanding of it—first as utterly simple, without any composition," including the most fundamental kind of composition, that of essence and existence. "In God there cannot be a distinction between essence and *esse* . . . but [rather] in the most determinate sense of subsisting and complete reality. God is, therefore, *ipsum esse subsistens*" (Te Velde 2006, 81). This is the most determinant sense of subsisting as a complete and absolutely perfect reality. Before discussing divine infinity in question 7, Thomas writes two questions on divine goodness: first on goodness itself and second on divine goodness.[13] This procedure follows from the recognition that a thing, inasmuch as it is perfect, is called good.

Question 7 begins with Thomas noting that a discussion of divine infinity is linked to "God's existence in all things; for God is everywhere, and in all things, inasmuch as He is boundless and infinite [*inquantum est incircumscriptibilis et infinitus*]"—Article 1: Whether God is infinite. Article 2: Can anything other than God be infinite through its essence? Article 3: Can anything be actually infinite in magnitude? Article 4: Can there be an actual infinite multitude of things?

In the first article, Thomas points to three objections to God's being infinite. The first concerns a standard objection already part of the Greek heritage: to be infinite is to be imperfect. The second objection comes from Aristotle's *Physics,* according to which finite and infinite are in the category of quantity, and since there is no quantity in God—it has already been shown that God is not a body—"it does not belong to Him to be infinite." The third objection observes that to be *here* rather than *there* and thus to be *this* rather than *that* is to be finite according to place and according to substance (*est finitum secundum substantiam*). God is surely different from any creature: "God is this, and not another." Thus God must be finite rather than infinite, or as Thomas puts the objection: "God is not infinite with respect to substance [*non est infinitus secundum substantiam*]." In the *sed contra*, Thomas quotes John Damascene, from *De fide orthodoxa* (1.4): "God is infinite and eternal, and uncircumscribable." Thomas goes on to write that "a thing is called infinite because it is not finite [*non est finitum*]." Although, Thomas says, all ancient thinkers recognized that the first principle from which all things flow is infinite because they thought that "things flow forth infinitely from the first principle [*considerantes res effluere a primo principio in infinitum*]," many did not understand what this first principle really was, and as a result they also erred concerning what it means for the first principle to be infinite.[14]

As Thomas continues his analysis, he notes that there are different senses by which we should understand that the infinite means not being limited (*non est finitum*). One can apply the notion of infinite both to matter and to form, which are for Thomas the coprinciples of all physical reality: "But matter is in a certain way limited by form, and form is in a certain way limited by matter [*Finitur autem quodammodo et materia per formam, et forma per materiam*]." Matter, by which Thomas means a potentiality for existing as something, is "perfected by form," that is, is made to be a concrete something: "Now matter is perfected by the form through which it is limited, and so infinity, as attributed to matter, has the nature of an imperfection; for it is, as it were, matter without form" (*ST* I, q. 7, a. 1).

Thomas then considers the case of "form," that by which something is actually the thing that it is: "By contrast, a form is not perfected by matter; instead, its scope is contracted by the matter. Hence, infinity, as attributed to a form not determined by matter, has the nature of perfection" (*ST* I, q. 7, a. 1). We see Thomas describing different senses of what can be called the "relative infinite"—two different senses of not being determined: the radical potentiality of matter and the undetermined yet perfect character of any form considered generically, separate from concrete, individual things. Here we have the relative infinity of imperfection and the relative infinity of perfection (Garrigou-Lagrange, n.d., 5).

In his *Writing on the "Sentences,"* Thomas uses the example of whiteness when discussing the relative infinity of form. Whiteness is a quality or form of a substance, but when we consider whiteness as a *principle* of determination of white things, in this sense, as distinct from a form existing in a particular concrete thing, the form of whiteness can be called infinite since, as a principle, it remains "undetermined in its nature to this or that particular white thing."[15] The relative infinity of such forms is an "intensive" infinity, not an "extensive" one—the latter refers to quantity.[16]

Moving from the relative infinity of created forms, Thomas writes:

> But, as is clear from what was said above [*ut ex superioribus patet*] (q. 4, a. 1), that which is the most formal of all things [*maxime formale omnium*] is *esse* itself. Therefore, since God's *esse* [*esse divinum*] is not received in anything [*non est esse receptum in aliquot*], and since, as was shown above [*et supra ostensum est*] (q. 3, a. 4), He is His own subsistent *esse* [*ipse sit suum esse subsistens*], it is clear that God Himself is infinite and perfect [*manifestum est quod ipse Deus sit infinitus et perfectus*]. (*ST* I, q. 7, a. 1)

As John Tomarchio, who offers one of the more sophisticated analyses of Thomas's discussion of divine infinity, puts it: "Having argued for the intensive infinity of a material form's quidditative perfection considered absolutely, Aquinas analogically extends this notion of intensive infinity to the perfection of existence considered absolutely, calling existence 'the most formal' of all perfections. He then goes further and analogically extends this notion of the infinity of existence considered absolutely to the actual existence of God" (Tomarchio 2002, 171).[17]

Tomarchio offers a good summary of the analogical analyses that inform Thomas's account:

> Aquinas moves analogically from the infinity of the principle of finite beings to the infinity of their ultimate cause. Having by degrees constructed this analogical concept of infinity and arrived at length

> by means of it to a meta-analogate, the unqualified infinity of divine subsisting existence, he grounds the complete meta-analogy on a relationship of causal similitude. If every effect in some way resembles its cause, and if the transcendent cause of all finite being is infinite, it should be no surprise that, in the image of their cause, all finite things are in a certain way infinite. (Tomarchio 2002, 186–87)

Note in the section from article 7, which I have been quoting, the simple phrase in Thomas's text, repeated twice, "as was shown above," or "as was said above." The "above" refers to questions 3 and 4 in which Thomas argued for the identity of existence and essence in God—a claim central to Thomistic philosophy and theology, a claim at the heart of Thomas's metaphysical understanding of the fundamental difference between Creator and creatures. It is, as Thomas says, on the basis of this understanding and the distinctions he makes concerning various senses of infinity, that he is able to conclude that God is infinite. Note also that Thomas concludes that God is infinite *and* perfect. Divine infinity is linked to perfection and perfection seen in an understanding of what it means to be.

For Thomas, God is perfect in the sense that He is beyond every creaturely mode of excellence.[18] This excellence or perfection concerns the kind or mode of being that a thing is. Citing a phrase from Pseudo-Dionysius, Thomas notes that "God does not exist in a certain way; He possesses, and this before all others, all being within Himself absolutely and limitlessly [*Deus non quodam modo est existens, sed simpliciter et incircumscriptive totum esse in seipso accepit et praeaccepit*]" (*S.c.G.* I, c. 28). God's perfection is not the aggregate of all creaturely perfections. God is not perfect with respect to some mode of perfection. He is "modelessly perfect" (Rocca 2004, 257).

Thomas's analysis includes making clear the difference between applying notions of infinity to the world of creatures, including incorporeal creatures, and what one means when speaking of divine infinity. In article 2 of question 7 of the *Summa*, he tells us that "things other than God can be relatively infinite, but not absolutely infinite [*quod aliquid praeter Deum potest esse infinitum secundum quid, sed non simplicter*]." Aware of the objection that angels do not have a material principle by which the form of an angel would be limited and hence would suggest that angelic form is infinite, perhaps absolutely considered, Thomas says that a created form subsisting in the way an angel's form subsists, separate from any matter, still is only relatively infinite.[19] Being created means that the creature, including an angel, "is not its own being"; "it follows that its being is received and contracted to a determinate nature. Hence it cannot be absolutely infinite." Only God is essentially, or absolutely, infinite. Furthermore, not even God, who possesses infinite power, can make anything to be absolutely infinite (*infinitum simpliciter*) (*ST* I, q. 7, a. 2, ad. 1). No creature, given what it means to be a creature, which in a fundamental sense means that its being (*esse*) "is received and contracted to a determinate nature"—no creature can be absolutely infinite—only the Creator. As Thomas writes:

> Therefore, with respect to the formality of existing [*ratio essendi*], there can be nothing infinite except for that thing in which is embraced all the perfection of existing, which can be varied in diverse things in infinite ways. And in this way God alone is infinite according to essence, for his existence is not limited to any determinate perfection, but embraces within itself every mode of perfection to which the formality of being [*ratio entitatis*] can extend. However this infinity can belong to no creature, for the existence of any creature is limited to the perfection of its proper species. (*De veritate* q. 29, a. 3)

As is evident in objection 3 in article 1 of question 7 from the *Summa theologiae*, one problem with the idea of an infinite God is how to understand the divine being as a concrete reality distinct from others, and yet at the same time preserve divine transcendence: God is beyond the very categories of created being. How can the infinite God be distinct from the created order of beings without at the same time becoming limited and finite as a distinct entity/reality?[20] God, precisely as the infinite act of being, not received in anything else, is by this fact individualized and distinct from those entities whose individuations occur with the reception of being, when, for example, a form is received by matter (Rocca 2004, 262). Infinity in creatures is linked to the distinction between being determined and being indeterminate.

God, however, is both infinite and determined. Because God is subsistent being itself, no addition, as it were, would be necessary to determine such being in one way or another. God's determination as subsistent being transcends the various kinds of determinations characteristic of created beings.[21] As Thomas observes in his *Commentary on the Book of Causes*, "The first cause is above being [*ens*] insofar as it is the infinite act of being itself [*ipsum esse infinitum*]" (*Super de causis* 6). "A creature's being, then, is made finite and limited by participating in the infinite act of being, which God *is* absolutely" (Rocca 2004, 260).

God is a something—*aliquid*—but by the absoluteness of his own *esse*: "The divine being is determined in itself and separated from all other things by the fact that no addition can be made to it."[22] The determination of the divine nature does not imply any limitation whatsoever. God falls outside the category of creaturely determinations. He is not one more being among beings; he is not the highest in the category of being. Thomas makes a similar point when distinguishing the Creator from all creatures. In speaking of Creator and creatures, Thomas habitually uses terms like *divisus* or *diversus*, but not *differens*. The last term would suggest only a difference, perhaps in terms of some wider category of likeness.[23] God is not defined or determined, as are all created realities. God is not determinant by way of limitation, but by way of distinction. For Thomas, God is a distinct reality precisely *through*, not despite, the divine infinity, which distinguishes and individualizes the divine nature without limiting or finitizing it (Rocca 2004, 262).

One final text to which I wish to refer comes from Thomas's *Commentary on the "Divine Names" of Pseudo-Dionysius* (*In De div. nom.*). The passage concerns the infinity of created being—and also an affirmation that this infinity is infinitely surpassed by divine being:

> Each thing, to the extent that it is finite and bounded [*terminatum*], to that extent has actual unity. But the one which is God is prior to all end [*finem*] and boundary and [prior to] their opposites, and is the cause of the boundedness [*terminationis*] of all things, and not only of existent things [*existentia*], but even of being [*esse*] itself. For created being [*ipsum esse creatum*] is not finite if it is compared to creatures, because it extends to all; if, nevertheless, it be compared to uncreated being [*esse increatum*], it is found to be deficient [i.e., to fall short], and [to be something] having the boundedness [*determinationem*] of its own intelligible character [stemming] from the forethought [*praecogitatione*] of the divine mind. (*In De div. nom.* 13, lect. 3.)

"Created being," taken formally, and thus in the infinite sense, corresponds to the Latin *ens commune*, or being as a universal. Still, that infinity is a mere product as compared to the "reality beyond reality" who is the Creator, the infinite beyond the infinity of mere created being.[24]

At the heart of Thomas's argument for diving infinity is the premise that "unreceived *esse*/being" is unlimited (*esse non est terminatum*). Thomas observes "that because things other than God have an *esse* that is received and participated, they do not possess it according to the total power of being" (*In De div. nom.* 5, lect. 1, n. 629). The view that act as such or, as in this case, that *esse* as such is not self-limiting, appears frequently in Thomas's writings. We have seen Thomas using it in his argument for divine infinity. He applies this analysis to "form" and then to "*esse*" (*In I Sent.* d. 43, q. 1, a. 1). As John Wippel points out, there is no demonstration of this principle in Thomas; it is likely that Thomas takes it to be self-evidently true. The "acceptance of this axiom [unreceived being is unlimited; or unreceived act is unlimited] presupposes a certain way of understanding *esse*, that is, as the actuality of all acts and the perfection of all perfections. . . . If this is one's understanding of *esse*, and it surely is Aquinas's, it will only be reasonable for him to conclude that *esse* is [itself absolutely unlimited]" (Wippel 2000, 172–73). For Thomas, actuality and perfection go together (*De potentia Dei* q. 7, a. 2, ad. 9).

Since Thomas's discussion of divine infinity occurs in the context of claims about what God is not, I should like to conclude by offering some reflections about this negative approach to God. God is not a purely negative transcendence. As Te Velde reminds us, we distinguish God as the cause of all things, as the Creator; thus, we must "correct," in a sense, the negative ascending movement of our thought toward God with the real descending movement from cause to effect. Recent scholars of Thomas, such as David Burrell and Denys Turner—and earlier A.-D. Sertillanges—have emphasized the apophaticism of Thomas's approach to God (White 2009, 258–59). But as Thierry-Dominique Humbrecht points out, we should speak of a "negative way" in Thomas's theological approach to God rather than referring to "negative theology." There is a positive aim of negation, expressed by returning the whole positive substance of the effect to the cause in a more excellent way: "God is not a negative transcendence but an excessive transcendence, which means that He is distinguished from all things by being all things in an excessive (unified, concentrated) way" (Te Velde 2006, 80).

Although Thomas does use terminology from Pseudo-Dionysius in discussing divine names, he interprets or recasts Pseudo-Dionysius in terms drawn from his own Aristotelian approach to knowledge (Rocca 2004, 304–5). When Thomas distinguishes his position on divine names from that of Maimonides, he writes:

> The idea of negation is always based upon an affirmation as evinced by the fact that every negative proposition is proved by an affirmative: wherefore unless the human mind knew something positively about God, it would be unable to deny anything about him. And it would know nothing if nothing that it affirmed about God were positively verified about him. Hence following Dionysius (*Divine Names*, c. 12), we must hold that these terms signify the divine essence, albeit defectively and imperfectly.[25] (*De potentia Dei* q. 7, a. 5)

Thomas Joseph White warns against an excessive apophaticism in discussing divine attributes such as infinity: "The mental negations by which one removes from one's ascriptions to God any significations denoting imperfections are themselves only a moment within the reflection on God as the primary cause of the world, and as preeminently perfect in an incomprehensible way. Such negations allow one to approach some understanding of God, who cannot be understood to exist in any kind of continuity with creatures as if he were one in a series of finite, participated beings" (White 2009, 272).

Here we see the connection between what it means for God to be Creator and to be infinite. As Thomas says: "Our intellect, since it knows God from creatures, in order to understand God, forms conceptions proportional to the perfection flowing from God to creatures, which perfections pre-exist in God unitedly and simply, whereas in creatures they are received, divided, and multiplied" (*ST* I, q. 13, a. 4). Still, God transcends all modes of signification derived from creatures.

Again, as White points out:

> Even negative reflection on the pure actuality of God not only presupposes awareness of divine causality, and therefore the positive resemblance of created effect to uncreated cause, but also *negates the limitations* that are proper to the modes of being that are charac-

> teristic of creatures. By negating these non-perfections (effectively a kind of negation of a negation), the mind makes more precise the *affirmation* of the super-eminent characteristics of the one God, and thus qualifies apophatically the epistemological mode of our knowledge of God. Such thinking, however, is meant to carry us forward toward a term that is utterly positive, yet transcendent of participated, finite being. Correspondingly, Aquinas makes perfectly clear that despite its imperfect, indirect character (lacking all *quidditative* knowledge of the divine essence), our naming of God signifies truly *what God is substantially in himself.* (White 2009, 273)

Terms such as "simplicity," "eternity," "infinity," "immutability," and "impassibility," according to White, "are in fact employed to qualify our truly positive designations that pertain to God as primary cause." This is the case because, although the various divine names "are formulated largely by way of negation . . . nevertheless, the *premise* of such negations is the *affirmation* of the pure actuality of God. . . . To remove composition, finitude, alteration, or suffering from God is to say something of his positive preeminence in its transcendence" (White 2009, 273; see Humbrecht 2005, 778). To speak of God as "simple" or "infinite," for example, is to make true judgments about God as He is in Himself.

NOTES

1. He admitted this position—that it was impossible (from a philosophical point of view) for God to communicate the power to create to a creature—as one of the possible positions in his *Writings on the "Sentences."*

2. David Bentley Hart observes that the understanding of the "distance" between God and the world was already a problem for Christianity in the centuries before the Council of Nicaea. Hart thinks that the great discovery of the Christian metaphysical tradition is the nature of transcendence, "transcendence understood not as a merely dialectical supremacy, and not as an ontic absence, but as the truly transcendent and therefore utterly immediate act of God, in his own infinity, giving being to beings" (Hart 2011, 264).

3. According to this theory, we are able to attribute to God, in an "eminent" mode, perfections that are found in creatures, inasmuch as God is their cause.

4. "As regards what God is, it is impossible to say what He is in His essence, so it is better to discuss Him by abstraction from all things whatsoever. For He does

not belong to the number of beings ["the class of existing things"] . . . not because He does not exist, but because He transcends all beings and being itself ["above existence itself"]" (*De fide orthodoxa* 1.9) (Schaff and Wace 1973, 4).

5. *De fide orthodoxa* 1.9 (Schaff and Wace 1973). Thomas cites this text from Damascene on several occasions. Thomas writes: *quoddam pelagos substantiae infinitum et indeterminatum* (*ST* I, q. 13, a. 11).

6. *Quod divina essentia in se nec ab homine nec ab angelo videbitur* (Denifle and Chatelain 1889, 170n128).

7. Thomas takes up this question in *ST* I, q. 12.

8. At the very opening of his *Confessions*, Augustine quotes from Psalm 147 (v. 5): *Magnus es, domine, et laudabilis valde: magna virtus tua, et sapientiae tuae non est numerus* [Great art thou, O Lord, and greatly to be praised; great is thy power, and thy wisdom is infinite] (Watts 1977, 3). How ought we to understand what it means for God to be great, for His power to be great, and for His wisdom not to be numerable (what Watts translates as "infinite")?

9. This point about Thomas's procedure is emphasized by both Sweeney and Davenport.

10. *Et ex hoc quod essentia est infinita, sequitur quod potentia eius inifinita sit* (*In I Sent.* d. 43, q. 1, a. 1) (Balas 1981, 93). See also Sweeney 1992, 432–37.

11. Sweeney's words, here slightly altered (Sweeney 1992, 432).

12. Also in *De veritate* (q. 2, a. 11), when referring to the infinite in the category of discrete quantity, Thomas speaks of a difference between the necessary and the contingent infinite.

13. According to Thomas, a thing inasmuch as it is perfect is called good. Ultimately, he will argue that God alone is good and other things are called good as they are caused by reason of the "similitude of divine goodness belonging to it" (*ST* I, q. 6, a. 4, sol.).

14. An example Thomas gives is those who thought the first principle was matter and consequently they thought it possessed a "material infinity [*infinitatem materialem*]" and thus some "infinite body [*corpus infinitum*]" was the first principle of things.

15. "Whiteness considered precisely as whiteness possesses all the perfection of whiteness and, thus, can be said to be infinite within the domain of whiteness. It is, however, not absolutely infinite, for whiteness is only one species of color and thereby is finite when considered precisely as color; similarly, whiteness is [in] only one of the four species of quality, which in turn is only one of the nine general modes of accidental predication; accordingly, it is also finite when viewed qua being" (Sweeney 1992, 435n57). "But the formality of any such form of whiteness can also be considered in itself and relative to more universal formalities, for example 'color'" (Tomarchio 2002, 168).

16. "Thomas thus moves between considering real relationships between principles of beings, namely the form and matter of things, to logical relationships among formalities, namely of 'whiteness' and of 'color.' That is to say, he moves from con-

sidering the indeterminacy of the color white relative to white things, to considering the indeterminacy of 'color' relative to 'whiteness.' Though moving from real to logical relations, he remains in the order of essence, nature, quiddity, or 'whatness'" (Tomarchio 2002, 168–69).

17. "God alone is infinite in all ways [all the ways he has described]. He is infinite subsisting existence, existence unlimited to any essence distinct from itself. There is no actual or real distinction between his existence and his essence; the distinction is purely logical, pertaining to our ways of conceiving of his infinite actuality. Likewise, there can be no distinction between an absolute consideration of his existence in itself and a consideration of it relative to something else. He is actually absolute existence. Conceived by us in the order of essence, he is called infinite in his essence because he actually and intensively includes in himself all essential perfections. Conceived of by us in the order of existence, he is called infinite in existence because he does not have existence but rather is existence subsistently and so unrestrictedly. Both in *what* he is and in *how he exists* he is infinite. Moreover, just as he does not *have* existence but is existence subsistently, likewise he is not distinguished from beings that *have existence* by himself having some limit relative to them, but rather by being infinitely what they are finitely, by being the supereminently and transcendently intensive plenitude of all their determinations. It is an infinity that is unqualified, absolute, actual, in the first moment of conception negative, but in the final analysis intensive. . . . Aquinas explicitly identifies 'infinite' as a divine attribute belonging to the *via negativa*, i.e., negative in signification: it denies creaturely limitations of God. . . . Aquinas's ascription of infinity to God undergoes a threefold analogical inflection. In the *via negativa*, it is used to deny creaturely limitations of God . . . ; in the *via causalitatis*, it is used to affirm that all creaturely perfections preexist in God virtually as their cause . . . ; and in the *via eminentiae*, it is used to affirm that God intensively and subsistently actualizes and transcends all the perfections to which the *ratio* of being extends, according to a uniquely supereminent divine mode" (Tomarchio 2002, 187).

18. It might seem, as Thomas says in *Summa contra Gentiles* (I, c. 28), that perfection cannot be attributed to God because "it does not seem that what is not made [*factum*] can be called perfect [*perfectum*]." But those things that are made—which are brought forth from potency to act—are said to be perfect in that they are completely made "at that moment when the potency is wholly reduced to act" so that it retains none of the potency relative to its completion; it is a completed being. "By a certain extension of the name . . . perfect is said not only of that which by way of becoming reaches a completed act, but also of that which, without any making whatever, is in complete act."

19. "Every creature is finite, simply speaking, inasmuch as its act of existence is not absolutely subsistent but is terminated in some nature to which it is communicated. Nothing, however, prevents some creature from being infinite relatively. Material creatures have an infinity on the side of matter, but they are finite on the part of the form, which is limited by the matter in which it is received. But created

immaterial substances are finite according to their act of existence, but infinite in that their forms are not received in anything. Thus we may speak of whiteness, existing separately, as infinite precisely as white, because it would not be contracted to any subject; its act of existence, however, would be finite, because it would be determined to a certain specific nature" (Aquinas, *ST* I, q. 50, a. 2, ad. 4).

20. The very fact that no being can be added to God distinguishes the Creator from every creature. See *Summa contra Gentiles* I, c. 26, and *De ente et essentia* 5.20–21.

21. Antoine Côté points out that Thomas (*In I Sent.* d. 8, q. 4, a. 1, ad. 1) writes of three types of determination that "permettent à l'être en puissance d'être un etrê déterminé (*ut sit aliquid*) fait valoir Thomas: (1) celle qui vient de l'addition d'une différence à un genre; (2) celle qui résulte de la reception d'une nature dans un sujet; (3) celle enfin qui survient par suite de l'addition d'un accident qui fait qu'un sujet existant est dit connaissant ou blanc. Aucun de ces modes ne pouvant convenir à Dieu, il s'ensuit que celui-ci est quelque chose de déterminé du seul fait de la propriété de ne pas recevoir d'addition" (Côté 2002, 122).

22. *ita etiam divinum esse est determinatum in se et ab omnibus aliis divisum, per hoc quod sibi nulla additio fieri potest* (Aquinas, *In I Sent.* d. 8, q. 4, a. 1, ad.).

23. "Car la difference qui existe entre deux étants créés implique la possession par chacun de perfections qui font défaut à l'autre, alors qu'en Dieu l'*esse* atteint à son plein épanouissement. Toute la perfection de l'être est intégralment réalisée sur le mode de l'éminence" (Côté 2002, 122).

24. This sentence and the reference comes from an unpublished essay by Lawrence Dewan, "Saint Thomas and Creation: Does God Create 'Reality'?" (February 5, 1998).

25. For an analysis of this text, see Te Velde 2006, 74; Humbrecht 2005, 151–68. The context of this general observation is the discussion of predicates such as good, wise, just, and so forth.

REFERENCES

Achtner, Wolfgang. 2011. "Infinity as a Transformative Concept in Science and Theology." In *Infinity: New Research Frontiers,* edited by Michael Heller and W. Hugh Woodin, 19–51. Cambridge: Cambridge University Press.

Aertsen, Jan. 1987. *Nature and Creature: Thomas Aquinas' Way of Thought.* Leiden: Brill.

Agostini, Igor. 2009. "La démonstration de l'infinité de Dieu et le principe de la limitation de l'acte par la puissance chez Thomas d'Aquin: Notes sur l'histoire de l'interprétation de la quaestio vii de la summa theologiae." *Les études philosophiques* 91 (4): 455–76.

Aquinas, Thomas. *Opera Omnia.* http://www.corpusthomisticum.org/iopera.html.

Balas, David L. 1981. "A Thomist View on Divine Infinity." *Proceedings of the American Catholic Philosophical Association* 55:91–98.

Baldner, Steven E., and William E. Carroll. 1997. *Aquinas on Creation.* Toronto: Pontifical Institute of Mediaeval Studies.

Basti, Gianfranco. 2013. "Infinity." In *Interdisciplinary Encyclopedia of Science and Religion.* http://www.disf.org.

Biard, Joël, and Jean Celeyrette. 2005. Introduction to *De la théologie aux mathématiques: L'infini au XIVe siècle*, edited by Joël Biard and Jean Celeyrette, 9–34. Paris: Les Belles Lettres.

Burns, Robert M. 1998a. "Divine Infinity in Thomas Aquinas: I. Philosophico-Theological Background." *Heythrop Journal* 39 (1): 57–69.

———. 1998b. "Divine Infinity in Thomas Aquinas: II. A Critical Analysis." *Heythrop Journal* 39 (2): 123–39.

Burrell, David B. 1986. *Knowing the Unknowable God: Ibn-Sina, Maimonides, Aquinas.* Notre Dame, IN: University of Notre Dame Press.

Bussotti, Paolo, and Christian Tapp. 2009. "The Influence of Spinoza's Concept of Infinity on Cantor's Set Theory." *Studies in History and Philosophy of Science* 40:25–35.

Clark, William N. 1952. "The Limitation of Act by Potency: Aristotelianism or Neoplatonism." *New Scholasticism* 26 (2): 167–94.

Côté, Antoine. 1992. "Note sur les sources de la doctrine de l'infinité divine chez Thomas d'Aquin." *Bulletin de Philosophie Médiévale* 34:197–214.

———. 1994. "Les grandes étapes de la découverte de l'infinité divine au XIIIe siècle." In *Actualité de la pensée médiévale*, edited by Jacques Follon and James McEvoy, 216–46. Louvain: Peeters.

———. 1995. "L'infinité divine dans l'Antiquité et au Moyen Âge." *Dialogue* 34 (1): 119–38.

———. 2002. *L'infinité divine dans la théologie médiévale (1220–1255).* Paris: J. Vrin.

Davenport, Anne A. 1999. *Measure of a Different Greatness: The Intensive Infinite, 1250–1650.* Leiden: Brill.

Denifle, Heinrich, and Emile Chatelain, eds. 1889. *Chartularium Universitatis Parisiensis I.* Paris: Delalain.

Follon, Jacques, and James McEvoy, eds. 1994. *Actualité de la pensée medievale.* Louvain: Peeters.

Garrigou-Lagrange, Reginald. n.d. *A Commentary on the First Part of St. Thomas' Theological Summa.* http://www.thesumma.info/one/one49.php.

Gravil, André. 2009. "Penser l'infinité: Thomas d'Aquin et Bonaventure face à la tradition des pères Grecs." *Les études philosophiques* 91 (4): 555–75.

Hart, David B. 2011. "Notes on the Concept of the Infinite in the History of Western Metaphysics." In *Infinity: New Research Frontiers*, edited by Michael Heller and W. Hugh Woodin, 255–74. Cambridge: Cambridge University Press.

Heller, Michael, and W. Hugh Woodin, eds. 2011. *Infinity: New Research Frontiers*. Cambridge: Cambridge University Press.

Humbrecht, Thierry-Dominique. 2005. *Théologie negative et noms divins chez Saint Thomas d'Aquin*. Paris: J. Vrin.

Kretzmann, Norman. 1977. *The Metaphysics of Theism: Aquinas' Natural Theology in Summa Contra Gentiles I*. Oxford: Clarendon.

———. 1999. *The Metaphysics of Creation: Aquinas' Natural Theology in Summa Contra Gentiles II*. Oxford: Clarendon.

Laborda, Miguel Pérez de. 2009. "La via remotionis nella Summa contra Gentiles." In *Studi sul pensiero di Tommaso d'Aquino—In occasione del XXX anniversario della S.I.T.A.*, edited by Lorella Congiunti and Graziano Perrillo, 181–210. Rome: Libreria Ateneo Salesiano.

LaMountain. George F. 1956. "The Concept of the Infinite in the Philosophy of Thomas Aquinas." *The Thomist* 19:312–38.

Pieper, Josef. 1999. *The Silence of St. Thomas*. South Bend, IN: St. Augustine's Press.

Rocca, Gregory. 2004. *Speaking the Incomprehensible God*. Washington, DC: Catholic University of America Press.

———. 2009. "Creatio ex nihilo and the Being of Creatures: God's Creative Act and the Transcendence-Immanence Distinction in the Works of Thomas Aquinas." In *Divine Transcendence and Immanence in the Works of Thomas Aquinas*, edited by Harm Goris, Herwi Rikhof, and Henk Schoot, 1–17. Leuven: Peeters.

Rolt, Clarence E., trans. 1977. *Dionysius the Aeropagite: On the Divine Names and the Mystical Theology*. Grand Rapids, MI: Christian Classics Ethereal Library.

Schaff, Philip, and Henry Wace, eds. 1973. "John of Damascus: An Exact Exposition of the Orthodox Faith." In *The Nicene and Post-Nicene Fathers*, 2nd series, vol. 9, edited by Philip Schaff and Henry Wace, 1–106. Grand Rapids, MI: Eerdmans.

Shapiro, Stewart. 2011. "Theology and the Actual Infinite: Burley and Cantor." *Theology and Science* 9 (1): 101–8.

Sweeney, Leo. 1992. *Divine Infinity in Greek and Medieval Thought*. New York: Peter Lang.

Tapp, Christian. 2011a. "Infinity in Mathematics and Theology." *Theology and Science* 9 (1): 91–100.

———. 2011b. "Eternity and Infinity." In *God, Eternity and Time*, edited by Christian Tapp and Edmund Runggaldier, 99–115. Burlington, VT: Ashgate.

Tapp, Christian, and Edmund Runggaldier, eds. 2011. *God, Eternity and Time*. Burlington, VT: Ashgate.

Te Velde, Rudi. 2006. *Aquinas on God. The "Divine Science" of the "Summa Theologiae."* Aldershot: Ashgate.

Tomarchio, John. 2002. "Aquinas's Concept of Infinity." *Journal of the History of Philosophy* 40 (2): 163–87.

Trottmann, Christian. 1995. *La vision béatifique: Des disputes scolastiques à sa définition par Benoît XII.* Rome: École Française de Rome.

Tugwell, Simon. 1988. *Albert and Thomas: Selected Writings.* New York: Paulist.

Turner, Denys. 2004. *Faith, Reason, and the Existence of God.* Cambridge: Cambridge University Press.

———. 2011. "A (Partially) Skeptical Response to Hart and Russell." In *Infinity: New Research Frontiers*, edited by Michael Heller and W. Hugh Woodin, 290–98. Cambridge: Cambridge University Press.

Villagrasa, Jesús. 2008. "Creazione e actus essendi: L'originalità della metafisica di Tommaso d'Aquino." In *Creazione e actus essendi: Originalità e interpretazioni della metafisica di Tommaso d'Aquino*, edited by Jesús Villagrasa, 83–137. Rome: Ateneo Regina Apostolorum.

Watts, William, trans. 1977. *Augustine's Confessions.* Cambridge, MA: Harvard University Press.

White, Thomas J. 2009. *Wisdom in the Face of Modernity.* Ave Maria, FL: Sapientia.

Wippel, John. 1998. "Aquinas and the Axiom That Unreceived Act Is Unlimited." *Review of Metaphysics* 51 (3): 533–64.

———. 2000. *The Metaphysical Thought of Thomas Aquinas: From Finite Being to Uncreated Being.* Washington, DC: Catholic University of America Press.

CHAPTER 5

Spinoza and Leibniz on the Absolute and Its Infinity

A Case Study

CHRISTINA SCHNEIDER

What is time? If no one asks me, I know; if I must explain it, I do not know.
—St. Augustine of Hippo[1]

PROLOGUE

A similar impasse to Augustine's presents itself when we are asked what is meant by "God's infinity" or by the "infinity of the Absolute." The impasse does not diminish at all if one looks at "God's infinity" within the metaphysical framework of Leibniz alone. It does not diminish even if one narrows down the issue further to the late Leibniz of the *Monadology*. God plays an important role within Leibniz's metaphysics,[2] and different connotations of "infinity" are present in Leibniz's theorizing about God.[3] Leibniz conceives of God as omnipotent, all-powerful, almighty, perfectly good, and eternal.[4] Behind all these *omni* notions, there lurk connotations of "infinity," just as they do in Leibniz's metaphysics, but they are not always made explicit by him.

As a mathematician and cofounder of the infinitesimal calculus, Leibniz was aware of infinity in a mathematical sense, and he had at least an intuitive concept of it that is not utterly alien to the rigorous mathematical concept of our epoch. As a metaphysician, however, he has problems with the concept. For Leibniz, and here he finds himself within the tradition, infinitesimals are not "real' or "actual," but they are "potential" or "ideal" (Leibniz, *Principes de la Nature et de la Grace,* in *Die philosophischen Schriften* [*GP*], 4:569): "Although mathematical meditations are ideal, that does not diminish their utility, since the actual things cannot escape their rules, and one can really say that the reality of phenomena that distinguishes them from dreams consists in this circumstance" (my translation).[5]

Nevertheless, Leibniz sets his mathematically spirited notion of infinity to work to escape the two "metaphysical labyrinths," as he calls them: the labyrinth of *freedom* and the labyrinth of *the continuum*. According to Leibniz, both problems are rooted in infinity and both may be resolved by taking infinity into account.[6] All this may be interesting in itself, and worthy of being explored when writing a monograph on Leibniz's metaphysics. For the present purpose—elucidating the "infinity of God" in Leibniz's metaphysics—it is of minor help.

Fortunately, there is no clue to Leibniz's late metaphysics independent of his theory of creation, and so also of his theory of God and its relation to "the world." This theory was intended to set his metaphysical framework apart from *Spinozism*. Looking at Spinoza's *Ethics*, right at the beginning, talk of infinity abounds. "Infinity" is deemed to "codefine" his notion of the Absolute and, by this, to set the stage for his metaphysical framework.

This leads to the question to be pursued in what follows: What implications might different connotations of infinity have for different notions of the Absolute and its relation "to the world"? I shall pursue this question by comparing Spinoza's attempt at an answer, on the one hand, and Leibniz's, on the other. As is well known, Spinoza's conception of the Absolute is often called *pantheistic*, and Leibniz's conception—explicitly formulated to be *nonpantheistic*—may be called *transcendent*.[7] This topic finds expression in my chapter title.

The case study is asymmetrical with respect to Spinoza and Leibniz. More emphasis is put on Leibniz. Spinoza's theory serves only as a starting point.[8] The case study is not exegetical in character. I attempt no deep and detailed historical and exegetical interpretation. It is systematic. The ideas of the two philosophers, as they are presented here, should serve as a platform intended to give substance to the claim that different concepts of infinity have implications for the concept of the Absolute and for its relation to "the world." The notion of God is especially important for considering Spinoza's framework.

PARADIGMS OF INFINITY: A COARSE-GRAINED TAXONOMY

The taxonomy to follow is not intended to be exhaustive. To clarify the concepts and their relations to the underlying theories much more work, historically and systematically, would be necessary. This, however, would be a topic of its own. For the present purpose, the coarse-grained indications may give orientation; they should help give some structure to the different uses of the word "infinity" in the subsequent discussion. Nevertheless, they indicate the role the different notions play in framing a notion of the Absolute.

Since the seminal work of Georg Cantor, there has been a clear notion of "infinity," or a mathematical theory of infinity. This theory, moreover, plays an important role within analytic philosophy. Being "extensional' in character, it may be doubted whether mathematical infinity may be adequately attributed to God, in any way. To put it loosely: speaking of God as an infinity, in the mathematical (extensional) sense, leads to the question, "An infinity of what?" This, in turn, seems to be in conflict with God's unity, another conceptual graveyard.

At the dawn of set theory, the mathematical notion of infinity was discussed with reference to and somehow "applied" to theological issues by Cantor himself (see Tapp 2005). The focus of the theological topics, however, has been the question of whether there is an *actual* infinity, opposing an old stance that entails that there is only *potential* infinity. Cantor conceived of his theory as showing that there can be *actual* infinity, as Tapp (2005) has shown.

There is another possible meaning of "infinity" in "the infinity of God" that is also mathematical in character, but is, at least prima facie and naïvely speaking, not about cardinality. It is of a geometric character: infinity as having no boundaries, being not embedded in a different, "greater," "more extended," region. This meaning of "infinite" is conducive to a concept of God as "having nothing outside God."

There are also meanings of "infinity" expressed by the "great-making" attributes or perfections, such as goodness, power, knowledge. They are attributed to God as pure perfections.

Within "pure perfections," one may distinguish two meanings. On the one hand, there is a meaning that may be called the "(pure-perfection-) *no-negation*-paradigm" or the "pure-perfection-*no-negation*-meaning." It is logical in that the definition of this perfection needs no correlative or codefining concept. It may be regarded as a conceptual or logical counterpart of the geometric paradigm. On the other hand, "pure perfections" attributed to the Absolute present themselves as the different *omni* notions: "all-goodness," "almightiness," "omniscience." The meaning is that these pure perfections are possessed by the Absolute in their highest degree. It is the "highest degree" of the pure perfection, or, more intelligibly, it is their unboundedness in degree. This is Leibniz's preferred paradigm, as may be seen in the opening paragraphs of an essay from 1686 (Leibniz, *GP*, 4:427) that treat of the meaning of "infinite" when infinity is attributed to God. One may call this connotation of infinity the "*pure perfection-omni* . . . meaning" or the "*pure perfection-omni* . . . paradigm."[9]

The different meanings associated with pure perfections, the no-negation paradigm on Spinoza's side and the different *omni* notions on Leibniz's side, make a difference to their attempts to solve problems about God. There is a further asymmetry between the two meanings of infinity: The pure-perfection/no-negation-paradigm is positively responsible for Spinoza's version of pantheism. The pure-perfection-omni . . . paradigm on Leibniz's side plays a role in his theory of creation, but not in his attempt to avoid "Spinozism." Leibniz's *not* conceiving of "the infinity of God" with respect to the no-negation-paradigm enables him to save a transcendence of sorts (but leaves him with another problem, as I shall show).

A GLANCE AT SPINOZA'S FRAMEWORK: *ETHICS*

The opening axioms and corollaries of Spinoza's *Ethics*, including uses of "infinity," play an important role in constituting his metaphysical framework. This metaphysical framework encapsulates at once his concept of the Absolute, and vice versa.

A note of caution about Spinoza's terminology: on the one hand, he uses expressions that point to conceptual analysis; on the other hand, he uses expressions that point to ontology. Both phraseologies are in parallel use, often intermingled. A contemporary reader might wonder whether Spinoza is engaged in conceptual analysis or doing ontology. However, the distinction is not drawn in Spinoza's thought, as his Sixth Axiom, which may be called a methodological axiom, reveals:

> A6: A true idea must agree with its object.

This means, roughly speaking, that conceptual language and ontological language express the same realm or, at least, realms that are "isomorphic." For present purposes, it seems advisable to interpret both languages in an ontological way.[10]

The Hallmark of Spinoza's Ontology—The First Axiom

> A1: Whatever is, is either *in* itself or *in* another.

Any understanding of the role this axiom plays in his ontology hinges on interpreting "in." A hint may be found in the corollary to proposition 6:

> Corollary to P6: For in Nature there is *nothing* except substances and their affections. (my emphasis)

This corollary formulates the categorical setting of Spinoza's metaphysical framework.

Definitions of Spinoza's "Finite"

> D2: That thing is said to be *finite in its own kind* that can be limited by another of the same kind. (my emphasis)

The central role that either finitude or infinity plays in Spinoza's metaphysics is indicated by the fact that his second definition is a putative clarification of "finite in its own kind." It is not "finite" that will be important, it is the negation of the finite, namely, the "infinite in its own kind." At face value, "*finite in its own kind*" and, correspondingly, "infinite in its own kind" allude to the geometric (boundary/no-boundary) distinction. It is rooted in the concept of "definition" that works with *genus*, *species*, and *differentia specifica*. By this, it foreshadows the pure-perfection-no-negation paradigm.

"Substance," "Attributes," and "Modes": The Definitions

> D3: By *substance* I understand what is *in* itself and is conceived through itself, that is, that whose concept does not require the concept of another thing, from which it must be formed. (my emphasis)

This is an instance of the conceptual and ontological double-talk: "being in itself" points to a strong ontological self-sufficiency that is echoed by a conceptual relation ("whose concept does not require the concept of another thing, from which it must be formed"). Anyway, this definition expresses *autonomy* driven to an extreme, as it implies that nothing that is created and not "self-created" can be a substance.

> D4: By *attribute* I understand what the *intellect perceives* of a substance, as *constituting its essence.* (my emphasis)

"Attribute," "constituting," and "essence" point to an ontological interplay. Attributes are responsible for a substance's being intellectually accessible. One may ask whether a "substance itself" is intellectually accessible or not, whether a substance is anything over and above its attributes/essence (and modes). If it is, it is not intellectually accessible, so it seems. Pursuing this topic, however, would lead our reflections too far astray (but see P1).

> D5: By *mode* I understand the affections of a substance, *or* that which is *in* another through which it is also conceived. (my emphasis)

"Modes," as it is usually interpreted, is Spinoza's version of "accidents." This interpretation is taken up below.

Spinoza's God

> D6: By God I understand a being *absolutely infinite*, that is a *substance* consisting of an *infinity* of attributes, of which each expresses an eternal and *infinite* essence. (my emphasis)

The use of the word "infinity" here is not very interesting. Prima facie, it expresses "cardinality."[11] More problematic and interesting is the use of "expresses an . . . infinite essence." Both together help elucidate "absolute infinite." The problem is how to understand "infinite" in "infinite essence." Spinoza defines "God" as an absolutely infinite being. Leaving, for a moment, the question of how to understand "absolutely infinite," so far nothing utterly new concerning substances, attributes, essences, and so on has been formulated:

- Accidents do not go from one substance to another (*non migratio*; see Spinoza's proposition P4[12]).
- Each substance has an essence.
- There are *only* substances and accidents (Spinoza's modes).
- Essences are internally variegated; by attributes.
- **And**: God is an *absolute infinite* being; conceived of as a *substance.*

Pantheism?

At this instant, one may ask two questions:

1. Why should that entail the existence of one and only one substance (namely, Spinoza's Absolute)?
2. Given an answer to the first question, why should this entail pantheism?

The answer to the first question is given by Spinoza himself in his explication of definition 6, and here, the meaning of "*infinite* essence" and "*absolutely infinite*" plays an important role:

> Exp. to D6: I say *absolutely* infinite, not *infinite in its own kind*; for if something is only infinite in its own kind, we can deny infinite attributes of it [NS: i.e., we can conceive infinite attributes which do

> not pertain to its nature], but if something is *absolutely infinite*, whatever expresses its *essence* and involves no *negation* pertains to its essence. (my emphasis)

"Infinite in its own kind" is made explicit using a definition of "infinity" as cardinality. "Absolute infinity" is explained by the no-negation-paradigm. A further tacit presupposition must be made explicit:

> For an attribute to "involve negation" means that its concept depends on the concept of an other, correlative or "co-defining," concept that is "responsible" for the "negation." That means: "Infinity" is understood by the *"pure-perfection/no-negation-paradigm."*

The consequences are plain:

1. The concept of the attribute that involves "negation" depends on another concept, or is "formed" by another concept. (Leibniz would say that the concepts of attributes are not "simple.")
2. The concept of the substance having the attribute "involving negation" depends on another concept.
3. Since substances are autonomous, by the third definition (D3), this cannot be the case.

It follows that there is only one substance. In his propositions 7 and 11, Spinoza argues that it pertains to a substance's essence to exist, and to exist necessarily. Spinoza's God therefore is the only substance and *exists necessarily.*

Concerning the second question, this follows: since there is not anything else but substances and modes, and since there is only one substance, anything else that is not *the* substance, God, is a mode of it.

Interpreting, further, the *in* as in "the predicate is *in* the subject" (construing Spinoza in a conceptual way) or similarly as in "accidents *in*here *in* substances" (construing Spinoza in an ontological way), one arrives at a pantheism of sorts. Conceiving of the Absolute as an autonomous and infinite being, and as a substance in a broadly Aristotelian sense, and using a certain connotation of "infinity," "infinite" as "pure-perfection/no-negation," entails a version of pantheism, Spinoza's version.

LEIBNIZ: THE STRUGGLE FOR TRANSCENDENCE

Spinoza is, in a sense, Leibniz's ghost. He is what Leibniz was afraid of being and saw himself as dangerously capable of becoming. The doctrine that God is in some way related to creatures as a whole to its parts, not as an extra item, was one that obsessed him: "If there has been no monads," he wrote to Bourguet, "Spinoza would have been right" (Wilson 1989, 89).[13]

The citation alluded to by Wilson is from a letter from Leibniz to Bourguet, December 1714:

> I do not know, Sir, how you can infer any Spinozism from it; that is being a little bit fast in drawing consequences. Quite to the contrary: It is precisely due to the monads that Spinozism is corrupted, because there are as many veritable substances, and so to speak, living subsisting mirrors of the universe or concentrated universes as there are monads instead of, as Spinoza says, only one single substance. He [Spinoza] would have been right, if there were no monads at all; in this case, beyond God, everything would be flowing and perishing in simple accidents or modifications, because there would not be any base of substances, which consists in the existence of monads, in the things. (Leibniz, *GP*, 3:575; my translation)[14]

Leibniz had reason to fear being accused of Spinozism. He also had reason to fear that the accusation was right. This quotation from his *Theodicée*, might reveal this:

> It is plain that this decision [to create the *best of all possible worlds*] does not change anything in the constitution of things and that it leaves them as they have been in the state of pure possibility; that means that it does not change anything: not with respect to their essence or nature and even not with respect to their accidents that are already perfectly represented in the idea of that possible world. (Leibniz, *GP*, 6:13; my translation)[15]

On the one hand, the danger has to do with Leibniz's theory of creation and principles that govern his overall architectonic. On the other hand, Leibniz's route of escape, his *theory of monads*, is also problematic.

Leibniz's Framework: The Absolute, Transcendence, and Ideas in God's Mind

Leibniz's pretheoretic notion of God is rather canonical. Leibniz's Absolute has personal traits: it is "thinking," judges, possesses will, and it is "*good.*" Moreover, Leibniz's Absolute is a creator and it is *transcendent.* As transcendent, it does not depend on anything except itself. So, Leibniz could be happy with Spinoza's third definition (D3) as a definition of "God," but *not* as a definition of "substance in general."[16] But Leibniz does not define "God" as Spinoza does.

"Infinity" plays an implicit role in Leibniz's theory of creation as perfection of knowledge, power, and goodness in the highest degree. It is the pure-perfection-notion with respect to the various *omni* notions. In the *Monadology* (paras. 36–41), Leibniz proves the existence of a necessary substance usually called "God" by the *principle of sufficient reason.* He infers that God has all pure perfections, in an unlimited and unbounded way. That transcendence poses a problem for Leibniz's theory of God, and His relation to the "created world" is revealed by the role ideas, possible worlds (and associated with that, complete concepts), play. I discuss both topics in the next subsection.

Ideas in God's Mind

"Ideas," for Leibniz, are the objects of thought. In modern terms one could say that ideas are the semantic content of whatever may have semantic content. Ideas are in the mind of the "thinker." A putative mind not containing ideas would not be able to think. It would not be a mind. His notion of *innate ideas*, even for humans, expresses this. Since God is *omniscient*, and omniscience presupposes thinking, ideas of whatever there is or could be, in some mode of being, are in God's mind.

Ideas are individual entities, they are not universals. Ideas may be "eternal" or necessary truths. Paradigmatic cases are ideas of logic and mathematics, but some ideas of ethics and metaphysics are also necessary truths. Ideas may also be contingent, as semantic contents of contingent things.

Ideas in God's mind form configurations of greater complexity—*possible worlds.* These configurations can be regarded, *cum grano salis*, as similar to maximally consistent sets of propositions.

Within possible worlds, there are subconfigurations—*complete concepts*. Complete concepts are the ideal version of individuals, or they are what is thinkable about individuals. Moreover, they express all and everything that can truly be said about that individual during its whole career, including all relations the individual stands in. And so, individuals in general (Leibniz sometimes calls them "substances") are *not* autonomous.

In Leibniz's theory of creation, different possible worlds come with different "degrees of goodness." One and only one of them is the "*best*." "Goodness" is an intrinsic feature of a whole possible world. Further, a world is not good because God judges it to be so; on the contrary, God judges it to be good, because it is good. God is infallible in judging, and this again is logically guaranteed by His omniscience.

By His omniscience, God knows which possible world is best. By His "all-goodness" he wills the best of all possible worlds to come into existence, and by His "omnipotence" God creates it.

God, however, has no power over ideas, possible worlds, and complete concepts, neither over their content and their "goodness" nor over how they may enter into configurations of higher complexity. As a consequence, God cannot make one possible world better or worse by adding or eliminating ideas or configurations within it. Nevertheless, ideas, possible worlds, and complete concepts belong to God's mind in a robust ontological sense.

The Danger of Spinozism

The story of the danger of Spinozism is easily told: God knows all possible worlds, all the ideas in His mind, all complete concepts, including the "goodness" of possible worlds, perfectly, by omniscience. God infallibly "judges" which possible world is the best. He knows which possible world is the best and, by His perfect goodness, "wills" the best of the possible worlds.

And then can God "create" this best possible world "outside" Him, outside His mind as an "extra item"? How can this ideal configuration be given an existence beyond God's mind?

As it stands, the danger of Spinozism is not averted. This is due to two metaprinciples of Leibniz's metaphysics. First, there is his *rational-*

ism, characterized as follows. Whatever there is in whatever mode of existence is perfectly "intellectually" transparent to God, perfectly *thinkable* by God. Second, there is Leibniz's ***principle of the identity of indiscernibles*** (referenced as "pii," in what follows). And so, no idea can be both in God's mind and elsewhere. There is no copy of anything. Remembering the quotation from *Theodicée* (Leibniz, *GP*, 6:131): "It is plain that this decision to create the *best of all possible worlds* does not change anything in the constitution of things, and that it leaves them as they have been in the state of pure possibility; that means that it does not change anything: not with respect to their essence or nature and not even with respect to their accidents that are already perfectly represented in the idea of that possible world" (my emphasis; my translation; see my note 15).

One arrives at a Spinozism of sorts. This is due to his architectonic (that is, *pace* his phraseology in several instances, not Aristotelian) and to the pii.[17]

Monads: A Route of Escape?

To escape Spinozism, there must be something "exterior" to the Absolute. However, this is inconsistent with there being infinity construed on the "geometric-paradigm." Leibniz thought that his monads could perform this function. But, in the light of Leibniz's principles, this move is not without problems, at least prima facie.

Monads as External to God's Mind

To avoid Spinozism, and to secure God's transcendence, monads must be external to Leibniz's God and therefore to His mind. This, however, implies that monads must be absolutely unthinkable, otherwise they would be thinkable by God and so ideas in His mind. The possibility that monads are copies of ideas or of configurations of ideas (of *complete concepts*) is banned by pii. This looks fatally like *prime matter* or *bare particulars*. Even God, omniscient as He is, can only know of the monads that they are external to Him and that they are a plurality.[18] Nothing "intrinsic" to them can be thought or known by God. This casts doubt on the reality of the entities Leibniz himself sometimes calls "substances." Even worse, whatever is created must be totally inexpressible, and so totally unintelligible; a drastic consequence for any rationalist.

Trapped between Idealism and Unintelligibility?

Leibniz conceives of the Absolute, not least to avoid Spinozism, as a transcendent being and a creator that is prima facie completely independent of its creatures. Leibniz attributes to the Absolute the properties denoted by his canonical *omni* notions. His metaphysical architectural framework, together with his idealistic presuppositions, and his principle of the identity of indiscernibles, entails a pantheism of sorts.

Leaving the *omni* notions as they are, and with this his implicitly assumed definition of infinity, there are two options for escape:

1. Leibniz could uphold his rationalistic stance: his thesis that whatever there is, in whatever mode of existence, must be intellectually perfectly transparent, at least for an *omni*scient being, God. And, at the same time, Leibniz must relax the principle of the identity of indiscernibles. This would allow him to conceive of "the created world" as being a copy of the best of all possible worlds. In this case, however, creation is in danger of being a Platonic parallel universe.
2. Leibniz could unrestrictedly uphold the principle of the identity of indiscernibles and relax his rationalistic stance, by admitting "unthinkable" entities of sorts, entities that are "unthinkable" even for an omniscient being, God. This seems to endow his metaphysics with a realm that is intellectually and semantically completely inaccessible.

Leibniz takes the second route with his monads. Is he left, by this move, with internally opaque and mysterious entities—opaque even for God?[19] Is this incoherent with God's omniscience?

Monads Again

Paraphrasing and interpreting the first paragraphs of the *Monadology* (*GP*, 6:607–23), one can make a long and intricate story short:[20]

- *Monads* are soul-like entities, mutually adapted to each other by their *perceptions* (through Leibniz's preestablished harmony). They mutually perceive, and so they represent each other. They perceive and represent themselves.

- In *perceptions*, monads "mirror each other." This is the representational aspect of perceptions.
- Monads are individuated by their *perceptions. Perceptions* are their *internal* qualities.

Monads fall into in three classes:

1. "Simple monads." They have no awareness of their perceptions.
2. "Souls," as Leibniz calls them. They have a certain sort of awareness of their perceptions. This "awareness," however, is beyond any conceptualization. A "soul," however, is not always aware of its perceptions. Animals are Leibniz's example and paradigm (cf. Kulstad 1991).
3. "Spirits," the highest sort of monads. They can conceptualize their perceptions.[21] The first and second sort of monads are in continuity with each other. Spirits, however, are set apart by their capacity to conceptualize. Each spirit has phases where it does not conceptualize and where it has no sort of awareness.

Leibniz's framework in a nutshell:

- There is not anything in the created world but *monads*, their *perceptions*, and *appetites.*[22]
- Perceptions are internally variegated first-person perspectives, mutually adapted to each other by the preestablished harmony.
- As first-person perspectives, they are transparent only to the monad whose perceptions they are.[23] If they can be conceptualized at all, then this is only by the monad whose perceptions they are and, even then, in an indirect way. Perceptions are not ideas and not authentically intellectually accessible by any other entity than the monad whose perceptions they are, not even by the Absolute.[24]

Being soul-like entities, Leibniz's monads are not internally poor. On the contrary, each monad has many *perceptions* but, at least prima facie, God is not omniscient with respect to them. Of course, God as God "created" them in strict correlation with the *complete concepts* in His mind. Leibniz calls his monads "active forces" (*vis activa*), meaning that they are their own sources of change, and so their own sources of exercising their appetites, that is, they strive from one perception to the next in complete correlation with their "individual law." God creates the different active forces in accordance with His plan. And so, God knows which complete concepts are correlated with which monads, which active forces, and so with the different respective perceptions and appetites. Due to the mutual fit of complete concepts, He knows that the different monads are harmoniously adapted to each other, that they exercise their active forces in harmony with each other. What God does not know is "what it is like to be" this or that individual monad in this or that stage of its history, to use Thomas Nagel's expression. These "what-it-is-like-to-be-nesses" are perceptions.[25]

Is this a restriction of God's omniscience? Is this, consequently, a shrinking of His infinity (on the pure-perfection/*omni* paradigm)? As things stand, it seems so. But this impression hinges on conceiving knowledge, and a fortiori omniscience, as only intellectual and conceptual, as being exclusively bound to third-person perspectives.

If, however, God or the Absolute had some other epistemic access to his creation, and so to monads and first-person perspectives, and thereby secured omniscience, another problem would show up: Does this access to the first-person perspective not entail that the Absolute is *in* its creation, *in* the individuals it creates?

SOME CONSEQUENCES

By presupposing the pure-perfection/no-negation paradigm of "infinity," Spinoza was driven to his pantheistic notion of God and His relation to the world.

Leibniz, fearing Spinozism, endorsed, at least implicitly, a different paradigm of "infinity": the pure-perfection/*omni* paradigm. The "danger of Spinozism" is rooted in his idealism, his rationalism, and the identity of indiscernibles. Given his overall framework, Leibniz's paradigm of infinity is inert to both the "danger of Spinozism" and his escape route: his theory of monads. His escape route, however, leads him to conceive of the created world in a *panexperientialistic* way.

This has consequences for Leibniz's concept of God as *omniscient*, an instance of his paradigm of infinity. If omniscience is to be preserved, the concept of God's knowledge has to be modified, or enlarged, to encompass more than third-person-perspective access to created beings. God must not only share with each and every monad their perceptions, the "what-it-is-like-to-be-nesses." He must, due to His *omni*science, have those perceptions in the identical way the respective monads have them. This leads to another danger, if it is a danger at all: *God as being in His creatures*. This consequence might not be intended by Leibniz. In his exchanges with Clarke, he accuses Newton in nearly every letter of conceiving of God as the "world-soul." Alluding to a metaphor of Pascal's, Leibniz expresses God's relation to creation in paragraph 13 of his *Principes de la Nature et de la Grace* as follows: "On a fort bien dit, *qu'il est comme centre partout; mais sa circomference n'est nulle part, tout lui etant present immediament, sans aucune eloignemment de ce Centre*."[26]

With this, a task that is in a way new but in a way old is urged upon us. Rethink the Absolute!

NOTES

1. *Quid est ergo tempus? Si nemo ex me quaerat, scio; si quaerenti explicare uelim, nescio* (Augustine, *Confessiones* 11.14.17).

2. This is the main topic of Adams 1994.

3. See, e.g., Leibniz, *Principes de la Nature et de la Grace* (*GP*, 6:598–606, paras. 7 to 9), or, concerning Anselm's ontological argument (*GP*, 4:354), and, finally, his *Monadology*, paras. 43–45 (*GP*, 6:614).

4. See his characterization of perfections in *Discours de métaphysique* (*GP*, 4:427–63).

5. "Ainsi quoyque les méditations Mathematiques soient idéales, cela ne diminue rien de leur utilité, parce que les choses actuelles ne sauroient s'écarter de leurs règles; et on peut dire en effect, que c'est en cela que consiste la réalité des phénomènes, qui les distingue des songes."

6. This is the topic of Leibniz's essay on freedom (Leibniz 1857, 178).

7. Of course, there are different uses of the word "transcendent." Here it is used to indicate that the Absolute is not only different to its creation but also somehow separated from it. A good taxonomy with respect to this word may be found in Culp 2008.

8. For the quotes, the English translation of Curley (1994) is used.

9. For an elaborate account of "perfections," see Brugger 1979 (chap. 123.6:92).

10. The numbers of axioms, propositions, and so on refer to the first chapter of *The Ethics*, "Of God."

11. Other interpreters are inclined to read it in the sense of "all," an advisable proposal in several respects.

12. P2: Two substances having different attributes have nothing in common with one another. And P4: Two or more distinct things are distinguished from one another, either by a difference in the attributes of the substances or by a difference in their affections.

13. It should be noted that Leibniz, in a rather early essay (from 1678; see *GP*, 1:139), formulates a partly polemical critique of Spinoza's *Ethics*. This critique, however, is mainly methodological in character.

14. "Je ne say, Monsieur, comment vous en pouvés tirer quelque Spinosisme; c'est aller un peu vite en consequences. Au contraire c'est justement par ces Monades que le Spinosisme est detruit, car il y a autant de substances veritables, et pour ainsi dire, de miroirs vivans de l'Univers tousjours subsistans, ou d'Univers concentrés, qu'il y a de Monades, au lieu que, selon Spinosa, il n'y a qu'une seule substance. Il auroit raison, s'il n'y avoit point de monades; alors tout, hors de Dieu, seroit passager et s'evanouiroit en simples accidens ou modifications, puisqu'il n'y auroit point la base des substances dans les choses, laquelle consiste dans l'existence des Monades."

15. "il est visible que ce decret [to create the best of all possible worlds] ne change rien dans la constitution des choses, et qu'il les laisse telles qu'elles étoient dans l'état de pure possibilité, c'est à dire qu'il ne change rien, ny dans leur essence ou nature, ny même dans leur accidens, representés déjà parfaitement dans l'idée de ce monde possible."

16. Incidentally, due to his strict determinism, Leibniz is not plagued by problems philosophers of our days have: God cannot be "surprised" by "free actions" of His creatures. There is no libertarian freedom. Each creature—including humans—simply exercises its individual law similar to computers exercising algorithms.

17. It must be noted that Leibniz's "ontology" in general and with respect to his Absolute is different from Spinoza's: Leibniz's Absolute is not a substance with accidents (modes); accidents cannot be "in" the Absolute due to its immutability. So, complete concepts or possible worlds are not "in" the Absolute like accidents are said to inhere in substances. A detailed study of the different "notions" of substance of the rationalists may be found in Woolhouse 1993. This is also a topic of Adams 1994.

18. Leibniz thinks that there are infinitely many of them; here "infinity" is understood as cardinality.

19. Of course, externally, as a placeholder of sorts with respect to Leibniz's overall metaphysical architectonic, as making up together a metaphysical region, they are expressible intellectually. They may be intellectually grasped with respect to generic features. But this does not hold for each single monad with its "internal" aspects, its "internal features."

20. A detailed interpretation and reconstruction may be found in Schneider 2001.

21. How this can work is an intricate issue that cannot be dealt with here, but cf. Schneider 2001.

22. "Appetite" is the principle of change; it is the succession of a monad's perceptions and so it leads to a monad's change and to "time." In the present context, appetite may be neglected.

23. Leibniz's unfortunate metaphor of monads being "without windows" points to this circumstance.

24. The connection of perceptions to ideas has to do with Leibniz's well-founded phenomena; this is a topic of its own and is left to the one side in the present context.

25. Here, Leibniz's unfortunate metaphor of "monads being windowless" fires back: if there are "no windows," no one can "look inside."

26. "One can really say that His is like a center that is everywhere but his periphery is nowhere. Everything is immediately present to him without any distance from this center" (my emphasis and my translation).

REFERENCES

Adams, Robert M. 1994. *Leibniz: Determinist, Theist, Idealist.* Oxford: Oxford University Press.

Augustine. 2008. *Confessions.* Translated by Henry Chadwick. Oxford: Oxford University Press.

Blumenstock, Konrad. 1983. *Spinoza, Opera: Werke II.* Darmstadt: Wissenschaftliche Buchgesellschaft.

Brugger, Walter. 1979. *Summe einer philosophischen Gotteslehre.* Munich: Johannes Berchmans.

Culp, John. 2008. "Panentheism." In *The Stanford Encyclopedia of Philosophy*, rev. Spring 2013 ed. http://plato.stanford.edu/archives/spr2013/entries/panentheism/.

Curley, Edwin. 1994. *A Spinoza Reader: The Ethics and Other Works.* Princeton, NJ: Princeton University Press.

Kulstad, Mark. 1991. *Leibniz on Apperception, Consciousness and Reflection.* Munich: Philosophia Verlag.

Leibniz, Gottfried Wilhelm. 1857. *Nouvelles Lettres et opuscules inédits de Leibniz.* Paris: Foucher de Careil.

———. 1875–90. *Die philosophischen Schriften* [*GP*]. 7 vols. Edited by Carl Immanuel Gerhardt. Berlin: Weidmann.

Rutherford, Donald. 1995. *Leibniz and the Rational Order of Nature.* Cambridge: Cambridge University Press.

Schneider, Christina. 2001. *Leibniz's Metaphysik: Ein formaler Zugang.* Munich: Philosophia Verlag.

Tapp, Christian. 2005. *Kardinalität und Kardinäle: Wissenschaftshistorische Aufarbeitung der Korrespondenz zwischen Georg Cantor und katholischen Theologen.* Stuttgart: Steiner.

Wilson, Catherine. 1989. *Leibniz's Metaphysics: A Historical and Comparative Study.* Princeton, NJ: Princeton University Press.

Woolhouse, Roger S. 1993. *Descartes, Spinoza, Leibniz: The Concept of Substance in Seventeenth-Century Metaphysics.* London: Routledge.

CHAPTER 6

Kant and the Infinity of Reason

RUBEN SCHNEIDER

STATUS QUAESTIONIS

According to a traditional interpretation, Kant rejects every attempt at constructing a metaphysical theory of God. In Kant's view, the divine attributes are purely "transcendental predicates,"[1] which have only a subjective validity (cf. *Critique of Pure Reason* [*CPR*], A 822, B 850, in Kant, *Akademie Ausgabe* [*AA*], 03:533.03), without any knowable application to an external, transsubjective reality. So why is it of interest to us, as analytic metaphysicians concerned with the *metaphysical* attribute of divine infinity, to look at Kant's transcendental idealism?

For Kant, the systematic place of the infinity of God is in the relation between our finite cognitive capacities and the object of the transcendental ideal (God) that infinitely transcends these finite capacities. According to Kant, our knowledge is, because of its limitedness, bound to sensation (*Empfindung*) and intuition (*Anschauung*) and therefore is restricted to appearances (*Erscheinungen*). Knowledge (*Wissen*), according to Kant, is always more than mere knowledge of an object's existence—it entails the classification of an object within the totality of our experience. If we have theoretical knowledge of an object, we are thereby able to specify its relation to any other objects of our knowledge. In this, we are aided by the categories, theoretical or a priori concepts, which are nevertheless restricted to objects empirically accessible, and by the transcendental schemata (cf. Schöndorf 2004, 59 and 74–75). In the case of God, there is no corresponding intuition and no corresponding schematism (57). Therefore, we cannot have any theoretical or a priori knowledge of God. The reality "behind" the idea of God seems to be completely inaccessible for us.

On the other hand, Kant claims that the regulative idea of God necessarily springs forth from our reason and is a necessary condition for our knowledge of the empirical world: the transcendental ideal makes possible the connection necessary for our knowledge of empirically existing objects. The idea is of God as the being of all beings, the *ens realissimum* and *omnitudo realitatis*, the absolute unconditioned totality that contains "the highest condition of the possibility of all that can be thought" (*CPR*, B 391, in *AA*, 03:258.17–18) guarantees the unity of reason and the unity of the universe and all things in general (cf. *CPR*, B 390f.). Here Kant is presented with a dilemma. The idea of God makes possible the connection of appearances necessary for every kind of knowledge. However, the idea of God does not contribute to the knowledge of particular objects as much as the understanding (*Verstand*) does, but only the idea of God can elevate the understanding to a knowledge of existing objects (cf. Schöndorf 2004, 60). But how can the idea of God make possible the connection of understanding to reality if this idea itself has no connection to reality whatsoever? Is it really true that the idea of God has no connection to reality? And does it truly follow that our reason is finite per se, according to Kant?

Our cognitive capacity, according to Kant, is finite and restricted to appearances. However, Kant clearly denies Berkeleyan idealism: the thing in itself is not only a pure "limiting concept" (*CPR*, B 310, in *AA*, 03:211.35) without connection to reality, but its existence is necessary for the system of transcendental idealism. It is the origin of the affection of our sensibility (*Affektion der Sinnlichkeit*), never to be given up by Kant (cf. *Prol.*, §§57, 350; Schöndorf 2004, 55). Furthermore, Kant ascribes to the objects of the transcendental ideas the status of being reserved for a thing in itself: in §25 of the "Transcendental Deduction" he claims that I know that I exist, that is, not only as a phenomenon but also as a thing in itself. In the chapter on the "Paralogisms," Kant does not oppose the existence of the soul but the attempt to grasp it in its essence. And the object of the idea of the world for Kant also is an existing thing in itself.[2] And the same applies to the idea of God (cf. Schöndorf 2004, 65). Kant's critique of rational theology cannot be sweepingly understood as a denial of every kind of knowledge of God but merely as a repudiation of the attempt in rationalistic metaphysics to grasp God in his attributes as they are in themselves. As an independent transcendental object that is unknown to us in its "inner predicates" (*CPR*, A 565, B 593, in *AA*, 03:382.04) and different from matter (cf. *CPR*, A 618, B 646, in *AA*, 03:412.17–20; cf. Natterer 2011, 132; cf. Kant, *Reflexionen* [*Refl.*], 6130, in *AA*, 18:463.14–15, and *Refl.*, 6173, in *AA*, 18:477.21–23), and the world, as a "something = X" (cf. *CRP*, A 104, in *AA*, 03:80.10), God is not subject to Kant's critique in his transsubjective existence ("trans-subjective" means "located beyond transcendental subjectivity" here). On the contrary, in this respect, Kant presupposes the existence of God, as he explicitly says in the appendix to the Transcendental Dialectic in *CPR* (cf. Wundt 1924, 261, 264; Adickes 1927, 83–99; Fischer and Hattrup 1990, 131; Coreth and Schöndorf 2000, 196–97; Schöndorf 1995, 175–95; Schöndorf 2004, 53–75; Düsing and Düsing 2002, 100; Natterer 2003, 544, 573; Byrne 2007):

> Thus if one asks (in respect of a transcendental theology) *first* whether there is anything different from the world which contains the ground of the world order and its connection according to universal laws, then the answer is: *Without a doubt*. For the world is a sum of appearances, and so there has to be some transcendental ground for it, i.e., a ground thinkable merely by the pure understanding. If the question is *second* whether this being is substance, of the greatest reality, necessary, etc., then I answer *that this question has no significance at all*. For the categories through which I attempt to frame a concept of such an object are of none but an empirical use, and they have no sense at all when they are not applied to objects of possible experience, i.e., to the world of sense. Outside this field they are mere titles for concepts, which one might allow, but through which one can also understand nothing. Finally, if the question is *third* whether we may not at least think this being different from the world in accordance with an *analogy* with objects of experience,

> then the answer is *by all means*, but only as object in the idea and not in reality, namely, only insofar as it is a substratum, unknown to us, of the systematic unity, order, and purposiveness of the world's arrangement, which reason has to make into a regulative principle of its investigation of nature. (*CPR*, A 694–96, B 723–25, in *AA*, 03:457.14–458.03)

And a few paragraphs later in the appendix, we read:

> But in this way (one will continue to ask) *can* we nevertheless assume a unique wise and all-powerful world author? *Without any doubt*; and not only that, but we *must* presuppose such a being. But then do we extend our cognition beyond the field of possible experience? *By no means*. For we have only presupposed a Something, of which we have no concept at all of what it is in itself (a merely transcendental object); but, in relation to the systematic and purposive order of the world's structure, which we must presuppose when we study nature, we have thought this being, which is unknown to us, *in accordance with the analogy* with an intelligence (an empirical concept), i.e., in regard to the ends and the perfection on which those ends are grounded, we have given it just those properties that could contain the ground for such a systematic unity in accordance with the conditions of reason. (*CPR*, A 698–99, B 725–26, in *AA*, 03:458.14–27)

In the appendix, Kant introduces a twofold object-relatedness of the transcendental ideas: first an indirect relation to the object(s) of intuition (to the *Gegenstand in der Idee*), and second an unspecified relation to the transcendental objects in their qualities in themselves, or toward the thing in itself, opaque at its core, respectively (the *Gegenstand schlechthin* or a "Something") (*CPR*, A 670, B 698, in *AA*, 03:442.34). Therefore, we have to distinguish between the sense and the reference of a transcendental idea:

> Without using the terminology of "sense" and "reference," Kant has in effect given us an account of the meaning of "God" that sharply distinguishes the sense and reference of that name. The Critical Philosopher's reconstruction of the meaning of "God" takes it that the

word is used to refer to the whatever-it-is that is the transcendent ground of the order in the world as it appears to us. (Byrne 2007, 64)

Kant's criticism is directed against the attempt to treat God as a mere object of understanding, as a being alongside other beings that can be grasped in its content univocally. His critique is not directed against the presupposition of the transsubjective existence of God as an, in itself, "unknown Something" that cannot be treated as an object among other objects. But what follows from this line of reasoning, especially for the concept of God and the concept of our own minds?

THE EXISTENCE OF TRANSCENDENTAL OBJECTS IN THEMSELVES

How, on the basis of transcendental idealism, can Kant speak in the above-quoted passages of the existence of the necessary primal ground *in itself*, if the category of existence (*Dasein*) is only applicable to the empirical realm? The same problem also emerges when Kant claims that our freedom, and the affection of our sensibility by the things in themselves, is a kind of causality (cf. Adickes 1924, 61–70). This is only possible if the actual semantic meaning[3] of the categories (their "transcendental content") (*CPR*, A 79, B 105, in *AA*, 03:92.23–24) does not coincide with their application to intuition. In this case, it would be possible to *think* experience-transcendent objects by means of the categories, even if one still does not recognize them (i.e., if one still does not represent them *in concreto*, or detect them with an intuition and a temporal schematism, a time schema) (cf. *Prol.*, §45, Orig. 133, in *AA*, 04: 332.12–19; *CPR*, B 167, in *AA*, 03:128n; cf. Adickes 1927, 49–52, 62). Otherwise, the problem of an experience-transcendent use of the categories, and the problem of knowledge of the existence of God, could not arise in the first place.

Unspecified applications of the categories are possible within "thought," but not within "cognition." This Kantian distinction between thought and cognition resembles a twofold meaning of the term "category" in *CPR*: Categories, on the one hand, can be understood as functions of synthesis that can only be exercised on the material of intuition (i.e., on the manifold of sensation). On the other hand, categories can be understood as content-based starting points and also results of the synthesis, which must be distinguished from the sum of perceptions subsumed under them (cf. Adickes 1927, 75, 84). As mere functions of synthesis, the categories have no "significance," that is, they have no reference to any object and are therefore "empty" forms of thought (cf. Adickes 1927, 80). But, as starting points and results of synthesis, the categories alone, without intuition and schemata, have an unspecified reference to a "transcendental object" (cf. *CPR*, A 247, B 304, in *AA*, 03:207.23–27). Kant talks about the transcendental object as the referent of the categories, especially in the chapter on the "Synthesis der Rekognition im Begriffe," in the Transcendental Deduction in *CPR* A, and in the chapter "Phaenomena und Noumena" (cf. *CPR*, A 104, in *AA* 04:80.07–12; *CPR*, A 109, in *AA*, 03:82.07–83.19). The term "transcendental object" does not appear in the B-Deduction of *CPR*, but in substance it is still there, since by virtue of the *synthesis intellectualis* there is already a general object-relatedness of pure understanding. This is a relation to a mere object-structure (to the "transcendental object") (cf. Lotz 1955, 144–54) on the level of pure understanding *before* the differentiation of the *synthesis speciosa* on the level of intuition. So there is an intentionality of pure consciousness (of the transcendental apperception), the object of which is wider than that of a schematized cognition (cf. *CPR*, B 137, in *AA*, 03:111.21–24). Each term that has an object of reference also has an intension. With their reference to the transcendental object *in abstracto* (i.e., with their "transcendental significance") (Kant *CPR*, B 305, in *AA*, 03:208.17–18), the categories therefore have a "transcendental content," which is not identical with the content of the respective time-schemata (at least in the case of the categories of relation and modality).[4] Only in this way is it possible to justify the use of the term "causality" in the case of freedom, and the possibility of thinking the idea of God "by analogy" through the categories of infinity, reality, substance, causality, and necessity.[5]

What is denoted by the term "transcendental object" is of decisive importance for the present question. In the literature, this is the subject of an extensive debate that cannot be reported in detail here. But one can state that, as we saw with the transcendental ideas, a twofold object-relatedness emerges: first, an indirect relation to the objects of perception in their qualities in themselves, and, second, an unspecified relation to the transcendental object or to the thing in itself, opaque at its core.[6] The transcendental object has two meanings. First, it is the merely epistemic concept of the Something that corresponds to the unity of apperception: objectivity is given by the necessary constancy and autonomy of representations, which is guaranteed by the necessary connection in judgments governed by the categories, and by power of the necessary unity of the transcendental self-consciousness (*CPR*, A 250–251, in *AA*, 04:163.32–164.06; cf. *Refl.*, 5642, in *AA*, 18:283.29–31, and *Refl.*, 5554, in *AA*, 18:230.15–231.07; cf. Adickes 1927, 99n1, 100–101; Adickes 1920, 618; Allison 1968, 176–84). As we have seen, transcendental idealism is explicitly not designed to be a version of Berkeleyan idealism, so the meaning of the term "transcendental object" has to be deepened semantically. The transcendental object, in a second sense, is the thing in itself corresponding the phenomena (cf. Allison 1968, 171–72; Klemme 1996, 256, 260–64; Wundt 1924, 208; cf. *CPR*, A 252–253, in *AA*, 04:164.28–165.07, in *AA*, 23, 49, E43/44; *CPR*, A 288, B 344, in *AA*, 03: 231.03–08; cf. *Refl.*, 5652, in *AA*, 18:305.11–12; and *Refl.*, 5654, in *AA*, 18: 312.08–31). The existence of this corresponding thing in itself was never doubted by Kant himself.[7]

In the terminology of Henry E. Allison, Kant distinguishes a *transcendental object for us* from the *transcendental object in itself*. "Transcendental object in itself" is semantically identical with "thing in itself," as the cause of phenomena, so the terms are mutually dependent (cf. Allison 1968, 183). So, by the a priori object-relatedness of the categories, thinking transcending the categories is possible without any determination of things in themselves. Without this possibility of perception-transcendent thought, the restriction of all cognition to the empirical sphere in transcendental idealism would be utterly meaningless: cognition through experience would be total. It would itself become the thing in itself. So we are able to *think* content beyond the bounds of intuition by means of the categories, but we do not know if the things in themselves are as we think them: understanding knows of the *existence* of things in themselves without being able to determine their content. So "thing in itself" gains a "negative extension" (*CPR*, A 256, B 312, in *AA*, 03:212.21–27; cf. Adickes 1920, 615–16). The thing in itself, or the transcendental object *in itself*, is the "noumenon in the negative sense," an unknown as regards content, "something in general outside of our sensibility" (*CPR*, B 306–307, in *AA* 03:209.20–31; cf. *CPR*, B 307–308, in *AA*, 03:210.03–12). In a positive sense, that is, with determined content, it is the object of an intellectual intuition that is sealed from us.[8] The objects of the ideas also fall under those noumena explicitly called "transcendental objects" (*BDG*, in *AA*, 02:158.12–25).

Kant presupposes God as a transcendental object "*in itself*." We do not know if we are actually referring to an object with such qualities when we use our concept of God as determined by the categories. Nevertheless, the object of the idea of God actually exists as an unknown content, "something = x," as an "unknown substratum" (*CPR*, A 697, B 725, in *AA*, 03:458.01) of the unity of the world in the sense of negative theology (cf. *Prol.*, §57, Orig. 170–71, in *AA*, 04:354.34–355.10, and *Prol.*, §59, Orig. 182, in *AA*, 04:361.16–28; cf. Heintel 1958, 17).

UNIVOCAL KNOWLEDGE OF GOD AND KANT'S CRITIQUE OF THE PROOFS OF THE EXISTENCE OF GOD

How does the thesis that Kant explicitly accepts the transsubjective existence of God in the sense of negative theology fit with his critique of the proofs of the existence of God? There he explicitly speaks of the "impossibility of a proof for the existence of God." I sketch out the thesis that Kant does *not* attack the theoretical proofs of the existence of God *across the board*, but only if they presuppose a content-rich concept of God as the *ens realissimum*. Kant's critique is aimed against the rationalist attempt to gain *univocal* knowledge of God, that is, to grasp God in his essence in itself.[9] Kant's critique is not aimed against the presupposition of an unknown primal ground, which is different from matter, and necessary for the world.

Kant's target is the putative proofs of the existence of God of rationalist school metaphysics. He had already criticized those during his precritical phase in *Einzig möglichen Beweisgrund* (1763). As Sala has shown, the contingency-proof in the works of Wolff rests on the premise that there is a necessary being and, in a second premise, on the identification of this being with God, in virtue of its attributes (cf. Sala 1990, 116–17, 299–305). Kant's critique is aimed against any attempt to univocally determinate the attributes of this necessary, primordial, being (cf. *Critique of the Power of Judgment* [*CPJ*], §90, in *AA*, 05:466.26–37; cf. *CPJ*, §91, in *AA*, 05:473.16–20). The existence of God as demanded by the postulates of practical reason presupposes the *content-rich* concept of God, that is, the existence of the necessary ground of the world *as* creator whose content is determined by will and reason (cf. *Critique of Practical Reason* [*CPrR*], A 226, in *AA*, 05:125.14–25). The headings in the critique for the proofs for the existence of God would therefore have to be read with this emphasis: it is not a treatise on the "impossibility of a proof of the *existence* of God" but on the "impossibility of a proof of the existence of *God*" (where "God" implies an idea of God with determinate content) (cf. Sala 1990, 136).

The central idea of the critique of the proofs of the existence of God in *CPR* is this: reason is not just compelled to form the idea of God as a necessary condition for the continuous determination of all concepts (cf. *Refl.*, 6017, in *AA*, 18:424.16–17) but also to find an ultimate resting point in the regress of such conditions. But the conclusion that there is an absolutely necessary being, and the pure negation of its contingency, does not bestow on that being any determinate content that would give us insight into its "inner necessity" (as *necessarium per se* or *vi essentiae suae existens*, respectively) (cf. Natterer 2003, 546; cf. *Refl.*, 4156, in *AA*, 17:437.12–19). So reason identifies this being with the *ens realissimum*. So no conceptual room is left to ask *why* unless this being is infinite in reality (*CPR*, A 584, B 612, in *AA*, 03:393.03–07; cf. *CPR*, A 585, B 613, in *AA*, 03:393.17–23; *CPR*, A 592–593, B 620–621, in *AA*, 03:397.28–32; *CPR*, A 586–87, B 614–15, in *AA*, 03:394.16–23; *Refl.*, 5939, in *AA*, 18:395.12–15; cf. Schmucker 1969, 42). However, even this identification is purely conceptual, and this is exactly what Kant criticizes. The identification is logically compelling only if the judgment "Every absolutely necessary being is at the same time the most real being" can be transformed via *conversio per accidens* into the judgement "Some most real beings are at the same time absolutely necessary beings" (and, because of the uniqueness of the most real being, into the judgement "every most real being is a necessary being") (cf. *CPR*, A 588, B 616, in *AA*, 03: 395.15–19; *CPR*, A 608–9, B 636–37, in *AA*, 03:406.33–407.09). This is the case if existence can be derived from the concept of God with determinate content. This is exactly the point of the ontological argument,[10] which is rejected by Kant for well-known reasons. For these reasons, most of Kant's precritical argumentation (cf. *Vorlesungen Metaphysik, V-MP-L1/Pölitz*, in *AA*, 28:278–287) targets a conclusion that presupposes a concept of God with determinate content. The cosmological argument, on the other hand, is structured by Kant (according to Wolff's version) by two steps:

(a) the conclusion to a necessary being

and

(b) the conclusion to the continuous determination of the necessary being. (cf. *CPR*, A 604–5, B 632–33, in *AA*, 03:404.30–34; *CPR*, A 605–6, B 633–34, in *AA*, 03:405.09–16)

With step (a) we only reach necessary existence, on the basis of which we are not able to deduce necessary existence from the essence of God. Any such deduction from the essence of God would make a posteriori arguments for the existence of God unnecessary and therefore lead us back to the ontological argument (cf. Schmucker 1969, 57–59; Sala 1990, 298–99). The basic problem for Kant, in this case, is that this again is an argument on the basis of a content-rich concept of God, of the *ens realissimum.*

Kant seems to reject the conclusion that there is a necessary being (regarding the world) in some passages.[11] This is most explicit in the "dialectical presumptions" in the context of the cosmological argument (cf. *CPR*, A 609–10, B 637–38, in *AA*, 03:407.21–408.04). Sala and Schmucker think that the first two principles are of critical origin (cf. Sala 1990, 307; Schmucker 1969, 67). But, according to Sala, they do not contribute to the actual critique of the cosmological argument. Sala thinks with this critical doctrine Kant denies the validity of the first step of the cosmological argument (cf. Sala 1990, 301–2, 352). For an assessment of this view that Kant denies any justification of the conclusion to an entity, with content, which is not yet grasped in its essence in itself, we have to look at Kant's precritical analysis of the contingency-argument as it was presented in the Wolff school. The decisive passages are found in the *Einzig möglichen Beweisgrund zu einer Demonstration des Daseins Gottes* from 1763 (*BDG*, A 193–97, in *AA*, 02:157.29–159.24; cf. Sala 1990, 177–78). The cosmological argument again is presented in two steps:

(1) The causal conclusion to an independent cause of the world

and

(2) the conclusion by logical analysis of its concept that this has to be God. (*BDG*, in *AA*, 02:157.29–157.32)

Kant further subdivides step (1):

1.1. First, he concludes there is an independent, first cause (which is not further determined in content) by reflections on causality (*BDG*, in *AA*, 02:157.35–158.04).
1.2. Then, by the principle of sufficient reason, Kant concludes that this first cause has to be the *ens necessarium*.[12]

The identification of the first independent cause with the *ens necessarium* is possible only if the first cause is *causa sui*. Additionally, it would have to contain the cause for its existence within itself, so it would be *vi essentiae suae existens*. But then one would have to attribute logical necessity to something that essentially exists necessarily, that is, one would have to accept that the negation of the existence of the *ens necessarium* would lead to a logical contradiction (*BDG*, in *AA*, 02:158.12–25). Existence therefore would be included in its concept: the existence of the necessary being would be a real (*sachhaltiges*) predicate (cf. Sala 1990, 179–80, 107–14). And this is exactly Kant's premise for the identification of the *ens necessarium* with the *ens realissimum* with maximally saturated content (cf. Sala 1990, 182–83). So Kant's problem is not about the inference to a first, independent cause of the world (cf. *BDG*, in *AA*, 02:157.35–158.04; cf. *V-MP/Volckmann*, in *AA*, 28:402), to which can be attributed a necessity relative to the world. His problem concerns a univocal insight into the "inner necessity" of the ultimate ground: "In Wolff's proof it is indeed possible to recognize the independence but not the (inner) necessity; for its existence is only necessary as far as the world is concerned. From necessity he concludes the *omnitudinem realitatum*. If this is correct he does not have to underlay an experience."[13]

The knowledge of God in relation to the objects that depend on him is exactly what was understood as knowledge of God by analogy by classical Scholasticism (cf. Sala 1990, 129n). And this was the only goal of the classical proofs for the existence of God. Their goal was never a univocal knowledge of him: for Thomas Aquinas there is a *relatio realis*, merely from the direction of the world toward God. The other way around, there only is a *relatio rationis* (cf. Aquinas, *ST* I, q. 13, a. 7, co.; cf. Ricken 2003, 295; *BDG,* in *AA*, 02:158.12–25). We can know God only *as creator* (i.e., in relation to the world).[14] In the critical philosophy, Kant similarly holds that we recognize God not in himself but only *in relation* to our morality (cf. *Refl.,* 6094, in *AA*, 18:449.30–32, and *Prol.*, §57, in *AA*, 04:357.08–16). Byrne states:

> [Kant] stands in a long tradition of Western thinkers who have denied that there can be direct, univocal property ascriptions to God. Many thinkers have reached the same conclusion as Kant: direct talk about God, as opposed to talk about God's acts or his relations to the world, comes up against a boundary of sense. . . . There is a form of concept empiricism behind and within Aquinas's discussion of the divine names in question 13 of the *prima pars* of the *Summa Theologiae.* (Byrne 2007, 64)

The same argument against the cosmological argument presented above can be found in Kant's critical phase, exemplary in the *Vorlesungsnachschrift V-MP/Volckmann* of 1784/85 (cf. also *V-Phil-Th/Pölitz*, in *AA*, 28:23–25). There Kant deals with the *Leibnizschen Beweiß a contingentia mundi* and elaborates:

> The proof is quite short: if something exists, then something exists necessarily, for that which exists, is either necessary in itself or accidental—in which latter case it must have a cause. If there is no necessary being one ends up with an ever-continuing series of *caußatis*, so there must be a highest cause which is no *causatum* and which therefore is the *ens necessarium*, which was the major *syllogismi*, the minor is *atqui* something exists because I am, therefore something exists necessarily.[15]

The component steps of the cosmological argument (1.1) and (1.2) are clear here. Kant's critique starts out with the attempt to determine the *absolute* necessity of this necessary being. The critique is equivalent to that of the precritical phase and to the critique elaborated in *CPR*:

> But the absolute necessity of something no man can know, even though Wolff wanted to provide examples for that in which process he deceived himself by adducing absolutely necessary judgements; but the judgement is only the relation of the predicate to the subject, therefore very different from the *ente neceßario*—for an *ens necessarium* is posited necessarily, it is impossible to suspend, with necessary judgements the opposite shall be impossible which consists in suspending the predicate without the object. In the second case I suspend the thing with all its predicates in which case I do not at all contradict myself.[16]

Kant's critique further aims at the determination of the *attributes* of the necessary being, which leads to the structure of the critique presented in *CPR*:

> Now one has to ask: what are the attributes of this being? The transcendental theology contains 2 proofs, the ontological where we infer the existence of highest reality from its concept, and the cosmological where we presuppose: there is something, and conclude from that, that something is absolutely necessary—and then one sets out to deduce the attributes of the *entis necessarii* from its concept. . . . But effectively there is no cosmological proof, but the cosmological is only a hidden ontological proof.[17]

Here we see that Kant does not present arguments against the cosmological argument in his critical phase different from those in his precritical phase. It therefore becomes clear that Kant's intention in rejecting the "dialectical presumptions" in *CPR* remains the same: he rejects the attempt to reach the *content-rich* idea of God by arguing for cosmological causal conclusions. This does not imply excluding the possibility of transcending to a sealed substratum with content, existing in itself as the independent primordial cause.[18] By means of the causal principle we are able to reach a definite *essence* only *within the empirical sphere*, but not the essence of an extramundane being.[19] Inspecting the causal chain, one does not reach an absolutely necessary first member. Therefore the solution of the fourth antinomy can only consist in an extramundane necessary being that is a *substantia noumenon*, which is empty of content (cf. *CPR*, A 559–65, B 587–93, in *AA*, 03:378.01–381.21; cf. Heimsoeth 1966/69, 387–89). Kant's critique targets the determinate *substantia noumenon* as *noumenon positivum*, but not the necessary primordial ground of the world as transcendental object = x (or *noumenon negativum*, respectively). Further passages in *CPR* show that Kant's critique is about our inability to infer the existence of a determinate necessary being with content, using the existence of the world as premise (cf. *CPR*, A 562, B 590, in *AA*, 03:380.01–03; *CPR*, A 565–66, B 593–94, in *AA*, 03:382.01–07; *CPR*, A 566, B 594, in *AA*, 03:382.08–21; *CPR*, A 635, B 663, in *AA*, 03:422.21–31; *CPR*, A 636–37, B 664–65, in *AA*, 03:423.22–32). The inner essence of a cause transcending experience cannot be reached by any inference from the existence of worldly causal chains. The "regulative principle" refers to a content-determinate concept of God. With respect to Him, the rule holds: philosophize about nature in such a way *as if* there were a necessary primal ground to it (cf. *CPR*, A 616–17, B 644–45, in *AA*, 03:411.20–31; cf. also *CPR*, A 564–65, B 592–93, in *AA*, 03:381.12–21).

Kant's final version of the texts on the transcendental ideal can therefore be interpreted as implying a closed conception: the critique of the "dialectical presumptions" relates only to an extrapolation of worldly causality, aspiring to be a univocal concept of the highest necessity. Also, a thorough analysis of the physico-theological proof shows that Kant's critique is only initiated when a univocal content-rich concept of God is presupposed.

THE AMBIGUITY OF KANT'S DUALISM AND HOW TRANSCENDENTAL IDEALISM POINTS BEYOND ITSELF

According to Kant, the extension of our cognition is limited by the boundaries of empirical intuition and by the boundaries of the constitutive functions of our "apparatus or capacity of cognition." We recognize only "appearances," not "things in themselves." But this leads to two fundamental "perspectives" or "standpoints" within *CPR*. First, there is a "transcendental" standpoint of purely immanent cognition, in which everything that goes beyond cognition grounded in the conditions of sensuality, and the limited functions of our ability to know, is made in principle problematic, incognizable.

Second, there is a "meta-transcendental" standpoint that putatively explains how the distinction between "thing in itself" and "appearance" can be drawn, as can be found in Schöndorf 1995, and I paraphrase: This tension between the "cognition-immanent" (or "transcendental") standpoint and the "cognition-transcendent" (or "meta-transcendental") standpoint leads to Hegel's fundamental criticism of Kantian transcendental idealism. Kant's meta-transcendental statements contain a presupposition that contradicts its own explicit limitation of our cognition to appearances, namely: the presupposition that we are capable of knowing that our cognition is, itself, structured in a transcendental-idealistic manner. This must be excluded due to the transcendental-idealistic restriction of our cognition to appearances (cf. here Hegel [1831] 1934, 2:230–32, 440–41, and Puntel 1981, 208–29).

A more precise explanation of the relation between the transcendental and meta-transcendental standpoint in *CPR* can be made with the help of the so-called *theoreticity-operator* (or *theoretical operator*), "T" (cf. Puntel 2006, 120–26; for the entire topic, see 481–504):

> Definition 1: for a proposition ϕ, T is an operator such that T(ϕ) is informally read as "it is the case that ϕ" (e.g., "it is the case that the earth orbits the sun"). Then T is the *theoretical operator*. If T(ϕ) builds a syntactically correct proposition, then ϕ is a declarative (theoretical) proposition (Puntel 2006, 122–23).

"Theoretical" is not used here in the same sense as Kant uses it (in distinction from "practical") but refers to the unlimited dimension of that which is the case *simpliciter*. T(ϕ) is thus not to be read as "it is *empirically* [or a priori or a posteriori and so forth] the case that ϕ." The scope of "theoretical," in the intended sense, includes all "elements" that make up a conception or theory (including "subjective capacity," and so forth). This theoretical dimension is expressed in theoretical propositions, those having the structure T(ϕ). The theoretical dimension expressed in theoretical propositions carries no reference to a subject's capacities, or any other factor external to the theoretical dimension, and is thus "unlimited" (cf. Puntel 2006, 122, 148–61).

Definition 2: T*TR* the *transcendentally limited theoretical operator* (abbreviated to *transcendental operator*) is such that T*TR*(ϕ) is informally read as "it is *from the perspective of transcendental subjectivity* the case that ϕ."

It is then valid:

1. If the theoretical operator, "T," is limited in an implicit or explicit manner, the unlimited theoretical operator T is nonetheless presupposed: Each philosophical standpoint, "S," (skeptical, pragmatic, hermeneutic, dialectical, transcendental, and so forth) is always articulated in the form T(S). Informally: "It is the case that S (is the correct philosophical standpoint)," or "It is *simpliciter* the case that S is valid."
2. The *transcendental* standpoint S*TR* in *CPR* contains all propositions of the form T*TR*ϕ, whereas the meta-transcendental level of the *CPR* is articulated in propositions of the form T(ψ): T(T*TR*(ϕ)) or T(S*TR*), respectively. The *unlimited* theoretical operator is thus always presupposed on the meta-transcendental level.
3. In what relation, then, do the transcendental and meta-transcendental levels stand? Is the unlimited operator T on the *explicit*, transcendental level of *CPR* reduced entirely to the limited operator T*TR* thus reducing the entire theoretical dimension to the limited transcendental dimension? This would lead to inconsistency between the two levels, as the meta-transcendental level is, at least implicitly, always already presupposed.
4. With the results of the examination in this chapter, the understanding of the operator T*TR* can be made more precise: T*TR* means, on the explicit level of *CPR*, not a *total* limiting of T to a purely transcendental-subjective, subject-immanent dimension but rather a limitation only in respect of *content* and *descriptive* cognition. The knowledge of the simple *existence* of the object of cognition in content-discernment that is structured along meta-transcendental lines is not excluded by T*TR*. (It offers itself, however, on the tran-

scendental level, merely as things in themselves of undetermined content or *transcendental objects* = x). Transcendental subjectivity refers from itself to that which is other to it, that which is not given with determinate content (cf. Puntel 1969, 335–38).

5. This presupposes, however, that there is meta-transcendental structured cognition with content, which is situated in the fully theoretical dimension of T: this is in Kant's thought (following Westphal 1968 and Schöndorf 1995) as cognition by the *intellectus archetypus* or *intuitive reason*, that is, the *infinite divine reason*, which is always already presupposed as existing by Kant. When Kant claims that one cannot determine the possibility of such reason, this can only refer to a cognition *with regard to the unequivocal content* of divine reason from the perspective of T*TR*. For, if intuitive reason is impossible, then this senseless speech would result in potentially discernable characteristics which are in themselves, and for which an accompanying actual discernment is impossible (i.e., a potentiality which can never be actualized).
6. The fundamental ambiguity in Kant's theoretical philosophy, the tension between the transcendental and meta-transcendental levels, can therefore already be found on the explicit level of *CPR*. The dualism between phenomena and noumena arises at the end as a rift, or gap, between finite and infinite manners of cognition, as a sort of "noetic dualism": "It is a noetic dualism, for it refers to two ways of knowing one world. And it is a theistic dualism, for the mode of knowledge which stands as the criterion of reality is that of a divine, intellectually intuiting, creative reason." (cf. *CPR*, A 252–53, in *AA*, 04:164.28–165.07)

It follows that the gap between finite and infinite manners of cognition must always be overcome at one point or another: "When God perceives a thing as it is in itself, then the assumption can be made that God also grants to human intellect a minimum of knowledge of independent reality which is necessary so that we can say that the thing in itself cannot be of the space-time manner of being of empirical objects" (Schöndorf 1995, 181).

From this reasoning, the following picture presents itself.[20] Infinite divine reason itself is somehow present in our finite minds. The sentences emerging at the meta-transcendental point of view articulate their "topic" in an *absolute* manner, without any recourse to the distinction between subject(ivity) and object(ivity) and without any limitation. In these theoretical sentences the thing in itself *articulates itself* and the dimension that finds its expression in this way is the very dimension at which Hegel aimed with *absolute thinking*. It is the dimension of Being as such, and as a whole, and our mind is coextensive with this all-encompassing dimension of totality (*anima, quae quodam modo est omnia*; Aquinas, *De veritate* q. 1, a. 1, co). This has crucial consequences for any conception of God that putatively does justice to the full capacities of the human mind in any way. This far exceeds anything one can find in Kant's philosophy, even implicitly and allusively: "God" as the truly Absolute only "enters the stage" in a philosophically adequate manner when the philosophical thinker has reached the "standpoint" of absolute thinking (cf. Puntel 2011, 11). Then "God" can no longer be an external, object-like being (*ens*, *Seiendes*), a being among other beings, but coincides with the totality of Being (*esse*, *Sein*) itself: "God" is the all-encompassing dimension of absolute-necessary Being itself (the *ipsum esse per se subsistens* [Aquinas, *ST* I, q. 4, a. 2], freely creating the contingent dimension of Being) and as such stands not in any kind of objectifying or external relation to created beings. (Of course, this can only be hinted at here.) God and the world are not separate subdimensions of Being, which would encompass both of them.[21] But the difference between God as Being in itself and the existence of the created world is posited *within* God as a difference from himself. God's transcendence therefore obtains within his total "auto-immanence."[22] The existence of contingent beings, therefore, is (to follow Aquinas) a *participation in Being itself*. In this way our finite mind can be seen as a *participation* in the absolute spirit of God: *Est Deus in nobis* (*Opus Postumum*, VII, in *AA*, 22:130).[23]

NOTES

For an earlier version of sections "The Existence of Transcendental Objects in Themselves" and "Univocal Knowledge of God," see Schneider 2013, and for parts of the last section, see Schneider 2011, 223–25. Abbreviations used for cited works of Kant: *AA* = *Akademie Ausgabe*; *BDG* = *Der einzigmögliche Beweisgrund zu einer Demonstration des Daseins Gottes* (in *AA*:02); *CPR* = *Critique of Pure Reason* (*AA*:03);

CPrR = *Critique of Practical Reason* (*AA*:05); *PND* = *Principiorum primorum cognitionis metaphysicae nova dilucidatio* (*AA*:01); *Prol.* = *Prolegomena zu einerjeden künftigen Metaphysik* (*AA*:04); *Refl.* = *Reflexionen* (*AA*:14–19); *OP* = *Opus Postumum* (*AA*:21 and 22); *V-MP/Volckmann* = *Metaphysik Volckmann* (*AA*:28); *V-MP-L1/Pölitz* = *Kant Metaphysik L1* (Pölitz) (*AA*:28); *V-MP/Herder* = *Nachträge Metaphysik Herder* (*AA*:28); *V-MP/Schön*=*Metaphysik von Schön Ontologie* (*AA*: 28).

1. ". . . the highest being remains for the merely speculative use of reason a mere but nevertheless *faultless ideal*, a concept which concludes and crowns the whole human cognition, whose objective reality cannot of course be proved on this path, but also cannot be refuted. . . . Necessity, infinity, unity, existence outside of the world (not as soul of the world), eternity without all conditions of time, omnipresence without all conditions of space, omnipotence, etc.: these are purely transcendental predicates, and hence a purified concept of them, which every theology needs so very badly, can be drawn only from transcendental theology" (Kant, *CPR*, A 641–42, B 669–70, in *AA*, 03:426.06–14).

2. For the topic of the existence of the world as a "transcendental object" or thing in itself respectively, see *CPR*, A 494, B 522, in *AA*, 03:341.03–06; *Refl.*, 5961, in *AA*, 18: 400.24–401.01; and *Refl.*, 5652, in *AA*, 18:305.11–12. For the existence of the soul or the transcendental apperception as a thing in itself, see *CPR*, A 682, B 710, in *AA*, 03:449.28–33; *CPR*, A 565, B 593, in *AA*, 03:382.01–02; *CPR*, in *AA*, 03:276n; *CPR*, B 157, in *AA*, 03:123.02–05; *CPR*, B 157, in *AA*, 03:123.10–11; *Refl.*, 5939, in *AA*, 18:395.16–17; *Refl.*, 5984, in *AA*, 18:416.04–06; and, especially, *Refl.*, 5661, in *AA*, 18:318f. Cf. Schmucker 1990, 352–53; Schöndorf 1995, 193.

3. The term "semantic meaning" is to be understood as the intension of a concept. Kant uses the term "significance" (*Bedeutung*) to express the "relation to an object" (cf. *CPR*, B 300, in *AA*, 03:205.20–21) (= objective significance, the term "transcendental significance" denotes the relation to the transcendental object).

4. Cf. Klemme 1996, 256–57 and 272. For example, the category of substance means independence and not infinite duration or endurance in time, respectively. The difference between transcendental content and content of the scheme in this case does not imply any equivocation: the schemes are mediations of the concept in its purity and of intuition, or, respectively, consequences in the case of an empirical realization (i.e., independence in empirical realization necessarily entails a duration in time). Klemme is of the opinion that the concept of the transcendental object in the B-edition of *CPR* disappears for the very reason that Kant wants to avoid the use of nonschematized categories in judgments. If there were categorical judgments that are not judgments of knowledge, according to Klemme especially, the difference between thinking and knowing and between general and transcendental logic would disappear (Klemme 1996, 272–76). But the fact alone that, for example, the postulates of practical reason are *theoretical* sentences that characterize their objects using the categories while practical reason simultaneously does not provide any sensations but only reasons seems to contradict this. This kind of sentence would be

impossible if the categories had content only if they relate to a manifold of sensations. The same applies to the *analogue schematism* of dialectics, which already means a mediation of the sensory world with its negation or its "beyond" (the noumenal world) respectively (cf. Specht 1952, 57).

5. *CPR*, A 697, B 724, in *AA*, 03:257.19–258.03; *CPrR*, in *AA*, 05:100.33–36. Cf. Adickes 1927, 90–91. For the concept of analogy in Kant's theoretical philosophy, see Specht 1952; Heintel 1954, 107–11; Heintel 1958, 9; Lakebrink 1960, 244–57; Puntel 1969, 303–50; Byrne 2007, 96–97.

6. For the idea of God, see, especially, *CPR*, A 565, B 593, in *AA*, 03:382.01–10.

7. Cf. Kemp Smith 1962, 404; Aquila 1979, 302–3; Langton 1998, 32; Langton 2004, 129–36; and Adickes 1924, 149–50: "Der Begriff des Noumenon, vom Standpunkt gewisser radikaler Prämissen und Konsequenzen der Tr.ph. aus betrachtet, [führt] uns auf keine wissenschaftlich gültige Weise über den Kreis des Bewußtseins hinaus und in das Gebiet des Transzendenten hinüber . . . weil wir keine Möglichkeit haben, dem Transzendenten gleichsam Auge in Auge gegenüberzutreten und es mit unseren Gedanken zu vergleichen, daß also das Noumenon für uns—als strenge, einseitige Erkenntnistheoretiker—immer nur ein bloßes Gedankenprodukt . . . bleibt, dessen objektive Realität zu beweisen oder gar sein Wesen zu erkennen wir nie imstande sein werden. Damit hat aber auch die Skepsis . . . ihr Ende erreicht: sie richtet sich nur gegen die theoretische . . . Erkennbarkeit der Dinge an sich, nicht gegen ihre Existenz." In this context, Klemme speaks of a categorically undetermined concept of existence, which points at existence in itself (cf. Klemme 1996, 257–58).

8. *CPR*, B 308–309, in *AA*, 03:210.21–34; cf. *Refl.*, 6048, in *AA*, 18:433.16–21. Cf. Westphal 1968, 127; cf. *OP*, in *AA*, 22, 26.26–29: "Der Unterschied der Begriffe von einem Dinge an sich und den in der Erscheinung ist nicht objectiv sondern blos subjectiv. Das Ding an sich (*ens per se*) ist nicht ein anderes Object, sondern eine andere Beziehung (*respectus*) der Vorstellung auf dasselbe Object."

9. Cf. Coreth and Schöndorf 2000, 196. Cf. Natterer 2003 (546, 544): "Kant . . . akzeptiert durchaus eine *analoge* Erkenntnis Gottes . . . aber nicht eine positive Erkenntnis Gottes in sich. . . . Nur auf diese positive Erkenntnis Gottes *im menschlichen Begriff* mit dem Anspruch logischer und mathematischer Stringenz bezieht sich die kantische Kritik der Gottesbeweise." Cf. Schöndorf 2004 (63–64): "Here we see at the heart of Kant's critique: His claim means that we are fundamentally unable to make any descriptive utterance about God, be the proof of his existence as it may. This critique is in accordance with Kant's claims about the metaphysical doctrine of soul in the chapter of the paralogism. Kant does not deny that every human being has a soul and that I know that I exist. . . . However, he polemicizes severely against any attempt to attribute any properties to the soul. This is also Kant's opinion with regard to the knowledge of God. Here too, Kant does not deny that we know that God exists, but he comes out against any characterisation of God." Cf. Schmucker 1969, 47, and Natterer 2011, 127–33.

10. Kant does not use "ontological argument" to refer to Anselm's proof in the *Proslogion*, but to Descartes's ontological proof in his Fifth Meditation, and the reformulation of this argument by Christian Wolff and Leibniz (cf. Coreth and Schöndorf 2000, 193; Schrimpf 1994, 60–61).

11. Cf. *CPR*, A 605, B 633, in *AA*, 03:404–5n; in *AA*, 15, 31–32; *V-MP/Herder*, in *AA*, 28:13; *V-MP/Volckmann*, in *AA*, 28:399–410; *CPR*, A 615, B 643, in *AA*, 03:410.32–35; *CPR*, A 604, B 632, in *AA*, 03:404.15–22. For the "patchwork" character of these passages, see Sala 1990, 301–2; Strawson 1981, 256–58; Schmucker 1990, 378–81; Kemp Smith 1962; Sala 1987, 153–69; Natterer 2003, 528.

12. *BDG*, in *AA*, 02:158.04–08: "Nun ist der zweite Schritt zu dem Satze, daß dieses unabhängige Ding schlechterdings nothwendig sei, schon viel weniger zuverlässig, da er vermittelst des Satzes vom zureichenden Grunde, der noch immer angefochten wird, geführt werden muß; allein ich trage kein Bedenken auch bis so weit alles zu unterschreiben." Cf. *Refl.*, 6033, in *AA*, 18:428.06f.; *PND*, in *AA*, 01:394.10–395.03; cf. Sala 1990, 178.

13. *Refl.*, 3812, in *AA*, 17:301.07–11: "Im *Wolfischen* Beweise kan man wohl die independentz, aber nicht die (innere) nothwendigkeit erkenen; denn sein Daseyn ist nur um der Welt willen nothwendig. Aus der nothwendigkeit schließt er die *omnitudinem realitatum*.—Wenn das richtig ist, so hatte er nicht nothig, eine erfahrung zu substruiren." Cf. Schmucker 1969, 110–11; *V-MP/Volckmann*, in *AA*, 28:455; *CPR*, A 324–25, B 381–82, in *AA*, 03:252.35–253.05; and *Refl.*, 3712, in *AA*, 17, 252.11–17.

14. Cf. Doyle 1984, 121–60, and Wennemann 1988, 119–29. Cf. Aquinas, *Scriptum super IV libros Sententiarum* IV, d. 49, q. 2, a. 1, ad. 3: Deus omnem formam intellectus nostri subterfugit: quia quamcumque formam intellectus noster concipiat, illa forma non pertingit ad rationem divinae essentiae; et ideo ipse non potest esse pervius intellectui nostro; sed in hoc eum perfectissime cognoscimus in statu viae quod scimus eum esse super omne id quod intellectus noster concipere potest; et sic ei quasi ignoto conjungimur.

15. *V-MP/Volckmann*, in *AA*, 28.1:456.9–17: "der Beweiß ist ganz kurz: wenn etwas existirt so existirt auch etwas nothwendiger Weise, denn das was existirt ist entweder selbst nothwendig oder zufällig, beym lezten muß es eine Ursache haben. Ohne ein nothwendiger Weise existirendes Wesen anzunehmen, hat man eine immerwährende Reihe von caußatis, es muß also eine oberste Ursache geben die kein caußatum ist, welches also ensneceßarium ist, dies war major syllogismi, minor ist, atqui es existirt etwas, denn ich bin, ergo existirt auch irgend etwas nothwendiger Weise."

16. *V-MP/Volckmann*, in *AA*, 28.1:456.17–27: "Allein die absolute Nothwendigkeit eines Dinges kan kein Mensch einsehen, wiewohl Wolff Beyspiele davon geben wollte, sich aber selber betrog indem er absolut nothwendige Urtheile anführte, das Urtheil ist aber blos die relation des Prädikats zum Subject, also vom ente neceßario sehr verschieden, denn ein ensneceßarium ist geseztnothwendiger Weise, man kans unmöglich aufheben, bey den nothwendigenUrtheilen soll das Gegentheil unmöglich seyn, welches in der Aufhebung eines Prädikats ohne das Object besteht,

im 2ten Fall aber hebe ich das Ding auf mit allen seinen Prädikaten, wo ich mich denn gar nicht wiederspreche."

17. *V-MP/Volckmann*, in *AA*, 28.1:456.27–457.13: "Nun frägt sich: welches sind die Eigenschaften dieses Wesens? Die transcendentale Theologie enthält 2 Beweise den ontologischen, wo wir aus dem Begriff der höchsten realitaet auf sein Daseyn schließen und den kosmologischen, wo wir voraus setzen: es ist was, und daraus schließen daß etwas absolut nothwendigsey und nun geht man darauf hinaus aus dem Begriff des entisneceßarii seine Eigenschaften zu folgern. . . . Im Grunde aber giebts keinen eigentlichen cosmologischen Beweiß, sondern der cosmologische ist nur ein verstekter ontologischer."

18. Cf. Schmucker 1990, 287–89; Natterer 2003, 519–20. Cf. Schmucker 1990, 288–89: "Ein intelligibles Absolutes . . . kann nie als unmittelbare Ursache von Erscheinungen als Erscheinungen gedacht werden, sondern nur als Ursache des den Erscheinungen zugrundeliegenden intelligiblen Substrats derselben." Cf. *Refl.*, 4135, in *AA*, 17:429.16–17: "Das Phänomenon von einem Dinge ist ein Product . . . unserer Sinnlichkeit. Gott ist Urheber der Dinge an sich"; and *Refl.*, 5981, in *AA*, 17:414.29–31; cf. Natterer 2003, 527.

19. Cf. *V-MP/Schön*, in *AA*, 28:500: Es ist "eine nothwendige Hypothesis, ein oberstes Wesen anzunehmen, denn sonst bekommen wir niemahls ein oberstes Glied in der Reihe und ohne dieses haben alle andre Glieder keine Haltung. Es ist mithin eine nothwendige Hypothesis, einen Urgrund anzunehmen, demohngeachtet kann uns unsre schwache Vernunft, ungeachtet der Nothwendigkeit dieser Hypothese keinen deutlichen Begriff davon geben." Cf. *CPR*, A 562, B 590, in *AA*, 03:380.01–03; *Prol.*, §45, Orig. 133, in *AA*, 04:332.12–16.

20. For the following, cf. Puntel 2011, 2009.

21. ". . . the absolutely necessary dimension of Being and the contingent dimension of Being are not (in the ordinary sense) two subdimensions of a yet more primordial dimension. Instead, they are the *explicated* primordial dimension of Being itself; this dimension discloses itself first as two-dimensional, and then as the absolute, necessary, free (personal) dimension of Being that creates the contingent dimension of Being and thereby contains it. The stages of the (self-) explication of the primordial dimension of Being are precisely that: stages in its (self-) explication. The primordial, universal dimension of Being is the *explicandum*, and the absolutely necessary dimension of Being as creating-absolute-that-has-created-the-contingent-world is the *explicans/explicatum* of the primordial, universal dimension of Being" (Puntel 2011, 261).

22. "'Immanence' connotes a correlation with something outer or other, but the prefix 'auto-' excludes anything outer or other. The prefix could, however, be taken to suggest that there must be something outer or other *to be excluded*. Yet, in the case of the primordial, universal dimension of Being, explicated as the-absolutely-necessary-dimension-of-Being-as-creating-the-contingent-dimension-of-Being, there simply is no outer or other to which that dimension could relate, even by negation. . . . [T]he

transcendence of God in relation to contingent beings occurs *within God* as the consequence of God's own free act. In that God, by creating, posits contingent beings within Being, God creates within God the internal distinction by means of which God transcends contingent beings" (Puntel 2011, 261–62).

23. For translations, I express my gratitude to Georg Maximilian Knauer, Philip Steward, and Stephen Henderson.

REFERENCES

Adickes, Erich. 1920. *Kants Opus Postumum*. Berlin: Pan.

———. 1924. *Kant und das Ding an sich*. Berlin: Pan.

———. 1927. *Kant und die Als-ob-Philosophie*. Stuttgart: Fr. Frommans.

Allison, Henry E. 1968. "Kant's Concept of the Transcendental Object." *Kantstudien* 59 (1–4): 165–86.

Andersen, Svend. 1983. *Ideal und Singularität*. Berlin: De Gruyter.

Aquila, Richard E. 1979. "Things in Themselves and Appearances: Intentionality and Reality in Kant." *Archiv für Geschichte der Philosophie* 61:293–307.

Aquinas, Thomas. 1882–. *Opera omnia iussu Leonis XIII edita cura et studio Fratrum Praedicatorum* [Editio Leonina]. Rome.

Baumgarten, Alexander Gottlieb. (1739) 2013. *Metaphysics: A Critical Translation with Kant*'s *Elucidations, Selected Notes, and Related Materials*. Edited by Courtney D. Fugate and John Hymers. London: Bloomsbury.

Bennett, Jonathan F. 1974. *Kant's Dialectic*. Cambridge: Cambridge University Press.

Byrne, Peter. 2007. *Kant on God*. Aldershot: Ashgate.

Coreth, Emerich, and Harald Schöndorf. 2000. *Philosophie des 17. und 18. Jahrhunderts*. Stuttgart: Kohlhammer.

Doyle, John P. 1984. "Prolegomena to a Study of Extrinsic Denomination in the Work of Francis Suarez, S.J." *Vivarium* 22 (2): 121–60.

Düsing, Edith, and Klaus Düsing. 2002. "Negative und positive Theologie bei Immanuel Kant: Kritik des ontologischen Gottesbeweises und Gottespostulats." In *Societas rationis: FS für Burkhard Tuschling zum 65. Geburtstag*, edited by Dieter Hünig, Gideon Stiening, and Ulrich Vogel, 85–118. Berlin: Duncker und Humbolt.

Fischer, Norbert, and Dieter Hattrup. 1990. *Metaphysik aus dem Anspruch des Anderen: Kant und Levina*s. Paderborn: Schöningh.

Hegel, G. W. F. (1831) 1934. *Wissenschaft der Logik*, vols. 1 and 2. Edited by Georg Lasson. Hamburg: Meiner.

Heimsoeth, Heinz. 1966/69. *Transzendentale Dialektik: Ein Kommentar zu Kants Kritik der reinen Vernunft*. Berlin: De Gruyter.

Heintel, Erich. 1954. "Kant und die Analogia entis." *Wissenschaft und Weltbild* 7 (2): 107–11.

———. 1958. *Hegel und die Analogia entis.* Bonn: Bouvier.

Horstmann, Rolf-Peter. 1998. "Der Anhang zur transzendentalen Dialektik (A642/B670–A704/B737)." In *Immanuel Kant: Kritik der reinen Vernunft*, edited by Georg Mohr und Marcus Willaschek, 525–45. Berlin: Akademie.

Kant, Immanuel. (1755–) 1910–. *Sämtliche Werke.* 29 vols. Edited by Preussische Akademie der Wissenschaften. Berlin: Reimer.

———. 1998. *Critique of Pure Reason.* Translated and edited by Paul Guyer and Allen W. Wood. Cambridge: Cambridge University Press.

Kemp Smith, Norman. 1962. *A Commentary to Kant's Critique of Pure Reason.* New York: Humanities.

Klemme, Heiner F. 1996. *Kants Philosophie des Subjekts: Systematische und entwicklungsgeschichtliche Untersuchungen zum Verhältnis von Selbstbewußtsein und Selbsterkenntnis.* Hamburg: Meiner.

Lakebrink, Bernhard. 1960. "Der Kantische Begriff der transzendentalen Analogie." *Philosophisches Jahrbuch* 68:244–57.

Langton, Rae. 1998. *Kantian Humility: Our Ignorance of Things in Themselves.* Oxford: Oxford University Press.

———. 2004. "Elusive Knowledge of Things in Themselves." *Australasian Journal of Philosophy* 82 (1): 129–36.

Lotz, Johannes B. 1955. *Kant und die Scholastik heute.* Pullach: Berchmanskolleg.

Natterer, Paul. 2003. *Systematischer Kommentar zur Kritik der reinen Vernunft: Interdisziplinäre Bilanz der Kantforschung seit 1945.* Berlin: De Gruyter.

———. 2011. *Philosophie der Transzendenz: Mit einem systematischen Abriss der kantischen philosophischen Theologie.* Norderstedt: Books on Demand.

Puntel, Lorenz B. 1969. *Analogie und Geschichtlichkeit.* Freiburg im Breisgau: Herder.

———. 1981. "Transzendentaler oder absoluter Idealismus." In *Kant oder Hegel? Über Formen der Begründung der Philosophie*, edited by Dieter Henrich, 198–229. Stuttgart: Klett-Cotta.

———. 2006. *Struktur und Sein: Ein Theorierahmen für eine systematische Philosophie.* Tübingen: Mohr Siebeck.

———. 2009. "The *Phenomenology of Spirit* and the Unity of Hegel's Philosophy: A Systematic Reappraisal." In *Still Reading Hegel: 200 Years after the Phenomenology of Spirit*, edited by Edmundo B. Pires, 171–97. Coimbra: Coimbra University Press.

———. 2011. *Being and God: A Systematic Approach in Confrontation with Martin Heidegger, Emmanuel Levinas and Jean-Luc Marion.* Evanston, IL: Northwestern University Press.

Ricken, Friedo. 2003. *Religionsphilosophie.* Stuttgart: Kohlhammer.

Sala, Giovanni B. 1987. "Bausteine zur Entwicklungsgeschichte der Kritik der reinen Vernunft." *Kantstudien* 78:153–69.

———. 1990. *Kant und die Frage nach Gott: Gottesbeweise und Gottesbeweiskritik in den Schriften Kants.* Berlin: De Gruyter.

Schmucker, Josef. 1969. *Das Problem der Kontingenz der Welt*. Freiburg im Breisgau: Herder.

———. 1990. *Das Weltproblem in Kants Kritik der reinen Vernunft: Kommentar und Strukturanalyse des ersten Buches und des zweiten Hauptstückes des zweiten Buches der transzendentalen Dialektik*. Bonn: Bouvier.

Schneider, Ruben. 2011. *Kant und die Existenz Gottes: Eine Analyse der ontologischen Implikationen in Kants Lehre vom transzendentalen Ideal*. Münster: LIT.

———. 2013. "Die transsubjektive Existenz Gottes bei Kant." In *Kant und die biblische Offenbarungsreligion*, edited by J. Sirovátka and D. Voprada, 29–44. Prague: Karls-Universität Karolinum Verlag.

Schöndorf, Harald. 1995. "Setzt Kants Philosophie die Existenz Gottes voraus?" *Kantstudien* 86:175–95

———. 2004. "What Does Kant Say about the Theoretical Recognition of God?" *Universitas* (Taipei) 357:53–75.

Schrimpf, Gangolf. 1994. "Anselm von Canterbury. Proslogion II—IV. Gottesbeweis oder Widerlegung des Toren?" *Fuldaer Hochschulschriften* 20.

Specht, Ernst K. 1952. "Der Analogiebegriff bei Kant und Hegel." *Kantstudien Ergänzungshefte* 66. Berlin: De Gruyter.

Strawson, Peter F. 1981. *Die Grenzen des Sinns: Ein Kommentar zu Kants Kritik der reinen Vernunft*. Königstein/T.: Hain.

Wennemann, D. J. 1988. "Saint Thomas' Doctrine of Extrinsic Denomination as Mediate Correspondence in Naming God ex tempore." *The Modern Schoolman* 65:119–29.

Westphal, Merold. 1968. "In Defense of the Thing in Itself." *Kantstudien* 59 (1): 118–41.

Wundt, Max. 1924. *Kant als Metaphysiker: Ein Beitrag zur Geschichte der deutschen Philosophie im 18. Jhd.* Stuttgart: Enke. (Nachdruck Hildesheim u.a.: Olms, 1984)

CHAPTER 7

Infinity and Spirit

How Hegel Integrates Science and Religion, and Nature and the Supernatural

ROBERT M. WALLACE

In this study I will outline how through his conception of infinity and the "spirit" that's structured by infinity, Hegel integrates science and religion, and nature and the supernatural more explicitly and effectively than any other well-known thinker has done. Though, in retrospect, one can see a similar integration at work in the entire broadly Platonic tradition of which Hegel is an important recent member.[1]

Through Hegel's account of infinity and "spirit" we see that science, religion, ethics, the arts, and philosophy are all necessary not points of view on but *aspects of* a single self-determining reality, whose traditional name is "God." This is the essential proposal of the latter two-thirds of Hegel's *Encyclopedia of the Philosophical Sciences* (1817–1830), on "Nature" and "Spirit," which derive their structure from his account of infinity in the first third of the *Encyclopedia* (on "Logic") and in his *Science of Logic* (1812–1831).[2] Once one understands science, religion, and so forth in this way, as aspects of the ultimate reality, it makes no sense to try to delegitimize one of them by appealing to another one. Since they belong together, each must be practiced in a way that respects the others.

Because we have inklings of the integration that Hegel achieves, he has an ongoing influence in widely disparate circles, even though there is very little agreement as to what exactly it is that he integrates or how he does it. I hope that by presenting an accomplishment which there is reason to impute to Hegel and which appears to have great cultural and intellectual significance, I will encourage further work on this broad issue, including, of course, further exegetical work on the form that this accomplishment takes in Hegel and in related thinkers.

AN ULTIMATE REALITY?

How can science, religion, ethics, the arts, and philosophy all be necessary aspects of an ultimate "reality"?

They all seek to practice an "ascent" above one's initial opinions, appetites, and emotions, to something that's truer, better than, or more beautiful than those initial opinions, appetites, and emotions. By "ascending" in this way, whether through truth, goodness, or beauty, we make ourselves more able to govern ourselves, rather than being governed by whatever external forces caused us to have our initial opinions, appetites, and emotions. Insofar as we govern ourselves, in this way, we become more "real," as ourselves and not merely as products of our environment, than we would otherwise be. Hegel points to this higher degree of reality with his doctrines that "the finite *is* only by going beyond itself," as the infinite, and that "the finite is not the real, rather the infinite is the real."[3] By which he means that when the finite "becomes infinite" by going beyond its limiting relations to its circumstances, it "is" as itself, and not merely as the product of those circumstances. We can call this more intensive reality "ultimate," because it includes the more familiar kinds of "reality" but goes beyond them in a way that seems to be definitive. Nothing could be more real than what by governing itself makes itself what it is.

By rising above external circumstances in this way, science, religion, ethics, the arts, and philosophy all help to constitute the ultimate reality. Let me emphasize, however, that to say that science and so forth help to constitute this reality is not to say that they, regarded as finite, "human" activities, are what's real, and the "ultimate reality" derives its reality from theirs. For in that case there wouldn't be anything particularly "ultimate" about this reality. Rather, "the finite is only by *going beyond* itself," as the infinite. Insofar as science and so forth pursue truth, goodness, and so forth, they go beyond the finite and (in ways that will become apparent as we go along) they are more than what we normally think of as "ours." And the result is that the ultimate reality that they constitute is also more than merely human or merely ours, and is, in fact, logically prior to (more fundamental and more real than) what is finite, merely human, and, in the ordinary sense, ours.

In what follows, I will try to clarify this apparently paradoxical proposal. To begin with, let's look at how science, religion, and so forth each contribute to something that deserves to be called the ultimate reality.

SCIENCE AND THE ARTS AS ASPECTS OF THE ULTIMATE REALITY

It's not difficult to see how science is an aspect of the ultimate reality that I've described. Insofar as science seeks the truth, as such, rather than merely to satisfy our preexisting appetites or confirm our preexisting opinions, it goes beyond those appetites and opinions and reflects something that seems more our own than they are. We can let particular appetite-satisfactions and opinions go while knowing that we ourselves are still intact. But if we were to let our pursuit of truth go, we would become automatons, no longer governing ourselves in a significant way but simply reacting (through appetites and opinions) to the world that created us and impinges on us, and thus no longer existing as "ourselves."[4] So our pursuit of truth expresses us ourselves, our self-government, more than externally induced appetites or opinions can do. And the same is true of the sciences, as particular ways in which we pursue the truth. In this way the sciences help to constitute something that's more fully itself, and more real *as* itself, than what would otherwise be present.

Thus the idea that science shows or presupposes that there is no higher or more ultimate reality is refuted by the practice of science itself. For by rising above our externally induced appetites and opinions, science helps to constitute something that's more self-governing, and thus more real as itself and in a clear sense more ultimate than what lacks science. And a world in which nothing pursued science or the truth as such would be less self-governing and less real as itself than the world in which they are pursued. Its contents would be determined by an apparently infinite regress of causes, without anything that causes or governs itself.

Next, the arts. Insofar as they take us beyond the satisfaction of bodily appetites or our needs for pride and the like, the arts seem to put us in a state that expresses "us" personally more than our bodily appetites and emotional needs are likely to. For the body and our emotional needs were presumably formed largely by prior bodies and by experiences that came from outside us. Whereas by taking us beyond the body's appetites and externally induced emotional needs, the arts enable us to be less dominated by external influences as such. This would explain the fact that we find excellent works of art not merely pleasant or entertaining, but, as we say, "inspiring." By freeing us, to some degree, from merely external influences, so that we can, as we say, be "creative" and "express *ourselves*," the arts enable us to be more fully ourselves and they thereby contribute to the reality that's real "as itself," by not being governed by what's other than it.

RELIGION AS AN ASPECT OF THE ULTIMATE REALITY

With regard to religion, you might wonder how it could contribute to our being fully ourselves. Doesn't it do the opposite, by directing us to be governed by something, such as a "God," that's other than us? I want to suggest that even in the Abrahamic religions, with their focus on a God who seems to be separate and set over against us, there is an important sense in which this God in fact does or can function to make us more fully ourselves.

It's well known that religions in general urge their followers to subordinate purely self-centered concerns to something that's higher or more inclusive. The moral teachings of Judaism, Christianity, and Islam certainly do this. Though they sometimes promise rewards and punishments after death, their most exalted and most admired teachings celebrate virtue itself as bringing us closest to God. The best-known and most admired saying of Rabia of Basra, the eighth-century Sufi saint, is that she wanted to "burn paradise and douse hell-fire, so that . . . God's servants will learn to see him without hope for reward or fear of punishment."[5]

There is still the issue of the authority that God seems to have in these religions, which sets God over against those who must merely obey. Here, turning to Christianity, I would point out how in the Christian scriptures Jesus is reported as saying that "the kingdom of God is within you" (Luke 17:21).[6] St. Paul is reported as approving the view that "in" God "we live and move and have our being" (Acts 17:28). And numerous early Christian writers wrote of the possibility of our "becoming God" (*theosis*) as something that was made possible by God's "becoming man."[7] These latter formulations are in fact preserved and repeated in the Roman Catholic catechism and Mass. Similar formulations can be found in Jewish and Islamic mystical writings and in Advaita Vedanta and Taoism.

None of these formulations encourage the common idea that God is simply a separate being, one that "exists independently of" humans. Nor does such an idea recommend itself if we want God to be *infinite*; for as Hegel points out, any being that's separate is ipso facto *finite*, limited by its relation to the other beings, from which it's separate. (That relation being the relation of "being separate from" those beings.) This is Hegel's critique of the "spurious infinity" (*schlechte Unendlichkeit*), which is conceived of as separate from the finite but is therefore limited by its relation to the finite, and thus is finite itself.[8] So Hegel, drawing on the "orthodox" texts that I mentioned, and followed later by modern theologians like Paul Tillich and Karl Rahner, seeks a formulation that will preserve God's transcendence while not making God a "separate being."[9]

HEGEL'S VERSION OF TRANSCENDENCE: BEYOND BUT NOT SEPARATE

The biggest obstacle to understanding Hegel's relation to religion is the widespread notion that if Hegel is indeed serious about God, his "God" is "immanent" rather than "transcendent," and this sets him apart from what people take to be "orthodox" religious thinking.[10] It's likely that writers who describe Hegel in this way think that a "transcendent" God would be a being that's separate from the world, as Hegel's God is not.[11] But as I've just pointed out, first, Hegel has a good reason to avoid thinking of God as a separate being from the world (namely, that a separate being is limited by what it's separate from, and thus it's not infinite), and second, the Christian and other theistic traditions are by no means unanimous in thinking of God as a separate being from the world. For (to cite the Christian doctrines again) if God were a separate being, we could hardly "become God," and it's difficult to imagine how God's "kingdom" could be "within us."

Nor is it clear that a God who is not a separate being is therefore "immanent." Hegel gives no systematic role to the terms "transcendence" and "immanence," but if we look for a central concern of his that corresponds to what we mean by "transcendence" (such as passing beyond limits, or surpassing the material universe), it would be the difference between the (truly) infinite and the finite.[12] The infinite transcends or (as Hegel puts it) "goes beyond" the finite in that it's real *as itself* and not just as the product of other things.[13] Plato and Hegel both evidently intend to conceive of a reality that's "transcendent" in something like this sense, without being a "separate" or "independently existing" being or beings.[14]

This intention makes it clear why Hegel is not, as is sometimes suggested, a pantheist. The vertical dimension whereby the infinite goes beyond the finite prevents "everything" from being equally "divine," as it's supposed to be in pantheism. Insofar as it's a mere collection of finite things, whether denumerable or otherwise, "everything" isn't the kind of infinity that Hegel is interested in, because it seems clear to him that only a *qualitative* infinity, which produces a different kind of being from finite beings, can invite worship and deserve to be called divine.

But one still naturally wants to know how B can go beyond A and be "more real as itself" than A is, without being a separate being from A. The answer is that this can be the case if B is A's own going beyond its finitude, by becoming infinite and real as itself.[15] A can go beyond its finitude through rational self-government or the pursuit of truth, such as I described earlier, in which A is guided by reason rather than by whatever external forces caused it to have the opinions and appetites that it started out with. If anything expresses A itself, rather than expressing externally induced opinions or appetites, it's A's pursuit of truth. When it's guided by itself in this way, A *as* B is real *as itself*, and in that sense it's more real than it was merely as the externally guided, unthinking A. But since B is A's own going beyond its finitude, in this way, B is not a separate being from A.

Presenting God in this way, as the self-surpassing (becoming fully real) of finite things rather than as a being that's separate from finite things, is Hegel's way of interpreting (among others) the teachings that "the kingdom of God is *within you*" and that in God "we live and move and *have our being*." The kingdom of God is within us in the sense that we're capable of rational self-government, and we have our being in this God in the sense that it's only through our self-government "in" this God that we achieve full reality, full being, as ourselves.

But we're still talking about God, and not merely about us, insofar as this full reality is always "above" a great part of what we, as human beings, are. The finite "*is* only by going *beyond itself*" (second emphasis mine): Hegel is not reducing God to us as the finite beings that we ordinarily take ourselves to be. Rather, he is elevating us (in part) to something beyond what we ordinarily take ourselves to be. This is why the apparent paradox that I mentioned, that the reality that's more real than us is constituted by our activities, is only apparently paradoxical. For this ultimate reality is constituted by activities in which we in fact go beyond our finite selves, to a higher degree of reality than we normally possess.

I'll say more about this in what follows. But we can already see how Hegel's version of transcendence identifies a core of truth in religion, which lends itself to integration with science, ethics, the arts, and philosophy. It lends itself to this because it takes religion to be promoting the surpassing of one's everyday finite self, rather than promoting submission to something that's separate from oneself. This core of truth no doubt contrasts with much conventional religious talk, but no advocate of religion is likely to deny that religion encourages its followers to surpass their everyday ways of thinking and functioning. Jesus (in Luke), St. Paul, Rabia, and Hegel are simply defining with increasing precision what would be the result of our doing that. They make it clear how one can speak meaningfully of an ultimate reality that's neither reducible to humans as such, nor, as Karl Rahner put it, a mere "member of the larger household of all reality," as it would be if it were an additional being, separate from and alongside humans and the "world" (Rahner 1978, 63).

THE PLATO/HEGEL "PHILOSOPHER'S GOD": LOVE, FAITH, PRAYER . . .

As for the common objection that religious believers will be left cold by a "philosopher's God," such as one finds in Platonism and in Hegel, several points need to be made.

First, this kind of God is characterized not only by the rational self-government or freedom that is manifest in rising above pre-given appetites and opinions, but also by an important kind of love. The reason for this love is made most explicit by Hegel in a variation on his critique of the supposed "infinity" that turns out to be rendered finite by being opposed to finite beings. Hegel points out that being separate from others is a way of being related to those others, so that being guided by one's separateness from others is a way of being guided *by those others* as others and, to that extent, not being guided by oneself.[16] So being guided by one's separateness from others detracts from one's self-government.

But "self-centered" people and gods are, precisely, guided by their separateness from others—they are concerned about themselves, and not (as they will tell you) concerned about those "others." And to that extent they are guided *by* (their relation to) *those others*, and they fail to be *self*-governed. So people and gods who are fully self-governed will not be self-centered. Rather, they will be, in effect, loving: they will treat others the same way they treat themselves. In this way, freedom as self-government translates into an important kind of love.[17] Of course, this also makes it clear how being truly oneself entails ethics, in which we are expected (broadly) to treat others as we treat ourselves.

Second, since the ultimate reality, which is real "as itself," is real in a way that everyday finite realities are not, one could see it as the core of truth in the idea of God's "creating" the world. By its presence in and influence on the world, the ultimate reality gives the world all of the "full" reality, reality "as itself," that the world possesses.

Third, our adherence to the ultimate reality that's composed of freedom and love, despite the attractions of self-centered appetites, opinions, and so forth, is equivalent to what traditional religion calls "faith." This is because our commitment to the ultimate reality is to something that goes beyond our "all-too-human" nature, and which from the point of view of that nature (that is, from the point of view of self-centered appetites and opinions) has no evident authority at all.

Furthermore, turning away from those self-centered appetites and opinions toward freedom and love is the equivalent of what's traditionally called "conversion." The aid that we receive, in this faith and conversion, from the freedom and love that are around us and hidden (as potential) within us, is equivalent to what's traditionally referred to as "grace" and "salvation." And our praise of this aid and our effort to be receptive to it are what we traditionally call "worship" and "prayer."

Critics often suggest that the Plato/Hegel God is not a "personal" God. But the Plato/Hegel God is in fact much more personal than we usually are, because, as Hegel tells us, it's "supremely free." Through its freedom/love, it nurtures the potential for "personhood" in everything, including us.[18]

In all of these ways, this "philosopher's God" and our dealings with it reproduce what we see in traditional religion. The only apparent difference is that Plato and Hegel present it all in a more analytical vocabulary. So it seems reasonable to suggest that what's most inspiring in traditional religious stories and concepts may be, precisely, the transcendent, free, loving, and supreme reality that Plato and Hegel show we're able to experience.

Plus, as I've explained, what Plato and Hegel describe has the advantage over the conventional conception of God as a separate being that Plato's and Hegel's God is truly infinite, which is to say, truly transcendent.

I realize that the view of Plato that I'm suggesting here may be unfamiliar. Few present-day commentators on Plato focus on the way in which Plato's account of the soul, in book 4 of the *Republic*, together with his account of cognitive "ascent" in books 6 and 7, shows how we can understand "transcendence" as a process that takes place within the world rather than simply in opposition to it. But this is the aspect of Platonism that Aristotle, Plotinus, and Spinoza all develop in various ways, and which Hegel in his turn conceptualizes through his account of true infinity and spirit.[19] All of these thinkers are very serious about transcendence, and they all seek to understand it as in some way taking place within reality or the world, rather than flatly in opposition to reality or the world. It's a dimension of ascent, rather than a dualistic divide. For, as Hegel in particular spells out, a dualistic divide prevents either of its components from truly transcending the other.

SCIENCE AND THE SCIENTIST, "OBJECT" AND "SUBJECT"

I must also acknowledge the natural response of admirers of science to what I have been saying about science's contribution to the reality that's fully itself and that's traditionally called "God." The problem is that science doesn't seem to recognize any such "ultimate reality" as I have been describing. If science doesn't *recognize* it, how can I say that science helps to *constitute* it? This puzzling state of affairs fuels the suspicions toward "metaphysics" and religion that one often encounters among people who admire the sciences.

The explanation of this puzzle is that beginning with the scientific revolution in the seventeenth century, modern science has made it its business to focus solely on "objects" and to ignore the possible significance of its own rational activity—of the "subject," as German idealists call it. The narrow focus on "objects" was initially intended as a practical way of maximizing the likelihood of rapid progress within a delimited area. Since then, however, it has come to be taken for granted, to such an extent that a scientist who suggests that her own rational activity deserves attention in its own right is likely to seem like an eccentric who is distracting attention from the only true reality: that of "objects." Science in practice systematically excludes itself, its own rational activity, from the realm of "objective realities" that it addresses.

When one puts it that way, it's obvious that such an exclusion can only be defended as a temporary expedient, not as an established truth about what's real. Surely an activity that claims to be fully rational must ultimately address itself, the "subject," as well as its "objects." And indeed this is just what the great modern philosophers have tried to do, on behalf of science.

Kant's way of addressing this issue, in his three *Critiques*, was to keep the subject separate from its objects. Science as he understood it was properly concerned only with objects, understood in a Newtonian mechanistic way, while the subject had "moral faith" in certain things about itself that mechanistic science could not know about the world as such. The subject had moral faith in its freedom, responsibility, immortality, and so forth. Kant's thoughts, in the third *Critique*, about the "regulative" role of teleology in understanding life, did not succeed in bridging the fundamental divide between object and subject, and knowledge and "faith," which he had thus created. There was still no way that one could have knowledge of oneself and of how one should act; one could only have practical faith. But if one's *ideal* is knowledge, then a "faith" that's contrasted with knowledge is bound to seem like a poor substitute for it. As a result of this unresolved dualism of knowledge versus faith, it seems clear that Kant did not successfully integrate science with ethics and religion.

One alternative, which is often adopted, would be to exalt some kind of "faith," as the key to everything, *over* knowledge. As an admirer of science, Kant wasn't tempted to do this, so he remained stuck with the problem of how to relate the two.

HEGEL'S PLATONIC SOLUTION

A third approach, which goes beyond Kant's uncomfortable dualism and beyond the exaltation of faith, was sketched by Hegel in his early essay "Faith and Knowledge" (1802) and systematically developed in his *Science of Logic* and his *Encyclopedia of the Philosophical Sciences*. Hegel was in effect returning to something like what seems to have been Plato's original solution to the problem. Hegel explains how knowledge and faith, and object and subject each involve the other. Rather than being belief in a separate and very powerful being, "faith," in Hegel's view, is one's *commitment to the pursuit of knowledge*—and through knowledge, of being oneself, and being real as oneself—as opposed to mere opinion, appetite-satisfaction, and failure to be oneself. The "subject" that exhibits this commitment is far from being merely "subjective" since, being real *as itself*, it has a more complete "reality" than mere "objects," as such, possess. Thus "faith" in this sense generates full reality, and gives rational access to it. Rather than being opposed to knowledge or reason, this faith is part and parcel of knowledge and reason.

In his major works, Hegel investigates the central issue of what it is to be oneself, and to be real as oneself. In his *Science of Logic*, he presents being and nothing as indistinguishable except when they present themselves as "becoming," or coming-into-being, and perishing.[20] For then they entail a "something" that comes into being or perishes, and which thus is determinate in some way (has a definite quality). The question then is this: Is the something determinate in itself (*an sich*) or through its relations to others (*Sein-für-anderes*)?

We might suppose that the something could be determinate in itself by being separate from others, as "finite" things are. But as I've suggested, this separation still connects it to the others, because it constitutes a relation between them. So separation and finitude don't give us something that's determinate entirely in itself. Rather, they give us what amounts to an infinite regress, of which no part is fully determinate by itself, and thus the "promise" of determinacy seems to be endlessly postponed.

The qualitative infinite, on the other hand, which involves no boundaries, promises to be determinate entirely in itself. As I mentioned in connection with the common conception of God as a separate being from us, it's important not to conceive of the "infinite" as another "being," separate from finite beings and thus, in fact, limited by its relation to them and not infinite. Hegel proposes that in order to avoid this outcome and come up with a *true* infinity, we must conceive of the infinite as the finite's own going beyond its finitude.[21] Then there is no border between the two, and we have before us what Hegel at one point calls "the fundamental principle of philosophy."[22] He calls it that because in it we finally have something that is what it is in itself, rather than through its relations to others, but which at the same time allows for the apparent multiplicity and reality of finite beings.

How can the finite "go beyond its finitude"? The finite can do this through something like rational self-government. We are sometimes able to be more self-determining, and thus to be what we are more "in ourselves," by being governed by our own thinking rather than by appetites or opinions that originated outside us. Hegel here is drawing on Kant's notion of rational autonomy, in which the autonomous moral agent is governed by its own rational nature rather than by inclinations that probably originated outside it. A finite something that goes beyond its finitude, Hegel suggests, is like a finite human being that goes beyond its initial appetites and opinions.[23] There is something within the something that guides it, and through which it's no longer "finite," no longer limited and determined by what's around it.

The first signal that such an inner guidance is possible is what Hegel calls (paraphrasing Kant and J. G. Fichte) "the ought" (*das Sollen*). Kant had made it clear that our inner guidance takes us beyond the realm of mere "fact" to a "morality" that his readers might be inclined to call a realm of "value." Likewise Hegel's "infinite" is not a "fact." Like the initial notion of something's being what it is "in itself" and not just through its relations to others, the infinite or the Ought is an *aspiration*. But it's not "only" an aspiration, since it's only through this aspiration that anything, including the universe, can be what it is entirely in itself, and not by reference to anything outside it. So that "value" and full reality are not (as we commonly suppose) separate domains, but are intimately entwined.

In this way the Ought and what follows it are the equivalent, in Hegel's presentation, of the Good, which Plato placed at the summit of reality. And indeed the Good itself appears in that traditional Platonic role at the end of Hegel's *Science of Logic*. It's only through the aspirations that are associated with the Good or the Ought that the soul, in Plato, or anything at all, in Hegel, can be what it is entirely in itself. As Plato's Good had enabled the soul to be unified and to function as "itself," the Ought and what follows it enable Hegel's something to be self-determined and "itself."[24] In this way, value plays an indispensable role in constituting what's fully real, in the sense of being real as itself. Here Hegel follows Plato and Aristotle in identifying purposes, and value in general, as an essential aspect of reality, rather than a separate domain, as they are in David Hume or in Kant. This follows from the focus, which Plato, Aristotle, and Hegel all share, on the issue of how something can be what it is in itself, and not merely through an endless sequence of determiners.

So where Hegel differs from Kant is that by showing how the finite fails to be what it is in itself, Hegel shows that only the (value-based) infinite is fully real, in the sense of being real as itself. Knowing this, through Hegel's exposition, and knowing through our experience the freedom that constitutes the infinite full reality, we *know* the infinite, our freedom, and the highest reality, rather than (as in Kant's account) merely having "practical faith" in them. This knowledge of the finite's relation to the infinite creates a path from the finite to the infinite, an intelligible process of "ascent," in contrast to the unbridgeable duality between theoretical knowledge and practical faith, which Kant had left us with.

We see this ascent from finite to infinite again later in Hegel's system as an ascent from Nature to Spirit. As the true infinity is the self-surpassing of the finite, so Spirit is the self-surpassing of Nature. And in each case, what propels this surpassing is our effort to be fully ourselves, and in that sense fully "real." So again we have an intelligible process of ascent, this time from Nature to Spirit.

By presenting this process of ascent from Nature to Spirit, Hegel responds to the standard charge of advocates of "naturalism," that because we have no systematic understanding of the relationship between the "natural" and the "supernatural," we should ignore the latter and focus only on the former. (Or we should "reduce" the latter to the former.) Following the example of Plato's analysis of ascent, in the sun, line, and cave allegories in the *Republic*, Hegel shows how natural beings such as ourselves can and do come to function in ways that can appropriately be described as "supernatural." This functioning merits such a description not because it belongs to a completely different "world" than nature, but because it's more self-determining or self-governing than such paradigmatic "natural" processes as those studied by physics. Rather than being two separate "worlds," the "natural" and the (properly understood) "supernatural" are lower and higher phases on a scale of increasing self-government and selfhood as such.[25]

HEGEL'S PHILOSOPHY OF NATURE

In the "philosophy of Nature" portion of his *Encyclopedia*, Hegel divides Nature into "mechanics," "physics," and "organics." "Organics," of course, has to do with life; this is what we call "biology." "Mechanics" has to do with the simple pushing and pulling of space, time, and matter—what we now call "physics." What Hegel himself calls "physics" is the intervening domain between the merely mechanical and the living, and has to do with the organization of matter into the four elements and the planet that's composed of them, which unfolds as light, electricity, and chemical processes. What he traces through this whole increasing complexity is Nature's increasing ability to organize itself into processes that have a "center" or a "self." The earth, life upon it, plant and animal species, and the functioning of individual organisms exhibit increasingly intense versions of this "self-ness." In all of this we can see the process of the finite's going beyond itself through increasing degrees of self-determination and thus of "infinity."

Hegel's philosophy of Nature is controversial not merely because of its apparently antiquated theory of "elements" and so forth, but more importantly precisely because of the way it focuses on "self-ness." "Self-ness" is not an everyday concern of modern physics, chemistry, biology, or neuroscience. On the contrary, scientists often seem to regard it as a mere by-product of processes that they seek to understand without reference to any "self."

Hegel is saying, however, that such an agenda ignores the scientist's own fundamental experience of seeking "self-ness" *in herself*, through the cognitive "ascent" that seeks to replace initial opinions and appetites with truth. Hegel is saying that neither space, time, matter, nor anything else can be more fundamental or better *known* than this essential activity of "ascent" in which the scientist, like every human being, is constantly engaged. So it's legitimate to examine space, time, matter, living things, and so forth, from the perspective of this issue of "self-ness," self-organization, and self-determination.

Indeed, it's *more* legitimate to examine them from this perspective than from any other. For self-ness (etc.) are by their very nature the ultimate reality, in reference to which every other candidate "reality" must be judged and understood. They are what is what it is by virtue of itself, rather than by virtue merely of its relations to other things, so that if we seek to understand reality *as such*, and not only in its myriad "manifestations," selfness is what we must examine first. Since it is what we ourselves are, and what the entire activity of investigation that we call "science" is, we know it through our mere awareness of our own activity, and thus it's not only more fully real but also better known by us than anything else. To bracket what we are and what we know best, and try to investigate only what we *aren't* and what we know *less well*, is to consign ourselves to ignorance of something than which we could never know anything more real or more fundamental.

AND HIS PHILOSOPHY OF SPIRIT

So Hegel presents "Spirit" as the reality that focuses most fully upon itself, inasmuch as Spirit asks—in line with the famous injunction of the Delphic oracle to "know thyself" (Hegel, *Encyclopedia*, §377)—what it, "Spirit," really is and thus how it can most successfully *be* what it really is.

Within "Spirit," Hegel unfolds first the familiar "subjective Spirit," composed of our theoretical and practical thinking. This would include the practice of science and other cognitive activities. Then Hegel examines an "objective Spirit," composed of property, morality, the family, the state, and history. In addition to our "inner" functioning, Hegel calls all of these "external" institutions "Spirit" also because they are ways in which our external, social world enables us to be free and self-determining in our dealings with one another. Enabling us to find various kinds of rational self-determination in the external world, they prevent that world from being a mere (irrational) obstacle to our self-determination.

But then a question arises: Which of these two kinds of freedom is primary—the internal one that's composed of our theoretical and practical thinking, or the external one that's composed of property, morality, the family, the state, and history?

Hegel's answer is that neither of them adequately embodies freedom, since being limited by each other, each one outside of and opposed to the other, they are both finite. To combine them and thus go beyond their limitations we need a new, more inclusive kind of reality, which will preserve what's free and fully real in each of them. Hegel calls this more inclusive reality "*absolute* Spirit" (where "absolute" means "freed"). We know this reality as the arts, religion, and philosophy. They preserve what's fully free and thus fully self-determining and fully real in subjective and in objective Spirit, and they omit the rest. So Hegel describes them as a "reconciliation." In fact, he describes them, for reasons that I'll explain, as "the Spirit's elevation to God."[26]

ABSOLUTE SPIRIT: ART

To begin (as Hegel does) with art, it's fully present in the "outer" world of the senses, but it also goes beyond that world by giving it the additional dimension that we call "aesthetic." In this additional dimension, we don't experience time, space, objects, and ourselves in the way that we do in the objective and subjective worlds. Instead we're held, entranced, by the aura of the artwork and what it does to us.

It's tempting to describe aesthetic experience as "merely subjective," merely "in the eye of the beholder." But insofar as we engage seriously with the arts, we know that this can't be correct. We can often reach agreement with other people about whether the art that we make or experience together is relatively shallow and contingent, or deeper and more compelling. That's the sense in which the arts, while being independent objects in the world, are also, as it were, "thoughtful." We can evaluate them in a way that resembles the way we evaluate thoughts. How "compelling" are they, for those who appreciate them? Because of this dimension of "thoughtfulness," works of art have a more intensive presence than what's merely objective and lacks anything like thought, while on the other hand their physicality gives them a more intensive presence than what's merely subjective.

And this is what gives the arts the magnetism that they have for us. They give us a glimpse of a more intensive "reality," which transcends the subjective/objective divide. And we're inspired by this reality because we feel that through the arts, we ourselves transcend that divide. Indeed, looking back at what's most free and most real in the accomplishments of subjective and objective Spirit, and of Nature before them, we can see all of this as, in an important sense, "art." What's fully "itself" and thus most real, including us, *is* "art" (or whatever art in its turn will turn out to be), because in it the work, its creator, and its appreciators (as it were) "create themselves."

ABSOLUTE SPIRIT: RELIGION AND PHILOSOPHY

Since the arts enable us to have this experience, it's no wonder that many of us make the arts into something rather like a religion for ourselves, in which we engage in something that's not very different from worship. But religion in the normal sense of the word goes one step further than this. We can see it as an effort to consolidate or "totalize" the magnetism that we experience in the arts.[27] Because of their immersion in sense experience, works of art are after all disparate, there are boundaries between them, and in that respect they fail to achieve the full freedom, the infinity, that we're searching for.[28] Archetypal religious figures, on the other hand, such as Jehovah, Osiris, Orpheus, the Buddha, and Jesus, overcome this disparateness and finitude by "representing," in various ways, the unity or infinity of everything. The "art religion" (as Hegel calls it) of the Homeric gods is gradually displaced by the more intense, "totalizing" religions of Orphism and its successors, because we are searching for the complete self-determination or freedom that the latter represent for us.

But because the world offers us many of these "unifying" figures and associated totalizing religions, it's unclear whether any of them can really unify our world. And beyond that problem there is the even more challenging problem of the division between these figures and *ourselves*. This ultimately prevents any of these figures from giving us full freedom or infinity. As Buddhists have quipped, "If you meet the Buddha on the road, kill him." The Buddha may inspire you to find infinity in yourself, but if you regard him as a figure that's separate from yourself, he's also an obstacle to your finding infinity. Likewise for Jesus and all the other "gods" insofar as they are assumed to be beings separate from ourselves.

What Hegel calls "philosophy" supersedes this final disunity and finitude by understanding the entire process that we see in the arts and religion, including the divisions between different gods and between the gods and ourselves, as part of the single process of the finite's surpassing of itself in true infinity. Sense-experience (in art) and representation (in religion) set us over against what we experience or what's represented to us. So although we came to art and religion for infinity, what we experience in them is still, in important respects, finite. Philosophy, on the other hand, understands us and artworks and gods as aspects of the *self-comprehending process* that is the self-surpassing of everything finite, including us, the artworks, and the gods.[29]

Within this process, everything is integrated with everything. This is the ultimate accomplishment of the rational love that I outlined in section 5. Finite, infinite, Nature, Spirit, you, me, subjectivity, objectivity, value, science, family, state, artworks, religion, gods—nothing is rejected, everything is integrated and subsumed, as Spirit "raises itself to" or surpasses itself as that which alone is completely infinite and thus wholly "itself."

The traditional term for what is completely infinite and wholly itself is, of course, "God." But now it's clear, as I've suggested, that though it's higher, this God can't be separate from ourselves. Rather, as we seek to be wholly ourselves by participating in the process that Hegel describes, we go beyond our finite bodies, our emotional needs, all separateness from each other and from everything else, and (as in the Roman Catholic catechism, which quotes St. Athanasius) we "become God."

As in Plato, this ascent is a matter of becoming (wholly) oneself, not of becoming something different.[30] But what one discovers about "oneself," in the process, and what one discovers about "God," is certainly not what common sense or conventional science expected. One's true self, it turns out, is the transcendent God.[31]

In this remarkable conception we see again the apparent paradox, that the merely finite—us—can constitute, by "going beyond itself," what's truly infinite. But now we have a more complete understanding of how this process does indeed go *beyond* us as we ordinarily conceive of ourselves, beyond all separateness from each other and from everything else.

RESPONSES TO HEGEL

That we could be truly beyond all separateness is not an easy thought to think. As Hegel says, "finitude is the most stubborn category of the understanding."[32] One thing that's clear, however, is that Heinrich Heine misunderstood Hegel completely in his oft-quoted comment that "I was young and proud, and it gratified my self-esteem to learn from Hegel that . . . it wasn't the Lord in heaven, but I myself here on earth who was God."[33] What Hegel is in fact describing is the opposite of an accomplishment of which one could feel "proud" as a separate individual. As far as that kind of pride is concerned, the "last," the one who has the least of it, shall indeed be "first."

Heine wasn't alone, in the generation after Hegel, in failing to grasp what Hegel had been driving at. Not recognizing the role of love in Hegel's ascent, Ludwig Feuerbach criticized it as merely "intellectual," and held up a counterideal of nonintellectual "love" that he hoped to find in the senses and in matter.[34] Karl Marx, focusing on the familiar misuses of religion, suspected that Hegel and religious traditions had conceived of "Spirit" as "higher" in order to sanctify the power of the ruling classes. Imposing an interpretation that had nothing to do with Hegel's intention, Adolf Trendelenburg objected that Hegel's true infinity didn't "follow from its premises."[35] Søren Kierkegaard caricatured Hegel's true infinity as a stick with which Hegel beat his opponents, and his concern for "system" as a psychological compulsion rather than the simple effort of thought to be as coherent as possible.

These critics all missed the point of Hegel's ascent, both in its continuity with the orthodox Christian doctrine of "becoming God" (*theosis*) and in its continuity with the Platonic tradition of rational transcendence as the source of unity and reality as oneself.

Despite repeated efforts, the Plato/Hegel view has not been well expounded since Hegel's time.[36] But there are reasons to think that the present situation in philosophy may make possible a new appreciation of what Plato and Hegel accomplished. In recent decades writers such as Charles Taylor, Gary Watson, Susan Wolf, John Martin Fischer, and Alfred Mele have developed conceptions of human rational self-government that resemble Plato's and Hegel's in their general approach.[37] Ethics and the arts are getting respectful attention; commentators on science are doing their best to clarify the nature and the limits of science's understanding of reality; and not everyone regards religion as inherently and in all respects irrational. Plato and Hegel dealt with all of these issues in a remarkably integrated and consequently powerful way. So when a better understanding of their response feeds into current discussions, a major illumination could occur.

When we appreciate Plato's and Hegel's view we see that science, religion, ethics, the arts, and philosophy are all aspects of the same "ascent," the same freedom, and the same freest and fullest reality or "person." And thus if science is indispensable, so are religion, ethics, the arts, philosophy, and the fullest reality or person. To deprive oneself of any of these, on the grounds of its supposed incompatibility with one or more of the others, is to render oneself finite and unfree.

NOTES

I would like to thank Tom Bennigson, Thomas Burns, and Alan Montefiore for very helpful comments on drafts of this chapter.

1. I give a good deal of additional textual support for my way of reading Hegel in Wallace 2005. The "broadly Platonic" tradition, toward which I can only gesture in this chapter, seeks to overcome materialism, mechanism, nominalism, relativism, and skepticism through a single systematic effort. See the description of "Ur-Platonism" in Gerson 2013 (10), and compare to Gerson 2005. Some leading members of the broadly Platonic tradition are Plato, Aristotle, Plotinus, Spinoza, and Hegel.

2. Again, more details are in Wallace 2005. I give a quick survey of other efforts to interpret or replace Hegel's philosophical theology in note 36, below.

3. Hegel 1989 (Miller trans.), 145, 149; G. W. F. Hegel, *Gesammelte Werke* (*GW*), vol. 21 (Hamburg: Meiner, 1985), 133 and 136; G. W. F. Hegel, *Theorie Werkausgabe* (*TWA*) (Frankfurt am Main: Suhrkamp, 1969–), 5:160, 164. In connection with this idea of a higher degree of reality, to which Hegel refers simply as "reality" (*Realität*) as such, please note that "real" here is not to be understood primarily in contrast to "illusory" or "imaginary" or the like. Rather, to be "real" is to be, as its Latin root *res* suggests, "thing-ish," that is, having an inherent unity of some kind, in contrast (say) to a mere aggregation of items. Thus A can be "more real than" B without this implying that B is illusory or imaginary; it merely implies that B is less organized or "itself," and more like an aggregate. Hegel proceeds from his introduction of "*Realität*" in the *Science of Logic* directly to the "something" (*Etwas*), which he describes as "relation to itself," and indeed as "the beginning of the Subject" (Hegel 1989, 115; *GW*, 21:103; *TWA*, 5:123). When he calls the "something" the "beginning of the Subject," here in the *Logic*'s initial "Doctrine of Being," Hegel is saying that through its "relation to itself," the something foreshadows what he describes in the *Logic*'s culminating "Subjective Logic" as the domain of "freedom" or self-government. So "reality," as preliminary to the "something," exhibits very much *in nuce* the "self-relation" and self-governing unity that we later find fully developed as the "Subject" and its freedom. That's another way in which, unlike the "reality" that's contrasted to "illusion" and so on, Hegel's "reality" can come in degrees. I explain in more detail in chapter 3 of Wallace 2005 and in the whole book how this "more intensive" (Hegel 1989, 137; *GW*, 21:125; *TWA*, 5:150) "reality" of infinite freedom is the theme of Hegel's philosophical system as a whole.

4. By contrasting us with automatons, I don't mean to take any position regarding determinism or libertarian free will, as such. I'm merely drawing attention to our need to take seriously our own rational functioning as enabling us to go beyond pre-given appetites and opinions. If we can't actually function in this way, we might as well abandon the idea that we can practice science or any other rational discipline.

5. Sells 1996, 151. Teachers like Rabia, who don't focus on an "afterlife" as such, don't reduce religion to mere morality, insofar as they are concerned with the fuller "reality" or God that is achieved through the "ascent" of which morality is one aspect. This is the way in which "mystical" traditions, which are concerned with the eternal present rather than with an "afterlife," are still fully "religious."

6. On the issue of how to translate this famous line in Luke, see Ramelli 2009, available online (March 2013).

7. For example, "The Word of God became man, that thou mayest learn from man how man can become God" (Clement of Alexandria, *Exhortation to the Heathen*, chap. 1, para. 871). For other examples, see the Wikipedia article "Divinization [Christian]," citing among many other sources the *Catechism of the Catholic Church*,

and for commentary, see Christensen and Wittung 2007. See also St. Augustine's famous saying: "You [that is, God] were more inward [to me] than my most inward part" (*Confessions* 3.6.11).

8. "It will be found that in the very act of keeping the infinite pure and aloof from the finite, the infinite is only made finite" (Hegel 1989, 137; *GW*, 21:124; *TWA*, 5:149). It's probably clear by now that as is usually true in theological discussions, the kind of "infinity" that Hegel is discussing here is a "qualitative" infinity rather than a mathematical or quantitative one. He discusses mathematical infinities in the second section ("Quantity") of the *Logic*'s "Doctrine of Being." The relation between the two types of infinity, as Hegel presents it, is too complex for me to discuss here.

9. See Tillich 1957, 6–7, and Rahner 1978, 63. One could also mention Alfred North Whitehead, Charles Hartshorne, Jürgen Moltmann, David Ray Griffin, and Philip Clayton, all of whom are usefully surveyed in Culp 2008 and Culp 2017.

10. "Hegel seemed to be denying any kind of transcendence (at least in a non-trivial sense) to God" (Pinkard 2002, 303). Compare Desmond (2003, 2) and Houlgate (2006, 435). As evidence of Hegel's rejection of transcendence, Moore (2012, 178) cites Hegel's *Encyclopedia* §38, in which I assume he's referring to Hegel's objection there to the notion of a "beyond" (*Jenseits*), which Hegel associates with the "Ought." But Hegel in fact approves of a certain kind of "beyond-ness," which is the way in which true infinity, in his words, "goes beyond" the finite. See note 13, below, for the key citation on this. I discuss this whole issue in section 3.17 (96–102) of Wallace 2005.

11. "Theism," as Charles Taylor puts it, "looks on the world as created by a God who is separate and independent of the universe," and "this cannot be accepted by Hegel" (Taylor 1975, 100).

12. Hegel does occasionally use the adjective "immanent," in roughly the sense that Kant gave to it (*Critique of Pure Reason*, A 296–97), as pertaining to the realm of what can be known. Unless one assumes (as Hegel does not) that God cannot be known, this kind of "immanence" has nothing directly to do with our notion of immanence as the opposite of ("orthodox") divine transcendence.

13. *Transcendere* means literally "to climb over, surmount, surpass," and Hegel explicitly describes the infinite as "going beyond" the finite: "The infinite is only as a going beyond [*als Hinausgehen über*] the finite" (Hegel 1989, 145 [Miller's translation actually says "transcending"!]; *TWA*, 5:160; *GW*, 21:133). So we have to be careful not to read too much into Hegel's objection to notions of a "*Jenseits*" (a "beyond"). In the next sentence but one after the sentence that I just quoted, Hegel spells out what it is that he really objects to: "The finite is not sublated by the infinite as by a power existing outside it; on the contrary, its infinity consists in sublating its own self." What he objects to is not the notion of going beyond, as such, but the notion that such going beyond involves or is brought about by "a power existing outside" the finite. That is, he objects to conceptions of the "beyond" as a separately existing being. So rather than rejecting "transcendence" as such, Hegel is presenting what amounts

to a revised conception of it, what we might call a "*true* transcendence." (As for being "real as itself," Hegel's whole discussion in the *Science of Logic* of the "something" and the "finite" is a discussion of how something can be fully "in itself" and thus "real" [Hegel 1989, 111–15; *GW*, 21:98–102; *TWA*, 5:118–22], and the upshot is that the infinite is what's "real" [Hegel 1989, 149; *GW*, 21:136; *TWA*, 5:164] and thus "in itself.")

14. Plato undermines the idea that God is an "independently existing being" when he makes it clear in the *Timaeus* that the "craftsman" who created the world had no choice but to create it, because he was "without jealousy" (29e). That is, God's nature requires God to create a world; so we can't coherently conceive of a God without a world; so the two don't "exist independently" of each other in the usual sense. Plato's conception of phenomena "participating in" transcendent Forms likewise suggests a closer relationship than the two "existing independently" of each other. Plato and Hegel both make it clear that it doesn't follow from X's not being a separate being from Y, that X is identical to Y. It may instead be the case that X "participates in" Y (Plato) or that Y is the "going beyond itself" of X (Hegel). I have to acknowledge that Plato's middle-period preoccupation with the "separation" (*chorismos*) of the Forms vis-à-vis what "participates" in them prefigures our conventional assumption that God is a "separate" being. But it's well known that Plato in his *Parmenides* criticizes this "separation" trenchantly—without showing any sign of abandoning his fundamental concern with "ascent." This is how the broad Platonic tradition (as I've called it) that includes Aristotle and Hegel gets under way.

15. To quote Hegel's formulation again: "The infinite *is* only as a going beyond the finite. . . . The finite is not sublated by the infinite as by a power existing outside it; on the contrary, its infinity consists in sublating its own self" (Hegel 1989, 146, emphasis in original; *TWA*, 5:160; *GW*, 21:133).

16. "Mutual repulsion and flight is not a liberation from what is repelled and fled from; the one as excluding still remains connected to what is excluded" (Hegel 1989, 175 [translation revised]; *GW*, 21:163; *TWA*, 5:196).

17. The free alternative to being guided by one's separateness from others is not to be guided by what we merely happen to share with others, but rather to be guided by our shared search for the True and the Good. So the love that Plato and Hegel advocate isn't indiscriminate promotion of whatever we all happen to want, but rather a fostering of rational freedom in each and all of us, which is a fostering that undoubtedly will often involve promoting the material conditions that enable such rational freedom to be actualized in us.

18. "Supremely free" (Hegel 1989, 841; *GW*, 12:251; *TWA*, 6:570). Hegel (1989, 824; *GW*, 12:236; *TWA*, 6:549) spells out "personality" as being "for itself" rather than "for" (dependent on) anything else, and being "practical" (and also theoretical or contemplative). Nurturing: "The universal . . . could also be called *free love* . . . for it bears itself toward what it is different from *as toward itself*" (Hegel 1989, 603; *GW*, 12:35; *TWA*, 6:277).

19. On Aristotle and Plotinus as broad Platonists, see Gerson 2005 and Gerson 2013.

20. Hegel's first major work, his *Phenomenology of Spirit* (1807), rotated around the same issue of subject and object, but complicated it in ways that the *Science of Logic* and *Encyclopedia* avoided. So it's easier to extract his fundamental thought from the *Logic* and *Encyclopedia* than from the *Phenomenology*.

21. See the text cited in note 15, above.

22. *Encyclopedia Logic*, §95, remark. One could suggest that insofar as there is a distinction between the infinite and the finite, there is still a separation and a "relation" between them. But Hegel's point is precisely that in this distinctness, they still depend upon each other (since "the infinite *is* only as a going beyond the finite" [see note 3, above]) for their identity. So the "relation" between them is not, like ordinary relations, "external" to their identity.

23. "At the name of the infinite, the heart and the mind or spirit [the *Gemüt* and the *Geist*] light up, for in the infinite the mind or spirit is not merely abstractly present to itself, but rises to its own self, to the light of its thinking, of its universality, of its *freedom*" (Hegel 1989, 138; *GW*, 21:125; *TWA*, 5:150).

24. Plato explains how the Good enables the soul to be unified and to function as "itself" in *Republic*, book 6 ("from having been many things, he becomes entirely one" [443d–e], through the functioning of his rational part) and books 6 and 7 (on reason's reliance on the Good). Hegel discusses the "Ought" in Hegel 1989 (131–36; *GW*, 21:118–23; *TWA*, 5:142–48), and the Good itself in Hegel 1989 (818–23; *GW*, 12:231–35; *TWA*, 6:541–48). Commentators often stress Hegel's criticisms of Kant's and Fichte's misleading conception of the "Ought" to such an extent that they neglect the "Ought's" key role, for Hegel, in indicating how the finite can in fact go beyond itself as the infinite. We have to ask why the "Ought" becomes an issue here at all, in the *Logic*'s conceptual development. If it were merely a "blind alley," it would be the only "blind alley" in the *Logic*. It's more plausible to understand it as Hegel's acknowledgment that Kant and Fichte genuinely seek, through the "Ought," to go beyond finitude, and that such a seeking indicates the *possibility* of what it unfortunately fails to achieve. The prominent role of the Good at the conclusion of the *Logic* and the alternation of ontological topics with practical ones in the *Encyclopedia* make it clear that the ontological implications that Hegel associates with the "Ought" introduce us to a fundamental principle of his system.

25. In this way, Hegel's "idealism" (as he calls it) does not assert, like George Berkeley's idealism, that all reality is ideas located in minds, or like Kant's idealism that important features are imposed on reality by minds. Rather, it shows how what most deserves to be called "real," because it's self-governing and thus is what it is by virtue of itself, is minds or "Spirit." The processes studied by physics are real in the sense that they can be studied objectively, but not in the sense that they are what they are by virtue of themselves. This is the gist of Hegel's definitive account of what he means by "idealism," in Hegel 1989 (154–56; *GW*, 21:142–43; *TWA*, 5:172–73).

26. "Reconciliation" (*Encyclopedia*, §§552R, 555, 561; *TWA*, 10:364, 367, 369). "Spirit's elevation to God" (*Encyclopedia*, §552R; *TWA*, 10:354; cf. §50, remark; *TWA*, 8:132).

27. In *Encyclopedia*, §572, Hegel lays out the sequence of art, religion, and philosophy as embodying (respectively) "intuition" (*Anschauung*), "representation" (*Vorstellung*), and "self-conscious thought," and he associates religion and "representation" with "totality" (*Totalität*).

28. This is a way of stating Plato's objection to the arts, in the *Republic*, that they are "images of images" and so forth. But it's clear from Plato's own manifestly artistic efforts (which he himself occasionally acknowledges as such) that although philosophy surpasses the arts in principle, it doesn't thereby render them dispensable. Hegel formalizes this state of affairs by presenting the arts as aspects of "absolute Spirit," ultimately subsumed but not abolished by philosophy, in accordance with his principles of "sublation" (*Aufhebung*) and true infinity.

29. "This knowledge is thus the concept, cognized by thought, of Art and Religion, in which the diverse elements in the content are cognized as necessary, and this necessary as free" (*Encyclopedia*, §572; *TWA*, 10:378).

30. On Plato's ascent as becoming oneself, see note 24, above.

31. When in the *Encyclopedia*'s treatment of Spirit everything finite goes beyond itself, as what Hegel now calls "the self-thinking Idea" (*Encyclopedia*, §574) and associates in his final quotation with Aristotle's God, it "goes back" to his *Logic* (*Encyclopedia*, §574), in which the culminating "Idea" was the origin of Nature and Spirit. Nature and Spirit always presupposed this infinity (the "Idea"), and now they have explicitly returned to it. Thus Nature and Spirit are the *epistrophe,* or turning back, that (following the traditional pattern first spelled out by Plotinus) reverses the *proodos*, "progression," or flowing out that occurs, in Hegel, in the *Logic*. Plotinus's *epistrophe* was modeled on Plato's descriptions of ascent in *Republic,* book 7 (etc.), and his *proodos* was modeled on Plato's creation story in the *Timaeus*. Hegel has shown us a way in which to understand these traditional concepts.

32. Hegel 1989, 129; *GW*, 21:117; *TWA*, 5:140.

33. Heine 1981, 47. Cf. Rosen 1969: "One may say that Hegel makes the suppression of nihilism dependent upon hybris, or the sanctioning of man's desire to be a god" (234).

34. It's sometimes suggested that Feuerbach's "anthropotheism" restates what was really going on in Hegel's philosophical theology. This, however, is a mistake, because Feuerbach didn't reproduce the vertical dimension of (Plato's and) Hegel's thinking, which corresponds to religion's "transcendence." This is why Feuerbach's various proposals have not inspired or attracted much of a following.

35. Trendelenburg 1870, 60. Hegel presents true infinity not as following from prior "premises" (which none of his transitions do), but as fulfilling the claims of being and reality to be determinate in themselves (*an sich*). See Wallace 2005, 59–80.

36. During the two centuries since Hegel, a series of writers have tried either to explain how the Plato/Hegel synthesis works or to state something similar in their

own way. In his *The World as Will and Representation* (1818), Arthur Schopenhauer laid out a duality of "will," on the one hand, and a blissful liberation from "will" (vol. 2, paras. 65–70), on the other. But because Schopenhauer didn't bring out the significance of our pursuit of the true and the good, or rational transcendence, there was no apparent path that could lead from "will" to the liberation that Schopenhauer described. Hence, no doubt, his pessimism. Later, Friedrich Nietzsche, Martin Heidegger, and Jean-Paul Sartre wrote about becoming oneself, authenticity, and freedom, but none of them noted how the pursuit of the true and the good can be crucial in this connection, by raising one above automatic responses to one's heritage or environment. Francis Herbert Bradley (1893) gave a version of Hegel that likewise neglected the role of rational transcendence in becoming fully oneself and thus provided no path that an individual could travel from "appearance" to mystical "reality." Like Schopenhauer and Bradley, Ludwig Wittgenstein's apparently positive allusions to "the mystical" in his *Tractatus Logico-Philosophicus* (1921) didn't connect it to our everyday experience of rational transcendence and thus they left obscure the role of this "mystical" in our lives. John Niemeyer Findlay and Wilfrid Sellars, in the middle of the century, and John McDowell (1994) and Iris Murdoch (1992) likewise didn't clarify the role of rational transcendence in (full) reality, and thus they weren't able to effectively overcome scientism's notion of "reality" as simply what's "objective." R. G. Collingwood came close to Hegel's project of integration in his *Speculum Mentis* (1924), but he too did not spell out the notion of rational transcendence as such. Alfred North Whitehead identified the Platonic rational transcendence in general terms in his *Religion in the Making* (1926) and *Process and Reality* (1929), but he didn't articulate it in everyday terms as freedom and love, so the concrete relevance of his account has remained obscure. Nor have commentators on Hegel from Findlay through Charles Taylor, H. S. Harris, Robert Pippin, Stephen Houlgate, Peter Hodgson, or (in Germany) Dieter Henrich or Walter Jaeschke brought out the centrality of rational transcendence in Hegel's system. So rational transcendence has not been effectively presented since Hegel's time, and Nietzsche's, Bertrand Russell's, and Heidegger's influential critiques of Platonism and Hegel have not been effectively countered.

37. I'm referring to Taylor's "Responsibility for Self" (1976, 281–99), and not to Taylor's *Hegel* (1975), in which he unfortunately did not identify or appreciate Hegel's contribution to the same Platonic train of thought about rational self-government that he (Taylor) was pursuing in "Responsibility for Self."

REFERENCES

Bradley, Francis H. 1893. *Appearance and Reality: A Metaphysical Essay.* London: Swan Sonnenschein.

Christensen, M. J., and J. A. Wittung, eds. 2007. *Partakers of the Divine Nature: The History and Development of Deification in the Christian Traditions.* Madison, NJ: Fairleigh Dickinson University Press.

Collingwood, Robin G. 1924. *Speculum Mentis, or the Map of Knowledge.* Oxford: Clarendon.
Culp, John. 2008 and 2017. "Panentheism." In *Stanford Encyclopedia of Philosophy.* https://plato.stanford.edu/entries/panentheism.
Desmond, William. 2003. *Hegel's God: A Counterfeit Double?* Aldershot: Ashgate.
Gerson, Lloyd P. 2005. *Aristotle and Other Platonists.* Ithaca, NY: Cornell University Press.
———. 2013. *From Plato to Platonism.* Ithaca, NY: Cornell University Press.
Hegel, Georg W. F. 1969. *Wissenschaft der Logik. I.* Frankfurt am Main: Suhrkamp.
———. 1969. *Wissenschaft der Logik. II.* Frankfurt am Main: Suhrkamp.
———. 1970. *Enzyklopädie der philosophischen Wissenschaften im Grundrisse 1830. I.* Frankfurt am Main: Suhrkamp.
———. 1970. *Enzyklopädie der philosophischen Wissenschaften im Grundrisse 1830. III.* Frankfurt am Main: Suhrkamp.
———. 1981. *Wissenschaft der Logik. Zweiter Band. Die subjektive Logik oder Lehre vom Begriff (1816).* Hamburg: Meiner.
———. 1984. *Wissenschaft der Logik. Erster Band. Die Lehre vom Sein (1832).* Hamburg: Meiner.
———. 1989. *Hegel's Science of Logic.* Translated by A. V. Miller. Atlantic Highlands, NJ: Humanities Press.
Heine, Heinrich. 1981. *Confessions.* Translated by P. Heinegg. Malibu, CA: Joseph Simon.
Houlgate, Stephen. 2006. *The Opening of Hegel's Logic: From Being to Infinity.* West Lafayette, IN: Purdue University Press.
McDowell, John. 1994. *Mind and World.* Cambridge, MA: Harvard University Press.
Moore, Adrian W. 2012. *The Evolution of Modern Metaphysics: Making Sense of Things.* Cambridge: Cambridge University Press.
Murdoch, Iris. 1992. *Metaphysics as a Guide to Morals.* London: Chatto & Windus.
Pinkard, Terry P. 2002. *German Philosophy, 1760–1860: The Legacy of Idealism.* Cambridge: Cambridge University Press.
Rahner, Karl. 1978. *Foundations of Christian Faith: An Introduction to the Idea of Christianity.* London: Darton, Longman & Todd.
Ramelli, Ilaria. 2009. "Luke 17:21: 'The kingdom of God is inside you.' The Ancient Syriac Versions in Support of the Correct Translation." *Journal of Syriac Studies* 12 (2): 259–86.
Rorty, A. O., ed. 1976. *The Identities of Persons.* Berkeley: University of California Press.
Rosen, Stanley. 1969. *Nihilism: A Philosophical Essay.* New Haven, CT: Yale University Press.
Sells, Michael A., ed. 1996. *Early Islamic Mysticism: Sufi, Qur'an, Mi'raj, Poetic and Theological Writings.* New York: Paulist.
Taylor, Charles. 1975. *Hegel.* Cambridge: Cambridge University Press.

———. 1976. "Responsibility for Self." In *The Identities of Persons*, edited by A. Oksenberg Rorty, 281–99. Berkeley: University of California Press.

Tillich, Paul. 1957. *Systematic Theology.* Chicago: University of Chicago Press.

Trendelenburg, Adolf. 1870. *Logische Untersuchungen.* Leipzig: Hirzel.

Wallace, Robert M. 2005. *Hegel's Philosophy of Reality, Freedom, and God.* Cambridge: Cambridge University Press.

CHAPTER 8

Bolzano's Concept of Divine Infinity

CHRISTIAN TAPP

Cuius regni non erit finis.
His kingdom will have no end.
—Nicene-Constantinopolitan Creed

Infinity is central to each of the three areas in which Bernard Bolzano (1781–1848) had expertise: mathematics, philosophy, and theology. The Bohemian priest dealt with all the respective concepts in these disciplines: quantitative infinity in mathematics, qualitative infinity in philosophy and theology, and some of their mutual interrelationships.

I focus on "infinite" used as a divine predicate. Since this use of "infinite" is related to a quantitative use in several ways, we will also dwell on these.

QUANTITATIVE AND DIVINE INFINITY IN THE *PARADOXIEN DES UNENDLICHEN*

In his late major work *Paradoxien des Unendlichen* (*PU*) (in English, *Paradoxes of the Infinite*),[1] Bolzano set out to resolve the most significant mathematical paradoxes of infinity. For him, infinity was a key concept for mathematics, physics, and metaphysics (*PU*, §1). Hence, he found it crucial to dissolve all kinds of paradoxical "*Schein*" (appearance) and to make it clear that the concept of infinity is perfectly rational.

To do so, Bolzano first presents a definition of infinite quantities: "I propose the name infinite multitude for one so constituted that every single finite multitude represents only a part of it" (*PU*, §9).

At first glance, this definition seems defective. It immediately raises two questions: What is a finite multitude? How could every finite multitude be a part of all infinite multitudes?

To begin with the second question: take, for example, the finite multitude that consists of me and Bolzano. Compare this two-element set with the set of natural numbers. To be sure, the two-element set of me and Bolzano is not part of the infinite set of natural numbers. So how could the set of natural numbers fulfill the definition of infinity, according to which every single finite multitude *must somehow be a part of* an infinite set? This puzzle is quite easily solved by interpreting more precisely the definition of an infinite multitude as a multitude of which "every single finite multitude represents only a part." If we take "represent" to mean that every finite multitude can be *mapped* onto a part of any infinite multitude, and not that every finite multitude *is* a part of it (in the sense of being a subset), the second problem is solved.

With respect to the first question (What is a *finite* multitude?), we should note that Bolzano proposes his definition at the end of a lengthier paragraph in which he talks about a series of individuals of a species *A*. He says that such a series may contain more or less numerous sets of terms:

> In particular, the series may contain so many terms that it cannot, compatibly with taking in and exhausting all the individuals of that species, be conceded to have a *last term*; a point which we shall handle in greater detail in the sequel. Assuming it for the present, I propose the name infinite multitude. (*PU*, §9; emphasis in original)

Hence, Bolzano introduced his incomplete definition of an infinite multitude in the context of a discussion about a series without a last term. Not having a last term, however, is a sufficient, but not necessary condition for being infinite. (Standard Cantorian set theory gives us lots of examples for its being nonnecessary. Take, for instance, the second infinite ordinal number $\omega + 1$. It is infinite but has, according to its standard representation, a last element, namely ω.)

Hence, I propose the following for making sense of Bolzano's definition of quantitative infinity:

1. A series is called "certainly infinite" iff it has no last term.
2. A series is called "finite" iff it can be mapped one-one onto a part of every certainly infinite series.
3. A series is called "infinite" iff every finite series "represents a part of it," that is, can be mapped one-one onto a part of it.

In this manner, we have captured the substance of Bolzano's definition and used two of his own suggestions to make the definition fully acceptable: the concept of a series without a last term, and the idea of representations as one-one mappings.

Today, it has become mathematically standard to define infinite sets as sets that allow a one-one mapping onto a proper subset of themselves. Astonishingly, Bolzano already noticed that infinite sets have this property and even that this property is characteristic of infinite sets (*PU*, §20). However, it was Richard Dedekind (1831–1916) who made this into a definition of infinity (Dedekind 1888, note on no. 64). In Dedekind's work, a variant of Bolzano's definition also survived in the form of the alternative definition that a set *S* can be called "infinite" if for every natural number *n*, *S* has at least *n* elements.[2]

So much for Bolzano's definition of quantitative infinity. In *Paradoxes of the Infinite*, he is quite successful in employing this concept to resolve all kinds of alleged paradoxes traditionally assumed to be generated by the concept of infinity.

But some philosophers claim that, besides this relatively precise notion, we also need another concept of infinity: a concept of *qualitative* infinity. Hegel, for instance, famously claimed that quantitative infinity is "bad infinity" and that we have to form a "dialectical" notion of "true infinity" (*wahrhafte Unendlichkeit*) that transcends the opposition of the finite and the infinite, in which the infinite was merely the negation of the finite (see Hegel [1830] 1970, §§93–95). Bolzano distinguished carefully between different Hegelian claims: (1) Hegel is completely right, says Bolzano, in criticizing the concept of potential infinity as "bad" infinity. What people call "potential infinity" is a property of growth processes, not of quantities. As the values of potentially infinite growth processes may always stay finite, potential infinity is a crude concept in Bolzano's eyes (*PU*, §11): "Infinite" shall be used as a predicate of quantities,[3] but with respect to quantities it does not make sense to call them infinite when they in fact always stay finite (§12). (2) However, Bolzano claims, Hegel is completely wrong to identify mathematical infinity with potential infinity. A limitlessly growing magnitude may always stay finite. In contrast, all the substantial mathematical uses of infinity are those of actual infinity. So whatever Hegel is talking about, it is not mathematical infinity. (3) Bolzano is very skeptical about an additional qualitative concept of infinity. However, he does not bluntly reject it as incoherent or empty. He just thinks it superfluous for the following reason: "What I refuse to admit is only this: that the philosopher knows any object to which he is entitled to attach the predicate of infinitude without having first established that in some respect or other that object exhibits infinite quantity, or at least infinite multitude" (§11).

What, then, would be the best touchstone for this claim? To prove it in the case of God!

> Now if I can once show that in God Himself, the Being Whom we regard as the most perfectly one, aspects can be found under which we see even in Him an infinite multitude, and if I can show that we attribute infinitude to Him under those aspects alone, then it will scarcely be necessary to go on and show that similar considerations lie at the bottom of all the other cases where the idea of the infinite holds good. (*PU*, §11)

So, God is a kind of test case for Bolzano: if, even in the case of God, his thesis about quantitative infinity being basic is true, then there is a very good reason to take it as being generally true. The same holds for the Hegelian doctrine of the All: if the All is all there is, such that outside of it there is nothing, then this All comprises an infinite multitude of entities, because it is an aggregate of all existing things, even nonactual things, such as absolute propositions and truths. Hence, even in the case of the All, it is unnecessary to deviate from Bolzano's account of quantitative infinity as primordial infinity. To coin a phrase: "No infinity without quantitative infinity!"

From the point of view of a philosopher of religion, this is a surprising thesis, since there are very few, if any, philosophical accounts of divine infinity that make use of quantitative concepts. But this is Bolzano's view of divine infinity, which we shall explore in this chapter.

Is Bolzano's thesis true? Is it the case that even in God we find quantitatively infinite aspects? Bolzano says yes:

> I say then: We call God infinite because we are compelled to admit in Him more than one kind of force possessing an infinite magnitude. Thus, we must attribute to Him a power of knowledge which is true omniscience, and which therefore comprises an infinite set of truths, to wit, all truths and so forth. (*PU*, §11)

This is a short statement of Bolzano's doctrine of divine infinity. We find it in a much more developed form in an earlier major work, the *Religionswissenschaft*.

DIVINE INFINITY IN THE *RELIGIONSWISSENSCHAFT*

The title of Bolzano's *Religionswissenschaft* (*RW*) (in English, *Science of Religion*) might be misleading. It is not a book of religious studies, but one about philosophical and revealed theology. Bolzano treats the divine attributes twice in this book: first in the context of natural theology (part I), then again in the context of revealed theology (part III). The two treatments are mutually consistent, though the revealed theology account is enriched by considerations about the "moral" and "real" value of each element of the doctrine and about its historical roots (which are, for the most part, biblical).

The Divine Attributes in Natural and Revealed Theology

When Bolzano treats the divine attributes in the context of natural theology, he scarcely mentions infinity.[4] This is completely different in revealed theology where he explicitly uses infinity as the central conceptual tool for explicating God's omni-perfection, and "infinity" even shows up in several section headings.[5] Bolzano's idea was that infinity is central for the theoretical development of the divine attributes. Even if he does not mention the word "infinity" in the first part, he uses it when he looks back to the first part in considering the "necessity of a revelation even for the most erudite people" (*RW*, I, §99). There he summarizes his earlier treatment of the divine attributes, saying that natural theology leads to the insight that a divine being exists and that it represents the epitome of all perfections insofar as it possesses "infinite power, wisdom, and holiness."[6] Natural theology, however, cannot reach much farther in envisioning the divine. In spelling out the reasons for this claim, Bolzano assigns infinity a crucial role:

> The first deficiency of this [the natural] religion reveals itself, if we want to envision the aforementioned properties of God somewhat more perspicuously. Because all divine powers and qualities (*Beschaffenheiten*) have a certain infinity, while our finite understanding struggles in conceiving the infinite, a humble person will become the more timid in her judgments about God, the more carefully she considers that it is the infinite being about which she dares to judge. (*RW*, I, §99)[7]

Hence infinity, even though not treated explicitly, was a key concept for Bolzano's natural theology.

Natural Theology (*RW*, part I)

Bolzano develops his natural theology based on the concept of God as an "unconditioned real" (*unbedingt Wirkliches*; *RW*, I, §66). This means: something is God if it is real and is not dependent upon anything. Using this notion, Bolzano presents a proof of God's existence (§67) and elaborates some of His properties: necessary existence (§69), substantiality (§70), the unconditioned nature of His powers (§71), immutability (§72), uniqueness (§73), and, finally, omni-perfection (§74).

Bolzano then considers the different powers of God (*RW*, I, §75). "Power" means "the potential to effect something." Effects can be internal, in the case of immanent powers, or external. Bolzano distinguishes four immanent powers—to think, to feel, to wish, and to will—and two external powers—to create, and to change. All these powers, which we know in ourselves, we also meet in God, except the power to wish, which would make no sense to ascribe to a being enjoying highest beatitude (§75, no. 6; see also §77, no. 1). The guiding principle for Bolzano's considerations is that God must have these perfections in the highest degree possible in themselves and in conjunction with the other perfections.

Let us now consider two highly instructive examples of Bolzano's treatment of divine properties, or powers: the power of knowledge (omniscience) and the external powers (omnipotence).

*God's Omniscience (*RW*, I, §76)*

An omni-perfect being must have the power to think, which must be as great as it can be, given its other powers. Because the power to think is not limited by other powers (instead, the other powers are limited by the power to think), it must be as great as it can be in itself: as the power to reach all the goals of thinking, to know all truths *an sich*. There are, however, infinitely many truths *an sich* (as Bolzano putatively proves elsewhere; see my next main section). Hence, God's power to know is infinite in the sense that it extends over infinitely many truths *an sich*. Note, however, that Bolzano does not explicitly talk about the infinity of God's power to know here, but that omniscience is conceived of as knowing *all* truths *an sich*. That these are infinitely many is not a *conceptual part of omniscience* as it is conceived in the first part of *Religionswissenschaft*.

God's Omnipotence

God's omnipotence is God's power to produce external effects (*RW*, I, §79). In us, such a power is variously limited by, for example, limits of our power to think, as we forget things or wrongly think them to be impossible for us, thus inhibiting the producing of certain external effects. In God, however, "all these limitations cease to apply" (§79, no. 4). Hence, Bolzano concludes, God's power to produce external effects extends to *all* that (1) is possible in itself, (2) is compatible with the moral law, and (3) requires a determining ground of its being.

Note, again, that Bolzano does not speak of infinity in this context. There is not even the obvious implication of an infinite set, like the set of all truths *an sich* in the case of omniscience. Interestingly, all this is completely different when Bolzano considers the divine attributes in the context of revealed theology, in the third part of *Religionswissenschaft*.

Revealed Theology (*RW*, part III)

In part III of *Religionswissenschaft*, where Bolzano develops a "catholic dogmatics," he explicitly addresses infinity when reconsidering the divine attributes treated in part I.

Bolzano begins by considering God's omni-perfection (*RW*, III, §43). All possible perfections are united in the highest possible degree in God's essence. This teaching of Christianity is so obvious that many philosophers have taken "all-perfect being" to be the definition of God. (Bolzano's definition is, instead, that God is an unconditioned real, as we have seen in *RW*, I.) However precise our concepts are, God remains incomprehensible to us in the sense that all teachings and theoretical insights about Him provide no account of how He is in Himself, but only a picture or a shadow image (III, §47). Bolzano says the "true reason" for God's incomprehensibility is that God has "totally *infinite powers*." For example: His power of knowing extends to all truths. Since there are *infinitely many truths*, to understand God completely one would have to understand *infinitely many truths*, and this is impossible for finite beings such as human beings. Hence for us, He is incomprehensible.

Bolzano calls God's power of understanding "God's infinite intellect" (*RW*, III, §52) and repeats the claim of the first part that God's power to think extends to whatever is knowable, that is, to all truths *an sich* (§52).[8] But this is puzzling. The quick transition from *an infinite capacity to know* to *knowing infinitely many truths* is not completely obvious. Having an infinite capacity to know (*RW*, III) and knowing everything knowable (*RW*, I and III) are conceptually distinct. So how can the transition be justified? According to Bolzano, there are infinitely many truths *an sich* to be known. (We will consider this in the next section.) That assumption is true. But it is not part of the concept of "knowing everything" that the extension of "everything" is an infinity of truths. Furthermore, the converse implication does not hold: from the fact that someone knows infinitely many truths, one cannot validly conclude that she knows everything. Imagine I know all true equations of natural numbers (0 = 0, 1 = 1, 2 = 2, etc.). Then I know infinitely many truths, but not all, by far.[9] We will come back to this puzzle.

On Bolzano's understanding, "God knows everything" (*RW*, III, §55) contributes much to the venerability of God. To illustrate what it means to "know everything," Bolzano digresses from his method of systematic argument: God knows all numbers, all combinations of numbers in sums of two, three, or four, and so on, in products and quotients. He knows the many different lines in geometry, the different angles, triangles, quadrangles, and so on, the orders of curved lines, the many different surfaces, bodies, spaces; the mechanical combinations of many different physical powers. In the empirical world, He knows the many stars, planets, comets; how many living and dead organic and inorganic beings there are on planet earth, how many atoms in each one of the thousands of mosquitoes flittering in the evening sun; and the relations between any two atoms in the whole universe (which God must arrange in the correct way in order to stabilize the cosmos as it is). If we consider all this, then, Bolzano says, "we must sink on our knees before the infinite, because we feel, what it means to be infinite" (§55). God's infinity is what requires us to worship and adore Him.

We have here a complete reversal of Heidegger's much later critical remark that one cannot sink on one's knees and pray, and sing, or dance, in front of a *causa sui*, which stands *exemplum pro specie* for strictly philosophical concepts of God (see Heidegger 1957, 70). For Bolzano, in contrast, it is precisely divine infinity which illustrates that the being which fulfills the formal concept of an unconditioned real is also worthy of the highest adoration and worship.

THE INFINITY OF TRUTHS

In our analysis of Bolzano's doctrine of divine infinity, we have been puzzled by the logical connection between knowing everything knowable and knowing infinitely much. Part of a solution consists in Bolzano's claim that the class of truths *an sich* is infinite. He offers proofs of this, for example, in the *Wissenschaftslehre* (*WL*) (in English, *Theory of Science*) (*WL*, I, p. 146, 147; *RW*, I, p. 35; *PU*, §13).

These proofs proceed inductively. According to the principle of complete induction, they have two parts: first, that there is at least one truth, and, second, that for every natural number n, if there are at least n truths, then there are at least $n + 1$ truths. By the definition above, it follows that the class of all truths is infinite.

There is at least one truth. This is proven by an indirect proof (*RW*, I, p. 35; *WL*, I, p. 145). Assume that it is not the case that there is a true proposition.[10] But, "[It is not the case that there is a true proposition]" is a proposition.[11] Hence it cannot be true, because we have assumed there is no true proposition. Therefore, by semantic descent, it is not the case that it is not the case that there is a true proposition, that is, our assumption was false and there is a true proposition.

The induction step from n true propositions to $n + 1$ is more interesting, but its idea is almost the same. I reproduce Edgar Morscher's semiformal reconstruction (Morscher 2011, 202):

> Assume the Inductive Hypothesis (IH): there are at least n true propositions, say $S_1, \ldots, S_n$. Then we have to prove: there are at least $n + 1$ true propositions. We do this by the following indirect proof, reducing the negation of our claim to the absurd:

1.	There are not at least $n + 1$ true propositions.	(assumption)
2.	There are no more than n true propositions.	(from 1 by PL)
3.	There is no true proposition except $S_1, \ldots, S_n$.	(from 2 and IH)
4.	[There is no true proposition except $S_1, \ldots, S_n$] is a proposition.	(premise)
5.	[There is no true proposition except $S_1, \ldots, S_n$] is true.	(from 3 by semantic ascent)
6.	[There is no true proposition except $S_1, \ldots, S_n$] is a true proposition.	(from 4 and 5)
7.	[There is no true proposition except $S_1, \ldots, S_n$] is different from each proposition of the $S_1, \ldots, S_n$.	(premise)
8.	There is a true proposition different from $S_1, \ldots, S_n$.	(from 6, 7; contradiction to 3)
9.	There are at least $n + 1$ true propositions.	(from 1, 3, 8 by indirect proof)

Although this proof is logically basic, it adduces two premises that require consideration. The first premise is that [There is no true proposition except $S_1, \ldots, S_n$] is a proposition (line 4). This is not to say much more than that "There are not more English sentences than $S_1, \ldots, S_n$" is an English sentence. The only special feature is Bolzano's distinction between linguistic sentences and sentences *an sich*. Linguistic sentences express sentences *an sich* if they fulfill certain semantic requirements (like being unambiguous; see Morscher 2011) that are fulfilled in the case of our first premise. The second premise is that the "new" proposition [There is no true proposition except $S_1, \ldots, S_n$] is distinct from each of the propositions $S_1, \ldots, S_n$ (line 7). Bolzano argues for this premise, for example, in *PU*, §13:

> For if we fix our attention upon any truth taken at random . . . and label it A, we find that the proposition conveyed by the words "A is true" is distinct from the proposition A itself, since it has the complete "proposition A" for its own subject.

The main objection to Bolzano's proof, and similar proofs (from Dedekind, Cantor, and others) is Frege's assertion that "p is true" does not express anything more than "p." But for his sentences *an sich*, Bolzano offers a clear-cut criterion of identity: propositions are identical if and only if they have the same subject-idea and the same predicate-idea. Therefore, "p is true" and "p" can be identical only if [p] is also the subject-idea of p and [is true] is also the predicate-idea of p, that is, only if p is the self-referential proposition [p is true]. In our case, in contrast, $S_1, \ldots, S_n$ are all part of the predicate-idea of p and we can quite safely assume that they are not part of the predicate-ideas of $S_1, \ldots, S_n$, respectively.

Given Bolzano's proofs that there exist infinitely many truths *an sich*, part of our problem is solved. It is now clear that a being that knows everything knowable, in fact knows infinitely many truths, namely, all the infinitely many truths *an sich*. Because the converse inference is invalid, however, there remains the problem of explaining how Bolzano could be justified in skipping so quickly from knowing infinitely much to knowing everything knowable.

A FORMAL BACKGROUND CONCEPT OF INFINITY

My thesis is that, in fact, there is no way to get from knowing infinitely much to knowing everything knowable. Hence, in order to interpret Bolzano in a coherent way, we need to ask whether our reconstruction of Bolzano's "infinite capacity of knowledge" as the capacity to know infinitely much is correct. This interpretation was suggested by Bolzano's (*PU*, §11) claim of the priority of quantitative infinity. Nevertheless it seems much more coherent to ascribe to Bolzano another, quite formal concept of infinity in the background. What concept is that?

The Formal Background Concept . . .

In *Wissenschaftslehre* (I, §87) Bolzano discusses how other philosophers have conceived infinity. He chooses four candidates and discusses them.

(1) The first candidate is *indeterminateness*: something is infinite insofar as it cannot be precisely specified or determined. Bolzano rejects this conception on the basis of several arguments. Let us consider one of them. Bolzano says that there is only one understanding of infinity as indeterminateness that makes sense; that infinite sets are not determinable by their proportion to finite sets. This is correct, says Bolzano, but indeterminateness in one respect does not mean determinateness per se. Hence one cannot say that infinite sets are indeterminate per se. He presents a knockdown counterexample: the line between two points in Euclidean space contains infinitely many points. The set of these points is therefore infinite, although it is precisely determined by the two terminal points. Hence it does not make sense to identify infinity and indeterminateness.

(2) The second candidate is *nonaugmentability*: something is infinite insofar as it cannot be augmented or increased. This definition is clearly false, says Bolzano, for one can increase the set of points on a line between two points (which is infinite) by simply extending the line beyond one of the terminal points.[12] Hence there are infinite sets that are clearly augmentable and, therefore, nonaugmentability cannot be a necessary condition for infinity.

(3) The third candidate is "*being larger than every possible set.*" But this definition is unfulfillable, because every set that might fall under it must, by definition, be larger than itself, which is impossible.

(4) The fourth candidate is "*having no end, limit, or bound.*" In Bolzano's eyes, this is the best attempt at defining "infinite." But it must be

corrected because X's being infinite does not mean that X has no limit in every respect. Again, he offers the counterexample of a line between two terminal points in Euclidean space. Having two terminal points, the line has limits in a certain respect.

Having no limits is the formal background concept I claim to be central for Bolzano's whole doctrine of infinity. Before I argue that it is indeed justified to ascribe this concept to Bolzano, even though he had some reservations concerning it, let us see how it fits into his thinking about infinity, mathematical and divine.

Bolzano's idea of mathematical infinity was inspired by the natural idea that counting (with constant speed, in finite time, and idealized beyond our creaturely constraints) will not come to an end. To count a set *s* means to present a part one-one mapping from the natural numbers onto *s*. Coming to an end in counting *s* means that all elements of *s* have been reached. A set *s* can then be called "finite" if all countings of *s* come to an end. A set *s* is infinite iff there is a one-one mapping of the natural numbers onto *s*. The connection between the precise mathematical definition of infinity and the general philosophical concept of "having no limits" is then as follows: a set is infinite (in the mathematical standard sense) iff it has no limit with respect to idealized counting.

Bolzano's distinction between "having no limits in a certain respect" and "having no limits in every respect" is helpful, especially when one considers mathematical theorizing about infinity after Bolzano. Cantorian transfinite ordinal numbers, for example, do not fulfill the criterion of "having no limits in every respect" because each one is limited by the next transfinite ordinal number. But they have no limits in a certain respect, namely, with respect to idealized counting of their members.

Is God infinite in the sense of having no limits whatsoever? On Bolzano's theory, we cannot say so without reservation. God's perfections are restricted to what is possible in itself, compatible with the moral law, and in need of a ground of its being. He therefore seems not unrestricted in every respect, unless such seeming restrictions do not really restrict divine power but rather exclude mere impossibilities from the extension of the powers ascribed to God in the form of the divine predicates.[13] Nevertheless, the concept of being without restrictions is helpful in developing

Christian doctrine. For example, if Christian faith calls God's intellect, wisdom, or will "infinite," it implies that certain restrictions, which we know from our human nature, do not apply in God's case.

. . . and Its Justification

Accepting this background concept of "having no limits" leads to terminological difficulties with the use of "being unlimited" and "being infinite" in today's mathematics. According to this use of language, there are infinite things that are limited. (For example, every transfinite ordinal number is an infinite number, but limited, for example, by the next transfinite ordinal number. Or, take Bolzano's set of points on a line that is infinite but limited.) Also, there are unlimited things which are finite (for example, the space between the curve $y = 1/x^2$ and the x-axis, taken from $x = 1$ to $+\infty$). But let us leave this aside here in favor of finding a coherent interpretation of Bolzano independently of the question of which terminology best expresses it.

I see two main reasons for parenthesizing Bolzano's reservations toward this concept and, instead, understanding him with the help of it. The first reason is that Bolzano's treatment of omnipotence strongly suggests this formal understanding. Bolzano asks how great our power to influence external things is. Our human powers over external things are already limited by the limitation of our other powers, such as our restricted cognitive capacities and our weak will. God, in contrast, does not suffer from such limitations. His will is perfect, and He can do whatever is possible and compatible with his other attributes (*RW*, III, §57). Hence, his external power is infinite in just the formal sense of being limitless.

The second reason in favor of my thesis is that the formal background conception helps to solve the second part of the puzzle of the quick transition from an infinite capacity of knowing to knowing everything knowable. The infinity of God's intellectual capacities does not *mean* that God knows infinitely many truths *an sich*, but that these capacities have no limit. An unlimited capacity for knowledge, however, includes the capacity to know everything knowable. For why should a truth that is knowable in principle be excluded from being known by a being with the unlimited capacity of knowledge?

CONCLUSIONS

Divine infinity is an integral part of Bolzano's concept of God. It is a test case for his broader claim that the concept of quantitative infinity is more basic than all other concepts of infinity; that all meaningful discourse about infinity presupposes quantitative infinity. This thesis from §11 of the *Paradoxes of the Infinite* must not be understood as reductivist, as though Bolzano would *only* accept quantitative infinity. He claims only that in each case in which we are entitled to use qualitative concepts of infinity, which are vaguer and more opaque, we must also be able to find aspects of the subject matter that exhibit the more precise quantitative infinity.

Ascribing to Bolzano a formal background conception of infinity as limitlessness or unrestrictedness, as suggested by his treatment of omnipotence, helps us to interpret his doctrine of God in a coherent way. However, what exactly it means to be unlimited, to suffer from no limitations in a certain respect, remains a topic for further philosophical analysis. Does every conception of "limits" require a metric or at least a total linear ordering? How can one align such a concept with the mathematical way of speaking, according to which there are unlimited finite magnitudes and also limited infinite magnitudes? Finally, what "respects" are to be considered, in the case of divine infinity, as unrestricted in certain, if not all, respects?

NOTES

An earlier version of this chapter was published in German (see Tapp 2011).

1. The standard English reference edition of *Paradoxien des Unendlichen* is Steele's edition (Bolzano 1950). For the German original text, there is a more recent edition (Bolzano 2012).

2. For Dedekind's famous definition of infinite chains, see Dedekind 1888 (71–73). Both definitions are equivalent over standard set theories.

3. Cf. the Aristotelian dictum the Scholastics used to phrase as *infinitum quantitatem sequitur*, "the infinite follows [= belongs to the category of] quantity" (Aquinas, *Summa contra Gentiles* 1. 1, c. 43).

4. Bolzano mentions infinity only in these contexts: the infinite regress in the causality proof for the existence of God that he critically analyzes in *RW*, I, §68; the description of the divine power of sensing (*Empfindungskraft*) as providing God with

beatitude of an infinitely high degree (*RW*, I, §77); the discussion of necessity to create infinitely many beatific creatures in the world, which is infinite with relation to space (§81, no. 2); and a cursory mention of God's infinite wisdom in the context of a discussion of the immortality of the soul (§85).

5. For example, *RW*, III, §52, has the heading "Die Lehre von Gottes unendlichem Verstande" ("The Doctrine of God's Infinite Understanding"), and "Die Lehre von Gottes unendlich vollkommenen Willen" ("The Doctrine of God's Infinitely Perfect Will") is part of the heading of *RW*, III, §57.

6. ". . . daß dieses Wesen als der Inbegriff aller Vollkommenheiten, und sonach begabt mit einer unendlichen Macht, Weisheit und Heiligkeit gedacht werden müsse" (*RW*, I, §99). This and the following English translations from the *Religionswissenschaft* are my own.

7. ". . . der erste Mangel in dieser Religion verräth sich, wenn wir uns die so eben erwähnten Eigenschaften Gottes noch etwas deutlicher vorstellen wollen. Weil nämlich alle göttlichen Kräfte und Beschaffenheiten eine gewisse Unendlichkeit haben, das Unendliche aber von unserem endlichen Verstände schwer aufgefaßt werden kann: so wird der bescheidene Mensch um desto schüchterner in seinen Urtheilen über Gott, je reiflicher er erwägt, daß es das unendliche Wesen sey, das er hier zu beurtheilen waget" (*RW*, I, §99).

8. God can even know our errors but, as Bolzano puts it, only "formally," not "materially." God can know that someone mistakenly believes the false proposition *p* but, in that case, God cannot know *p* itself. See also *RW*, I, §76.

9. *RW*, III, §53, offers a hint toward an explanation of this tension. There, Bolzano quotes Psalm 139 at some length to demonstrate that his doctrine is in accordance with scripture. The most important verses are 17–18a (KJV): "How precious also are Your thoughts to me, O God! How great is the sum of them! If I should count them, they would be more in number than the sand."

To be sure, the number of grains of sand is a speaking metaphor for an incredible large number. Verse 17b, however, calls their number "great," while Bolzano's version calls it "infinite" (my translation, see *RW*, III, §53). This close connection between infinity and greatness also helps to explain the fact that Bolzano calls God's intellectual powers "infinite."

10. To stay short and to facilitate understanding, I simply use "proposition" for Bolzano's "Satz an sich."

11. In the Bolzano literature "[p]" is used to denote the *Satz an sich* expressed by the linguistic sentence p.

12. Georg Cantor later made augmentability the main criterion of distinction between the transfinite and the absolute infinite. The transfinite shares with the finite the property of being augmentable, while the absolutely infinite is not augmentable. What exactly Cantor meant by this is a matter of discussion, however. See Tapp 2014.

13. Thomas Aquinas explicitly excludes inconsistent descriptions of states of affairs from the descriptions that are covered by his doctrine of divine omnipotence.

Dealing with the question of whether this would not mean a restriction of divine power, he says that "whatever implies contradiction does not come within the scope of divine omnipotence, because it cannot have the aspect of possibility. Hence it is better to say that such things cannot be done, than that God cannot do them" (*Summa theologiae* I, q. 25, a. 3, c.a.).

REFERENCES

Bolzano, Bernard. 1950. *Paradoxes of the Infinite*. Translated from the German of the Posthumous Edition by Franz Prihonsky and furnished with a Historical Introduction by Donald A. Steele. New Haven, CT: Yale University Press.

———. 2012. *Paradoxien des Unendlichen* (mit einer Einleitung und Anmerkungen herausgegeben von Christian Tapp). Philosophische Bibliothek, vol. 630. Hamburg: Meiner.

Dedekind, Richard. 1888. *Was sind und was sollen die Zahlen?* Braunschweig: Vieweg.

Hegel, Georg Wilhelm Friedrich. (1830) 1970. *Enzyklopädie der philosophischen Wissenschaften im Grundrisse*. Edited by Eva Moldenhauer and Karl Markus Michel. Frankfurt: Suhrkamp.

Heidegger, Martin. 1957. *Identität und Differenz*. Pfullingen: Neske.

Morscher, Edgar. 2011. "FESI." In *Bernard Bolzanos bessere Welt: Akten der Internationalen Tagung Salzburg, 27. und 28. Mai 2010*, edited by Kurt F. Strasser, 197–217. Brno: Marek.

Tapp, Christian. 2011. "Beobachtungen zur Lehre von der Unendlichkeit Gottes bei Bernard Bolzano." In *Bernard Bolzanos bessere Welt: Akten der Internationalen Tagung Salzburg, 27. und 28. Mai 2010*, edited by Kurt F. Strasser, 173–96. Brno: Marek.

———. 2014. "Absolute Infinity—A Bridge between Mathematics and Theology?" In *Foundational Adventures: Essays in Honor of Harvey M. Friedman*, edited by Neil Tennant, 77–90. London: College Publications.

CHAPTER 9

Cantor and the Infinity of God

BRUCE A. HEDMAN

INFINITY AND TRADITIONAL CHRISTIAN THEOLOGY

While on sabbatical leave in 1988, I had a conversation with the Very Reverend Thomas F. Torrance at the Center on Theological Inquiry, Princeton, New Jersey, in which he expressed doubts whether infinity could exist in a contingent universe. As a mathematician trained in the transfinite theory of Georg Cantor, I felt sure that it must. After he and I corresponded extensively, Torrance (2002) later was convinced to write that Cantor "regarded a completed set as an infinite magnitude, but he distinguished it as transfinite in contrast to the absolute infinity of God.[1] He held that mathematics has to do with a form of rationality which God has imposed upon both the human mind and the universe. It is a created harmony between them that gives the universe its rational unity" (Torrance 2002, 74).

Of course, Torrance was originally expressing one viewpoint in traditional Christian theology that "infinity" was an attribute characteristic of God alone. Thomas Aquinas (1945; *ST* I, q. 7, a. 2) defended the uniqueness of God's infinity: "Things other than God can be relatively infinite, but not absolutely infinite [*non simpliciter infinitum*]," which suggests that things other than God can be infinite, but only in a certain respect (Tapp 2016, 97–101). A quantity is *relatively* infinite, if it is simply unbounded. The *actually* infinite contains within itself already an infinite magnitude. Before Cantor most mathematicians spoke of infinity as only relative, not actual, only as unbounded. For example, in 1831 Carl Friedrich Gauss (1860, 2:269) wrote: "The infinite is only a *façon de parler* in which one properly speaks of limits."

We are deeply indebted to Joseph Warren Dauben for tracing how Cantor's deep Christian convictions intentionally shaped his theory of infinity and its mathematical formulations. Dauben wrote: "The theological side of Cantor's set theory, though perhaps irrelevant for understanding its mathematical content, is nevertheless essential for the full understanding of his theory and the developments he gave it" (Dauben 1979, 291).

CANTOR'S CHRISTIAN BACKGROUND

Cantor's paternal grandparents belonged to the Sephardic Jewish community in Copenhagen, and they emigrated to Russia during the Napoleonic Wars. Apparently, his paternal grandfather, Jakob Cantor, converted to Christianity, as he gave Christian saints' names to his children. Cantor's father, Georg Woldemar Cantor, was educated in the Lutheran mission in Saint Petersburg. He married Maria Anna Böhm, who, upon their marriage, converted from Roman Catholicism to Lutheranism. Their six children, of whom Georg (1845–1918) was the eldest, were baptized in the Evangelical Lutheran Church in Saint Petersburg. They gave their children disciplined religious instruction, and at age fifteen Georg was confirmed in the Lutheran Church. Georg's correspondence with his father shows that both were devout Lutherans.

In 1856 the family moved to Germany, and in 1860 Cantor graduated from the Realschule in Darmstadt with distinction in mathematics. In 1867 Cantor submitted his doctoral dissertation to the University of Berlin in number theory, and then joined the faculty of the University of Halle in 1869, where he was to spend his entire career. There he was made a full professor in 1879 at the remarkably young age of thirty-four.

ACADEMIA REJECTS CANTOR'S IDEAS

From his university days Cantor felt deeply that he had a calling from God to study philosophy and mathematics, rather than follow more lucrative pursuits. Later, Cantor believed that the transfinites had been revealed to him by God and that he was called to spread the truth about God's creation for the benefit of both the church and the world (Dauben 1979, 291). This faith sustained him during the decades when the mathematical establishment rejected his concept of the transfinites. In particular, Leopold Kronecker, his former professor in Berlin, absolutely rejected Cantor's approach to set theory. Kronecker was an early proponent of "constructionism," which allowed the existence of an entity only if an example could be explicitly constructed, thus disallowing the "law of the excluded middle," which proved proposition A by showing not-A to be contradictory. Kronecker regarded Cantor as a "corruptor of the youth" (Dauben 1977, 89n15) for teaching his ideas to the upcoming generation. As Kronecker was an influential member of the Berlin Academy of Sciences, he effectively blocked Cantor's professional advancement. The Swedish mathematician Gösta Mittag-Leffler on philosophical grounds refused to publish in his journal, *Acta Mathematica,* an article by Cantor, writing that "it was about a hundred years too soon" (Dauben 1979, 138).

Cantor had no support from his colleagues at the University of Halle. Philosophically he was the only faculty member who was not a materialist, determinist, or positivist. He denied that the universe was eternal and unbounded, and stressed that transfinite numbers did not imply otherwise. On the contrary, in a private correspondence to K. F. Heman in 1887, Cantor promised to show that in fact transfinites could disprove the eternity of time, space, and matter, though he never wrote out these arguments (Dauben 1979, 360). He challenged the existence of objective or absolute time decades before Einstein (108).

SUPPORT FROM ROMAN CATHOLIC SCHOLARS

In these early years, the only support Cantor received was from Roman Catholic scholars. These were stimulated to investigate Cantor's theories closely by the papal encyclical *Aeterni patris* of Leo XIII in 1879 (Leo XIII 1903). Leo XIII challenged Catholic intellectuals to engage with modern science both to clarify only apparent difficulties with orthodoxy and to guide science by the principles of Christian philosophy. But this support from Roman Catholic scholars became more forthcoming only when Cantor made a certain distinction.

Cantor introduced the mathematical concept of a "completed set." The natural numbers **N** = {1, 2, 3, 4 . . .}, for example, Cantor considered together as a set in themselves, as a completed infinite magnitude, as a *Ding für sich*. Taken as a whole, the set of natural numbers was defined by Cantor as the first transfinite number, which he denoted by the lowercase omega, ω, in deliberate distinction from the infinity symbol, ∞, which John Wallis (Barrow 2008, 340) had introduced in 1655 to mean simply "unbounded." As an intellectual precedent for this step Cantor appealed to Plato's *Philebus*: "All things that are even said to be consist of a one and a many, and have in their nature a conjunction of limit and unlimitedness" (Plato 1961, 1092).

More importantly, an argument often quoted by Cantor comes from Augustine's *City of God*: "All infinity is in some ineffable way made finite to God, for it is comprehended by his knowledge" (Augustine 1979, 2.238).

But Hellenistic cosmology equated God with the world's soul and the world as God's body. In opposition to this, traditional Christian theology maintained the uniqueness of God's infinity. Christian orthodoxy was particularly suspicious after Spinoza's monistic philosophy of substance, which thought of God as the infinite self-generating substance (*natura naturans*) that generates the world (*natura naturata*). Any concrete, temporal infinity was suspect. To be accepted by Roman Catholic scholars Cantor had to make a distinction.

The first theological paper to address Cantor's transfinites was written by a neo-Thomist, Fr. Constantin Gutberlet (1886), who argued that Cantor's completed set of infinite magnitude was consistent with God's unique infinity. He criticized Cantor, though, for believing that actual infinity existed in the created order. Whereas Cantor did not believe that the universe was infinite in either duration or extent, he did follow Gottfried Leibniz in positing an infinite number of elementary particles ("monads"), so that the transfinites existed in the physical universe. In a letter dated 1886 to Cardinal Johannes Franzelin, Gutberlet's teacher, Cantor made the distinction between Absolute Infinity, eternal and uncreated, reserved for God and his attributes, and the *Transfinitum* (the transfinite numbers), as created in the physical universe and in the mind of man (Dauben 1979, 145). Cardinal Franzelin was a leading Jesuit philosopher and papal theologian to the First Vatican Council. He approved Cantor's distinction as removing any threat to orthodoxy, which Cantor took as an imprimatur for his work. Cantor (1887), in fact, was proud to quote Franzelin's letter: "Thus the two concepts of the Absolute-Infinite and the Actual-Infinite in the created world or in the *Transfinitum* are essentially different, so that in comparing the two one must only describe the former as *properly infinite*, the latter as improperly and equivocally infinite. When conceived in this way, so far as I can see at present, there is no danger to religious truths in your concept of the *Transfinitum*" (83).

Cantor distinguished three levels of existence: (1) in the mind of God (*Intellectus Divinum*); (2) in the mind of man (*in abstracto*); and (3) in the physical universe (*in concreto*). From the beginning of his work Cantor was keenly aware of the paradoxes inherent in the ideas of "the set of everything" and "the set of all sets." These belonged to Absolute Infinity, which is beyond mathematical formulation and which is comprehended only in the *Intellectus Divinum*. But Cantor believed that God had put into the human mind the concept of number, both finite and transfinite, *in abstracto*, and often argued that their existence in the *Intellectus Divinum* is the basis for their existence in the human mind (Dauben 1979, 146).

Furthermore, Cantor believed that he had a duty to the Church to save it from the error of insisting on a finite creation while science pursued infinity *in abstracto*. This error would mar the faith it sought to foster. Cantor appealed to the principle set forth by Aquinas in *Summa contra Gentiles*: "For error concerning creatures, by subjecting them to causes other than God, spills over into false opinion about God, and takes men's minds away from Him, to whom faith seeks to lead them" (1955, *S.c.G.* II, c. 3, para. 6).

Like a modern Galileo, Cantor hoped to aid the Church in understanding the world science was unfolding (Dauben 1979, 232).

CANTOR'S ORDINAL NUMBERS

Cantor's mathematics of infinity found its pinnacle expression in his *Grundlagen einer allgemeinen Mannigfaltigkeitslehre* (1883). Here he summarized his earlier work, and introduced a new field of inquiry into mathematics. Concerning the new theory of the infinite Cantor presented in the *Grundlagen*, Dauben (1979) wrote:

> Whatever the disappointments Cantor was to suffer, his transfinite set theory represented a revolution in the history of mathematics. Not a revolution in the sense of returning to earlier starting points, but more a revolution in the sense of overthrowing older, established prejudices against the infinite in any actual, completed form. Consequently, Cantor's transfinite numbers were to prove no less revolutionary for philosophers and theologians who were concerned with the problem of infinity. (118)

Cantor introduced the lowercase omega symbol, ω, as the first transfinite number larger than all the finite ordinal numbers, that is, the natural numbers $\mathbf{N} = \{1, 2, 3, 4 \ldots\}$. From ω Cantor described an ever-increasing sequence of "ordinal numbers" in the following way. Using modern notation (Halmos 1960, 76), let $\omega + 1$ as the union of the set ω with itself, that is, $\omega + 1 = \omega \cup \{\omega\}$. Note that ω is an element of $\omega + 1$, that is, $\omega \in \omega + 1$. So ω is contained in $\omega + 1$, but not vice versa. Similarly, construct $\omega + 2$ as the union of $\omega + 1$ with itself, that is, $\omega + 2 = \omega + 1 \cup \{\omega + 1\}$. Thus, Cantor built up a hierarchy of ordinal numbers ω, $\omega + 1$, $\omega + 2$, $\omega + 3 \ldots$ Regard this set of ordinals $\{\omega, \omega + 1, \omega + 2, \omega + 3 \ldots\}$ as a completed set in itself, and denote it by $\omega \cdot 2$. Complete the set $\{\omega \cdot 2, \omega \cdot 2 + 1, \omega \cdot 2 + 2, \omega \cdot 2 + 3 \ldots\}$, and denote it by ω. Next consider $\{\omega, \omega \cdot 2, \omega \cdot 3, \omega \cdot 4, \ldots\}$ as a completed set, and denote it by $\omega \cdot 3$. Next consider $\{\omega^2, \omega^2 \cdot 2, \omega^2 \cdot 3, \omega^2 \cdot 4 \ldots\}$ as a completed set, and denote it by ω^3. Consider $\{\omega, \omega^2, \omega^3, \omega^4 \ldots\}$ as a completed set, and denote it by ω^ω.

Cantor discussed an arithmetic of ordinal numbers, along the lines of $(\omega + 1) + 1 = \omega + 2$, but more importantly he offered a precise definition of the size of an ordinal number. This expanding hierarchy of ordinal numbers is becoming increasingly inclusive, as lower ordinals are contained in higher. However, they are in fact not getting larger. Cantor used the commonsense idea of one-to-one correspondence from daily experience to define the size of ordinals. Take a set of five sheep and the five fingers of one's hand. We say both are of size five, because there is a one-to-one correspondence between the sheep and the fingers; one can associate each finger with exactly one sheep and each sheep with exactly one finger. A function $f(x)$ is called a "one-to-one correspondence" between two sets A and B, if for every element a in A there is one and only one element b in B such that $f(a) = b$. Cantor defined two sets A and B to be the same "cardinality," if there was a one-to-one correspondence between them. Interestingly, all of the above ordinals are of the same cardinality, that is, one can find a one-to-one correspondence $f(x)$ between any two of them. Cantor denoted this cardinality of the natural numbers **N** = {1, 2, 3, 4 . . .} with the first letter of the Hebrew alphabet, aleph, with a subscript zero, that is, $\aleph_0$, pronounced in German "aleph-null" and in English "aleph-naught."

CANTOR'S CARDINAL NUMBERS

Now the question arises whether there are any cardinalities larger than $\aleph_0$? Let the set of all positive rational numbers be denoted by **Q**, that is, **Q** = {p/q, where p ∈ **N** and q ∈ **N**}. Perhaps one might think that because **Q** is a "dense" set it might have a larger cardinality. A "dense set" of numbers has the property that between any two of its elements there is another element. The set of natural numbers **N** is not dense, because there is no natural number between 4 and 5, say. However, **Q** is dense, because between any two rational numbers there is another rational number, their average. However, Cantor showed, using his now-famous "diagonalization argument," that there is a one-to-one correspondence between the natural numbers **N** and the positive rational numbers **Q**.

Arrange all rational numbers in the following table:

1/1	1/2	1/3	1/4	1/5	. . .
2/1	2/2	2/3	2/4	2/5	. . .
3/1	3/2	3/3	3/4	3/5	. . .
4/1	4/2	4/3	4/4	4/5	. . .
5/1	5/2	5/3	5/4	5/5	. . .
⋮	⋮	⋮	⋮	⋮	. . .

Clearly, all positive rational numbers will occur somewhere in the table, that is, the rational number p/q occurs in the pth row and the qth column. Now pair the natural numbers **N** onto the positive rational numbers **Q** along the diagonals in the following way:

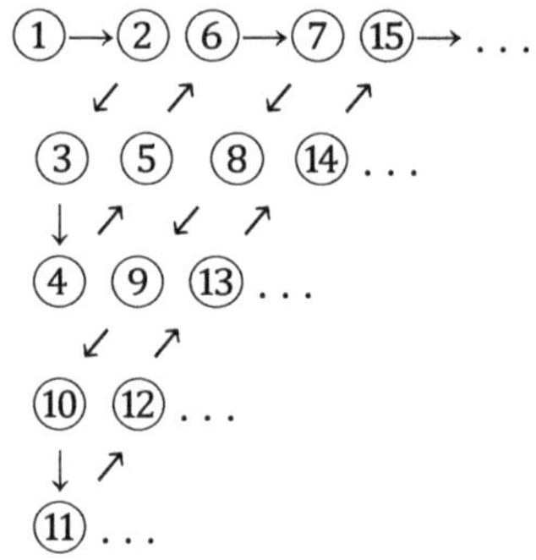

Clearly, every natural number is paired with a positive rational number, and vice versa, and this gives a one-to-one correspondence from **N** onto **Q**. So **N** and **Q** have the same cardinality $\aleph_0$, which is called "countably infinite." It is easily shown that the set of all rational numbers is of the same cardinality as **Q**.

So the question arises: Is there a cardinality larger than $\aleph_0$? Cantor found that the answer is in the affirmative, and gave the following argument. Call B a "subset" of A, if every element of B is also an element of A, that is, if x ∈ B then x ∈ A. Given a set A, define the "power set of A," notated $\mathcal{P}$(A), as the set of all subsets of A. For example, if A = {1, 2, 3} then there are eight elements in $\mathcal{P}$(A), specifically $\mathcal{P}$(A) = {∅, {1}, {2}, {3}, {1, 2}, {1, 3}, {2, 3}, {1, 2, 3}}, where it is understood that the empty set ∅ and A itself are subsets of A. Even if A is an infinite set Cantor discovered that the cardinality of $\mathcal{P}$(A) is always greater than the cardinality of A itself, written $\mathcal{P}$(A) >> A. Here is his argument.

Assume the contrary, that $\mathcal{P}$(A) and A have the same cardinality. (Here Kronecker and the constructionists reject using the Law of the Excluded Middle.) Then there is a one-to-one correspondence $f(x)$ between P(A) and A, that is, to every subset S of A there corresponds a unique element s of A so that f(S) = s. We want to distinguish whether or not s is itself an element of S, whether s ∈ S or s ∉ S. Let T be the set of all elements s of A for which s = f(S) is not contained in S, that is, T = {s ∈ A: s = f(S) ∉ S}. Since T is itself a subset of A, there is an element t of A such that f(T) = t. Either t is an element of T, or it is not, that is, either t ∈ T or t ∉ T. On the one hand, if t is an element of T, then T is the set of all elements of A whose s = f(S) is not in S, that is t ∉ T. On the other hand, if t is not an element of T, then, since T is the set of all elements s such that s is not an element of f(S), then t is an element of T, that is, t ∈ T. This contradiction implies that there can be no such one-to-one correspondence *f*(*x*) between $\mathcal{P}$(A) and A, and the cardinality of $\mathcal{P}$(A) must be greater than the cardinality of A itself, that is, $\mathcal{P}(A) >> A$.

Cantor (Dauben 1979, 194) designated the cardinality of $\mathcal{P}(\mathbf{N})$ using aleph with a subscript 1, that is, $\aleph_1$, so that $\aleph_1 >> \aleph_0$. He observed that the power set of a power set would be of larger cardinality still, that is, $\mathcal{P}(\mathcal{P}(\mathbf{N})) = \mathcal{P}^2(\mathrm{N}) >> \mathcal{P}(\mathrm{N})$, which he designated by $\aleph_2$. Denote the cardinality of $\mathcal{P}^n(\mathbf{N})$ by $\aleph_n$, and so Cantor found a set of increasing cardinal numbers $\aleph_0 << \aleph_1 << \aleph_2 << \ldots \aleph_n << \ldots$ In modern notation (Halmos 1960, 102), $\aleph_1$ denotes the least uncountable cardinal number, so Cantor's conviction that $\mathcal{P}(\mathbf{N}) = \aleph_1$ is the *continuum hypothesis.*

THE *CONTINUUM* HYPOTHESIS

In the seventeenth century, Descartes introduced the term "real numbers" to denote the lengths of all possible segments on the number line. Equivalently, the set of all real numbers $\mathbb{R}$ is the set of all decimal representations, finite or infinite, that is, $r = a_0 \cdot a_1a_2a_3 \ldots$, where $a_0 \in \mathbf{N}$ and $a_i \in \{0, 1, 2, 3, 4, 5, 6, 7, 8, 9\}$. More precisely, $r = \sum_i^\infty = \frac{a_i}{010^i}$. Since Cantor showed that **Q** was only countably infinite, the question follows whether $\mathbb{R}$ is also countably infinite. Cantor discovered the answer in the negative through the following argument.

Assume that $\mathbb{R}$ is countably infinite. Since the open interval of real numbers between 0 and 1 is a subset of $\mathbb{R}$, its cardinality is also countably infinite. Call this set $\mathbb{H}$, that is, $\mathbb{H} = \{r \in \mathbb{R}: 0 < r < 1\}$. Then there is a one-to-one correspondence *f*(*x*) between $\mathbb{H}$ and **N**, where for every $n \in \mathbf{N}$, $f(n) = r \in \mathbb{H}$. Write this correspondence in ascending order:

$$f(1) = r_1 = 0 \cdot a_0\, a_1\, a_2\, a_3 \ldots$$
$$f(2) = r_2 = 0 \cdot b_0\, b_1\, b_2\, b_3 \ldots$$
$$f(3) = r_3 = 0 \cdot c_0\, c_1\, c_2\, c_3 \ldots$$
$$\vdots$$
$$f(i) = r_i = 0 \cdot i_0\, i_1\, i_2\, i_3 \ldots$$
$$\vdots$$

Now form a new decimal expansion whose n^th^ decimal place differs from the nth decimal place of $f(n) = 0 \cdot n_0\, n_1\, n_2\, n_3 \ldots$ That is, we form a new decimal expansion $r = 0 \cdot r_0\, r_1\, r_2\, r_3 \ldots$ where, say, we let $r_n = 2$ if $n_n \neq 2$, or let $r_n = 3$ if $n_n = 2$. Thus, r differs from every decimal expansion in the above table, and there is no $n \in \mathbf{N}$ such that $f(n) = r \in \mathbb{H}$. This contradiction implies that $\mathbb{R}$ cannot be countably infinite.

Since $\mathbb{R}$ is uncountably infinite, the question arises whether the cardinality of $\mathbb{R}$ is $\aleph_1$. Could the cardinality of $\mathbb{R}$ be even larger $\aleph_1$? Cantor states his firm conviction in his *Grundlagen* that the answer is no, that the cardinality of $\mathbb{R}$ is $\aleph_1$, and he spent years in vain trying to prove it. The real number line had been called the *continuum*, and so the phrase *continuum hypothesis* came to describe Cantor's conjecture that the cardinality of $\mathbb{R}$ is $\aleph_1$. His dogged conviction about the nature of the *continuum* may explain Cantor's philosophical position with Leibniz in defending the existence of an infinite number of elementary particles, the "monads," in the physical universe, so that transfinites could exist *in concreto.*

ACCEPTANCE OF CANTOR'S SET THEORY

On August 8, 1900, at the International Congress of Mathematicians at the Sorbonne in Paris, David Hilbert (1902) posed the first ten of his now-famous twenty-three unsolved problems in mathematics. These were largely to set the agenda for mathematical research in the twentieth century. The first of his unsolved problems was Cantor's continuum hypothesis. Hilbert was an early and influential advocate of Cantor's set theory and transfinite numbers. Both believed that set theory could be used to express the foundations of mathematics, but both also knew that there were paradoxes to be avoided.

Cantor used the Hebrew letter Taw, ת, to denote the set of all cardinal numbers $\{\aleph_0, \aleph_1, \aleph_2 \ldots \aleph_n \ldots\}$. Yet, he was aware that the set of all cardinal numbers could not contain its own power set, $\mathcal{P}$(ת). These paradoxical constructions, such as ת, "the set of everything," and "the set of all sets," Cantor called "Absolute Infinity," which he believed to be beyond all mathematical determination and to be comprehended only by the mind of God.

In 1901, while struggling with Cantor's concept of cardinality, Bertrand Russell (1903, 101) came up with what has become known as "Russell's paradox." Let R be the set of all sets that do not contain themselves, that is, $R = \{S: S \notin S\}$. For example, if T is the set of all the squares contained in the plane, then clearly T is not an element of itself, that is, $T \notin T$. But consider the complement of T, that is, T^C, all sets that are not the set of all squares in the plane. Then, $T^C \in T^C$. Now consider whether R is, or is not, an element of itself. If $R \in R$, then by R's definition, $R \notin R$. But if $R \notin R$, then by R's definition it should be, and $R \in R$.

Russell's paradox shows that a naïve approach to set theory is inadequate. In 1908, Ernst Zermelo proposed the first system of axioms for set theory. But in 1921, Abraham Fraenkel pointed out that the Zermelo axioms could not generate power sets, which most theorists of the time accepted. In 1922, Fraenkel proposed appending Zermelo's axioms with two more, which, with the Axiom of Choice, constitute the Zermelo–Fraenkel axiomatic system, abbreviated ZFC, which today is the standard axiomatic system used by most set theorists. In 1940, Kurt Gödel showed that the continuum hypothesis cannot be disproven by ZFC, and in 1963, Paul Cohen proved that the continuum hypothesis cannot be proven from this system of axioms. Thus, Cohen established the independence of the continuum hypothesis from ZFC, for which work he received the Fields Medal in 1966. Thus, Cantor's prize conjecture turned out to be far richer than he ever imagined. In fact, Paul Cohen (1966, 151) has suggested that the continuum may be larger than all alephs.

CANTOR'S SET THEORY REVOLUTIONIZED SCIENCE

In 1931, Gödel published "On Formally Undecidable Propositions," which showed that in any axiomatic system large enough to contain the arithmetic of natural numbers there is a theorem that cannot be proven from these axioms. For more than fifty years, Frege, Russell, Whitehead, Hilbert, and others had tried to find an axiomatic system sufficient for mathematics and to prove its consistency. Gödel thus proved that their goal was unattainable. In classical mathematics the open question motivating these axiomatics was whether Euclid's Fifth Postulate could be derived from his other four. This "Parallel Postulate" is equivalent to the angles of a triangle adding to 180°. In 1829, Nikolai Ivanovich Lobachevsky published a geometrical system that assumed this sum to be less than 180°, and in 1831 János Bolyai assumed this sum to be greater than 180°. These alternative axioms led to hyperbolic geometry and elliptical geometry, respectively. But were these three geometries consistent? Gödel's result allowed only comparative consistency, that Euclidean geometry was as consistent as hyperbolic and elliptical, but that their absolute consistency could not be proven. Thus, Euclid's Fifth Postulate was "undecidable" in terms of his previous four.

In a letter from 1885, the Swedish mathematician Gösta Mittag-Leffler (Dauben 1979, 138) wrote to Cantor that his transfinites were as revolutionary as these non-Euclidean geometries. With the unparalleled success of Newtonian mechanics in predicting natural phenomena, the eighteenth century had come to view the universe as explainable by a single axiomatic system. "Classical science" had come to regard the universe as noncontingent. But the late nineteenth century into the twentieth saw scientific discoveries that revealed the universe to be far more complex. William Berkson (1974) chronicled in detail how the revolutionary concept of "fields," beginning with Michael Faraday and culminating in the equations of James Clerk Maxwell, explained electromagnetic phenomena without Newtonian "particle on particle" interaction. Einstein's theory of relativity revolutionized Newton's notions of space as a "container" and of time as absolute. Cantor's theory of sets and transfinite numbers was the font of another stream of thought that revolutionized classical science, as he stimulated work by Hilbert, Russell, Zermelo, Gödel, and Cohen.

A CONTINGENT UNIVERSE

Allow me to borrow from myself some definitions:

> To say that the universe is "contingent" means that it need not be the way it is. Its particular space-time structure is not a necessary consequence of its existence. Alternately, an "incontingent" universe[2] would possess a necessary structure; such a world would be uniquely determined by just the requirement of self-consistency. A contingent universe does not contain within itself a sufficient explanation of itself. For an incontingent universe one conceivably could find a single, consistent theory from which one could deduce uniquely the structure of the world, including the numerical values of all constants of nature. A contingent universe will be termed "open"; an incontingent universe will be termed "closed." (Hedman 1989, 99)

The ancient world largely thought of the universe as noncontingent, and so attempted to reason about the world a priori. Contingence is a medieval doctrine, which was implied by God's sovereignty in the *creatio ex nihilo* (Aquinas 1945, *ST* III, q. 46, a. 2). A contingent universe necessarily must have a beginning, and Aquinas held that the very idea that the world did not always exist could be known only by revelation and not reason: "The reason for this is that the newness of the world cannot be demonstrated from the world itself. For the principle of demonstration is the essence of a thing" (III, q. 46, a. 2). Thomas F. Torrance (1981a) argued that the concept of the contingence of the universe played a crucial role in the development of experimental science during the Enlightenment, as scientists recognized that to answer questions about nature required asking nature itself. Isaac Newton denied that "the world exists by necessity, and by the same necessity follows the laws proposed" (Barrow 1988, 323).

But Newtonians were soon to interpret his laws as determining the entire structure of the "closed" universe. Such mechanists thought of the universe as closed, self-contained, and self-explanatory. Questions about origin and purpose, which contingence raises, were not even considered legitimate by the mechanists. Those who favor a "closed" system believe in the "eternity of the universe," which Jaki (1980, 108) called "the first article of the secular faith." But such a philosophy must struggle to account for the role of the mind in such a mechanistic, deterministic universe. Why should mental, mathematical abstractions correlate with the physical world? This question spurred the developments advanced by Locke, Berkeley, and Hume in the eighteenth century.

Immanuel Kant finally closed the Newtonian system in upon itself by arguing that the mind does not experience the physical world itself, the "noumena," but only its own sense perceptions, the "phenomena." Scientific laws are not statements about physical reality itself, but are only the mind's own ordering of its sense perceptions. The mind has innate patterns, the "a priori," according to which it orders its perceptions. Kant considered Euclidean geometry, for example, as a priori knowledge. In Kant classical science found its final, self-contained, noncontingent character, which closed it off from not only metaphysics but from physical reality itself.

Torrance (1981b, 11–39) explained in detail how this Kantian interpretation floundered on the scientific revolutions of the late nineteenth and early twentieth centuries. If science is no more than the mind's projection onto its sensory input, how could such a thoroughgoing scientific revolution arise? Non-Euclidean geometries, Cantor's transfinites, Maxwell's fields, and Einstein's space–time continuum are beyond Kantian categories. Modern science has moved away from a positivist toward a realist interpretation of the reciprocity between the mind and the physical universe. Thomas Kuhn (1962) has detailed the progressive nature of scientific revolutions. Hermann von Helmholtz held firmly to his conviction that the universe could be explained entirely in terms of Newtonian mechanics. Einstein replied to him: "The view appears dull and naïve to a twentieth-century physicist. It would frighten him to think that the great adventure of research could so soon be finished, and an unexciting if infallible picture of the universe established for all time" (Einstein and Infeld 1938, 58).

Cantor was proudly the only member of Halle's faculty who was not a Kantian. Cantor wrote (Dauben 1979, 292–94) that his motive in studying transfinite cardinalities was to address certain applications in chemistry, optics, and biology. Cantor was explicitly opposed to any Kantian interpretation of science, and maintained passionately that the transfinites were not his mind's own invention, but were suggested to him through physical considerations. Cantor distinguished between the *in abstracto* and the *in concreto* existence of all numbers, finite and transfinite. But Cantor believed these dual realities were always found together. This correspondence between the physical and ideal aspects of numbers, Cantor believed, came from "a unity in the universe itself" (Dauben 1979, 132). Hedman (1993) wrote:

> The counter-intuitive, even paradoxical, properties of the transfinites argue against a Kantian *a priori*, and suggest that physical reality has impinged upon the mind from outside itself. I cite Cantor's interpretation of his own work as an example of what I call "incarnational mathematics." There is a created rationality embedded in our minds and within the physical universe. The mathematician possesses not merely a mind, but a mind embedded, "incarnated," if you will, into the physical world through his body. I believe history shows that the most productive mathematics has been suggested by physical considerations. After a period of abstract development, mathematics has often been refocused by physical applications of its abstractions. (12)

CANTOR'S FAITH IN THE *INTELLECTUS DIVINUM*

Dauben held that Cantor's religious views molded his creative process in discovering the transfinites. He wrote, "One is tempted to wonder if this view of God's role in ensuring the reality and existence of Cantor's *Transfinitum* was responsible for his discovery of the contradictory nature of that very concept" (Dauben 1979, 349). The mathematical entities Cantor studied, which existed apart from any mathematical system, were themselves grounded in a reality beyond this universe. Cantor relied on the Divine Intellect as "the certain repository of the *Transfinitum*" (349). This Absolute Infinity was the ground for the transfinites, but was itself mathematically indeterminable.

NOTES

I wish to thank Christian Tapp for several excellent suggestions and references in this chapter.

1. Tapp (2011, 96) aptly points out that here Torrance oversimplifies. Cantor distinguished the transfinite not only from the absolute infinity of God, but also from the absolute infinity of "proper classes."

2. Perhaps "noncontingent" would have been a phrase more widely used than "incontingent."

REFERENCES

Aquinas, Thomas. 1945. *Summa Theologicae*. Edited by Anton Pegis. New York: Random House.

———. 1955. *Summa Contra Gentiles*. Edited by Anton Pegis, James F. Anderson, Vernon J. Bourke, and Charles J. O'Neill. New York: Hannover House

Augustine. 1979. *City of God*. In *The Nicene and Post-Nicene Fathers: Series I*, edited by Philip Schaff. Grand Rapids, MI: Eerdmans

Barrow, John D. 1988. *The World within the World*. London: Oxford University Press.

———. 2008. *Cosmic Imagery: Key Images in the History of Science*. New York. W. W. Norton.

Berkson, William. 1974. *Fields of Force: The Development of a World View from Faraday to Einstein*. London: Routledge and K. Paul.

Cantor, Georg. 1883. *Grundlagen einer allgemeinen Mannigfaltigkeitslehre: Ein mathematisch-philosophischer Versuch in der Lehre des Unendlichen*. Leipzig: B. G. Teubner

———. 1887. "Mitteilungen zur Lehre vom Transfiniten." *Zeitschrift für Philosophie und philosophische Kritik* 91:81–125.

Cohen, Paul J. 1963. "The Independence of the Continuum Hypothesis." *Proceedings of the National Academy of Sciences of the United States of America* 50 (6): 1143–48.

———. 1966. *Set Theory and the Continuum Hypothesis*. New York: W. A. Benjamin.

Dauben, Joseph Warren. 1977. "Georg Cantor and Pope Leo XIII: Mathematics, Theology, and the Infinite." *Journal of the History of Ideas* 38 (1): 85–108.

———. 1979. *Georg Cantor: His Mathematics and Philosophy of the Infinite*. Cambridge, MA: Harvard University Press.

Einstein, Albert, and Leopold Infeld. 1938. *The Evolution of Physics*. New York: Simon & Schuster.

Fraenkel, Abraham. 1922. "Zu den Grundlagen der Cantor-Zermeloschen Mengenlehre." *Mathematische Annalen* 86:230–37.

Gauss, Carl F. 1860. *Briefwechsel zwischen C. F. Gauss und H. C. Schumacher*. Edited by C. A. F. Peters. Altona: Gustav Esch.

Gödel, Kurt. 1931. "Über formal unentscheidbare Sätze der *Principia Mathematica* und verwandter Systeme, I." *Monatshefte für Mathematik und Physik* 38:173–98.

———. 1940. *The Consistency of the Continuum-Hypothesis*. Princeton, NJ: Princeton University Press.

Gutberlet, Constantin. 1886. "Das Problem des Unendlichen." *Zeitschrift für Philosophie und philosophiche Kritik* 88:179–223.

Halmos, Paul. 1960. *Naïve Set Theory*. New York: Van Nostrand Reinhold.

Hedman, Bruce. 1989. "Mathematics, Cosmology, and the Contingent Universe." *Perspectives on Science and Christian Faith* 41 (2): 99–103.

———. 1993. "Cantor's Concept of Infinity: Implications of Infinity for Contingence." *Perspectives on Science and Christian Faith* 45 (1): 8–16.

Hilbert, David. 1902. "Mathematical Problems." *Bulletin of the American Mathematical Society* 8 (10): 437–79.

Jaki, Stanley L. 1980. *Cosmos and Creator*. Edinburgh: Scottish Academic Press.

Kuhn, Thomas. 1962. *The Structure of Scientific Revolutions*. Chicago: University of Chicago Press.

Leo XIII. 1903. *The Great Encyclical Letters of Pope Leo XIII*. Edited by J. J. Wynne. New York: Benziger.

Plato. 1961. *Philebus*. In *The Collected Works of Plato*, edited by Edith Hamilton and Huntington Cairns. Princeton, NJ: Princeton University Press.

Russell, Bertrand. 1903. *The Principles of Mathematics*. Cambridge: Cambridge University Press.

Tapp, Christian. 2011. "Infinity in Mathematics and Theology." *Theology and Science* 9 (1): 91–100.

———. 2016. "Chapter 4: Infinity in Aquinas' Doctrine of God." In *Analytically Oriented Thomism*, edited by Miroslaw Szatkowski, 93–115. Neuenkirchen-Seelscheid: Editiones Scholasticae.

Torrance, Thomas F. 1981a. *Divine and Contingent Order*. New York: Oxford University Press.
———. 1981b. *Christian Theology and Scientific Culture*. New York: Oxford University Press.
———. 2002. *Theological and Natural Science*. Eugene, OR: Wipf and Stock.
Zermelo, Ernst. 1908. "Untersuchungen über die Grundlagen der Mengenlehre I." *Mathematische Annalen* 65:261–81.

PART II

Systematic Approaches to the Infinity of God

CHAPTER 10

God Almighty

Divine Power and Authority in the Biblical and Patristic Periods

BERNHARD LANG

Power and authority must not be confused. The difference is immediately apparent from examples. Power is exemplified by the soldier who, in physical man-to-man combat, prevails over his enemy, and by the shaman who is able to ease someone's ailments by his healing touch. The notion of authority calls up different images, for instance that of a captain giving orders to the crew of a ship, or the commander of an army. Here physical power is located at the lower end, authority at the top of the chain of command. In our modern conceptual world, authority seems nobler and enjoys more prestige than mere physical power. Some people, such as scholars in the academic world, might enjoy much prestige and wield authority, but they have little or no power.

The power and authority of the biblical god, or any other god of whom we hear in ancient documents, transcends human power and authority. Traditional religious language leads us to believe that at least the biblical god commands infinite power and exercises unlimited authority. Indeed, a god who has infinite *power* may be called "almighty." A deity that has *authority and control* over everything may also be called "almighty." But the two notions must be distinguished: the first refers to a god's unlimited ability to do whatever he wishes, the second to a god's sovereign capacity to rule over everything, the entire universe. Both notions of almightiness apply to the monotheistic god of the Bible: he is the sovereign ruler over everything, but he may also be said to have unlimited power. The two notions go well together. The distinction between the two concepts, power and authority, is occasionally blurred, but we will use it in the interest of clarity. But our case should not be overstated. The limitations of the distinction should be borne in mind: this distinction is not drawn in the biblical and related texts themselves, but is imposed by the modern interpreter who thus risks forcing the ancient evidence into modern conceptual categories that do not fully apply. Yet, the conceptual distinction promises to be helpful.

I present my thought in this chapter in two sections. The *first section*, while presupposing the biblical god's supreme sovereignty, explores how the ancient Israelites envisaged or described their god's exercise of power. This section reviews a substantial body of biblical texts, arguing that, in the final phase of biblical history, truly unlimited power was ascribed to Israel's god. The *second section* comments on Israel's god as the sovereign ruler over everything. In this case, we review no biblical texts. Instead, we explore in a terminological study how Israel's god came to be called "almighty." The underlying Greek noun, *ho pantokrátôr*, means "all-ruler" in the sense of "ruler of all," "ruler of everything." The corresponding Latin word, the adjective *omnipotens*, translates as "all-ruling" in the sense of "ruling over everything."

THE EXPERIENCE OF DIVINE POWER IN THE BIBLICAL PERIOD

This section addresses the question of whether the biblical god was considered to be "almighty" (or omnipotent) in the sense of being able to act with infinite power. Almightiness, in this context, may be defined as the unrestricted faculty to achieve anything that the god wants to do or that his devotees feel he should do. "Our god is in the heavens; he does all that he pleases," says the Psalmist (Ps. 115:3). Other relevant general statements can be found in well-crafted, no doubt stereotyped polar expressions in Old Testament poetry: "Yahweh kills and brings to life; he brings down to Sheol and raises up. Yahweh makes poor and makes rich;

he brings low and exalts" (1 Sam. 2:6–7). The poets like to put this assertion into their god's own mouth: "I kill and I make alive, I wound and I heal.—I form light and create darkness, I make well-being and create calamity" (Deut. 32:39; Isa. 45:7). "He does according to his will among the host of heaven and among the inhabitants of the earth, and none can stay his hand or say to him, What have you done?" (Dan. 4:32). But although such passages sum up the general feeling well, they tell us little about the details of Yahweh's mighty acts. For this reason, we will take a closer look at the contexts in which Israel's god exercises his power.

There are three relevant contexts that correspond to the three roles or functions Yahweh was credited with by his early devotees. (1) He protected each individual, caring and providing for him personally. (2) He served as the national god for the entire Israelite people, blessing the king with his gifts of wisdom and valor, and granting victory in war. (3) He created and controlled the universe, favoring those who love justice, and punishing those who violate the divine order (Lang 2002; Lang 2008). There is some evidence in the biblical record that the ancient Hebrews had three different names for the various roles or functions of their god: the personal god was called "Shaddai," the national deity "Yahweh" or "Yahu," and the Creator "El" or "Elyon." However, the name of Yahweh was also used as a general term for Israel's god, irrespective of his perceived function.

We may visualize the ancient Israelite's experience of god in the form of three concentric circles (fig. 10.1). These stand for the individual (at the center) who is surrounded by the nation (middle ring) and, finally, by the wider world. Each of these three spaces is supervised and cared for by a deity: the personal god is responsible for the individual, the national god for the Israelite people, and the universal god for the wider world. Thus, no space was left empty, and the world was full of gods.

We will take a closer look at the three divine functions to assess how in each case god's power was understood.

The Personal God

Personal piety, in the ancient Near East and in Egypt, focused on a god or goddess chosen as a personal protector, whom scholars call "the personal god." A wife did not seem to have her own personal god, but she shared

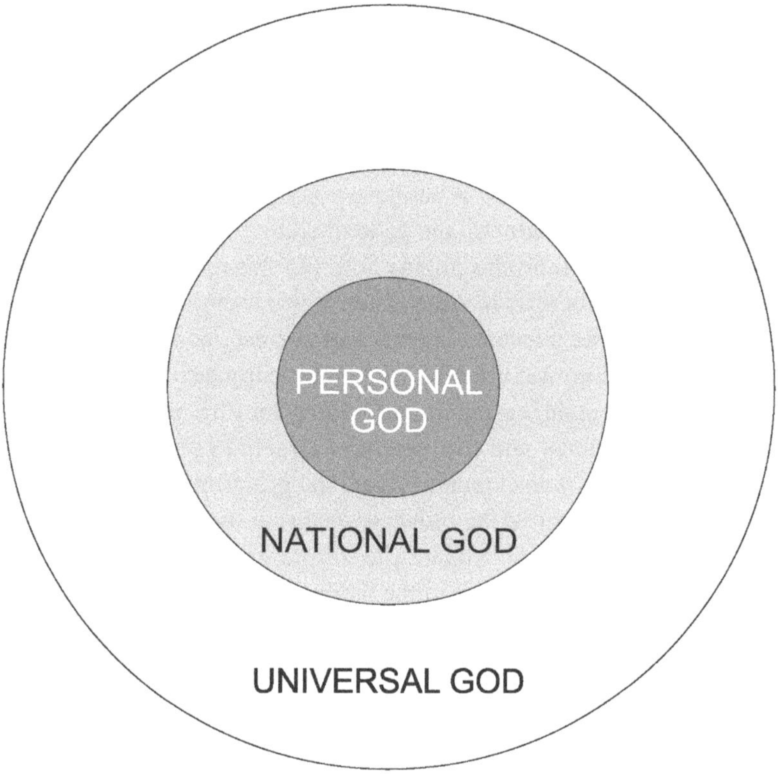

Figure 10.1. *A spatial model of the biblical god.* Ancient Israelite individuals thought of themselves as encompassed by the divine. The three concentric spaces of this mental map denote the individual and his personal god, the nation and its national god, and the universe dominated by the universal god. As monotheism developed, the three deities were identified as aspects or roles of the one god, and "God" became his personal name.

that of her husband, just as young children shared that of their father. In the polytheistic context, any god or goddess, irrespective of his or her particular competence, could be chosen as one's personal god. The personal god was believed to accompany his or her client from cradle to grave, to offer help and protection, at least as long as he or she is not offended by neglect or the violation of some ritual taboo or ethical law. The believer referred to his personal deity as "my god," an expression familiar to any reader of the Bible. Because of the monotheistic selection or reworking of the writings included in the biblical canon, there was only one god that an individual could choose as "his" god—none other than Yahweh. Many psalms echo how individuals felt about this deity's role as personal god: "The Lord is my shepherd, I shall not want. He makes me lie down in green pastures. He leads me beside still waters. He restores my soul," reads the beginning of the most famous of the psalms, Psalm 23. When in distress, the psalmist feels abandoned by his personal god, and asks: "My god, my god, why have you forsaken me? Why are you far from saving me?" (Ps. 22:1). But when all is well, the individual rejoices: "I remember you upon my bed, and meditate on you in the watches of the night; for you have been my help, and in the shadow of your wings I will sing for joy. My soul clings to you; your right hand upholds me" (Ps. 63:6–8).

Interestingly, animals also have a personal god. Most biblical texts emphasize the personal god's care for humans, but some deal more specifically with this deity's care for the animal world (in religious studies conceptualized as "Lord of the Animals"). But it is really the same god who cares for both humans and animals by blessing them with food, fertility, and health. This idea is skillfully developed in the book of Job, which I read as a kind of mystery or initiation drama.

This is the story: Job, a wealthy and well-respected non-Israelite who lives in the land of Uz, becomes the subject of a debate at the heavenly court. To God, Job is the most upright of all men. Satan has doubts. Job's piety, he suggests, might reflect this man's gratitude for the divine blessing that made him the wealthiest man. But will he stay pious when deprived of his wealth? Will he curse God when he has lost his health? Satan obtains permission to test the man, and now Job is visited by many a misfortune. Thieves take his cattle, donkeys, and camels. A flash of lightning hits his outbuildings, killing sheep, goats, and the servants. A hurricane destroys the house where Job's children are celebrating, and all of them are killed. To complete the series of disasters, Job is disfigured by a severe skin disease. All of this puts an end to Job's former life; struggling to exist, he suffers in the liminal space between life and death.

A period of testing follows. "Curse God and die," his wife suggests. But Job remains firm in his trust in God. Three friends who visit seek to comfort him, and to offer their view of his situation. They confront him with traditional wisdom: since misfortune is the result of some crime,

Job should examine himself and admit to being guilty. Job, however, protests his innocence. By his heroic insistence on his innocence, Job has stood the test, or so the reader must feel. Then the story reaches its climax. Prepared through loss and pain for higher wisdom, he is granted a glimpse of how God rules the universe. A voice of thunder sounds. The Creator, who appears in majesty and addresses Job, describes the world that he has put in order and on which he has bestowed his blessings. God founded the earth, confined the wild seas, allocated a place to each star in the heavens, sends rain to water the earth, and cares for the animals: lion, raven, wild goat, ass and ox, ostrich, horse, hawk, and eagle. He describes graphically the power of Behemoth and Leviathan, *alias* hippopotamus and whale, transfigured into unconquerable mythical beasts. Overwhelmed by God's monologue, Job, in humility, desists from his accusations of God, surrenders himself, and admits his ignorance. Now he no longer stands at the threshold of death, but at the threshold of life, to which he is restored. Job regains his health and his possessions with flocks doubled in size. More children are born to him: seven sons and three daughters. As a pastoralist, he lacks nothing, and his happiness is complete. Former friends return and join in his celebrations. Job lives for another 140 years, and then dies in peace, full of days.

We may recapitulate Job's story in terms of the notion of the personal god. Job's personal god initially served him well, making him rich, healthy, and respected, just as one would expect. But one day, the personal god abandoned his protégé, for no obvious reason, and Job is left to his misfortune. But Job remains a faithful servant of his personal god, who eventually restores him to health and wealth.

Literary critics have consistently pointed out that the book of Job, in the words of Northrop Frye, "is technically a comedy by virtue of its happy ending" (Frye 1983, 196). When literary critics speak about comedy, they make sure that no one misunderstands this to mean merely a funny story or play. The book of Job includes all the basic ingredients of an authentic comedy: the protagonist's crisis or conflict as a result of his or her being involved in the larger conflict among seemingly inalterably opposed elements in the world; the hero's subsequent initial triumph; and, finally, some sort of festive reconciliation in a world that allows even the sinner, and more generally evil, to exist. As William Whedbee (1998) explains, "The comic vision does not necessarily eliminate evil and death; it is not incorrigibly and naively optimistic; it does not shut its eyes to the dark, jagged edges of life in this world." Incorporating rather than ignoring the haunting riddles of life, comedy "simply affirms that a harmonious, prosperous society is still most desirable in the midst of a world of pain and death" (260–61).

A prosperous society, yes, but the fate of the individual must not be neglected. In fact, it merits close consideration, as in the book of Job, in which the protagonist's health and wealth are eventually restored by his personal god, who is also the Creator.

According to the dictum of Hendrikus Berkhof, a Christian theologian, "a purely powerless god is no God" (Berkhof 1986, 147). The author or authors of the book of Job would certainly have agreed with the dictum. Clearly, the god of Job, and, more generally, the personal god, is a very powerful deity. Yet he does not always respond to prayer. Thus the prayer of Jesus to be spared seizure and crucifixion remains unanswered, despite the appeal to divine omnipotence: "Abba, Father, all things are possible for you. Remove this cup from me" (Mark 14:36).

The National God

The personal god cares for each human individual, but the tribal or national god protects the entire people or nation of Israel. Two more or less elaborated myths tell how the connection between Yahweh and his people originated. According to the book of Genesis, Yahweh elected and blessed Abraham and his family, promising him descendants and land. In this myth, Abraham's grandson Jacob, renamed Israel, is the decisive ancestor. His twelve sons were the ancestors of the twelve tribes of Israel. The book of Exodus, however, tells a different story. It tells of the revelation to Moses of Yahweh, and the giving of his law, religious and civic, to the people led by Moses. The giving of the law was sealed by a formally concluded covenant that established a mutual bond of loyalty between Yahweh and the Israelites, as told in Exodus. The covenantal law famously stipulates the exclusivity of the relationship, expressed in Yahweh's fundamental commandment: "You shall have no other gods before me" (or "besides me"; Exod. 20:3). This is the first of the Ten Commandments Yahweh gave to the Israelites.

One of the national god's main functions was to support his people in times of war. He accompanies the king's army on the battlefield and gives the enemy into Israel's hands. This notion is echoed in one of the attributes recognized by devotees of the national god: he is described as "warlike." Unfortunately, the relevant expression is traditionally mistranslated as "jealous," thus portraying Israel's god as a "jealous god." This rendering is mistaken and should be abandoned, for the term actually refers to Yahweh's warlike character (Lang 2011). He is prone to anger, war, and violence. The Bible offers the following definitions: "Yahweh, the Irascible [*qannâ*] is his name" (Exod. 34:14), and "Yahweh is a warrior, ["the Irascible"] is his name" (Exod. 15:3, reconstructed).

One biblical book illustrates most clearly the ancient Israelites' notion of the national or war god: the book of Joshua. According to the story told in this book, the early Israelites were a warlike tribal people that originally lived in the desert lands east of the River Jordan. But the Israelite tribes did not wish to stay in the desert; instead, they remembered the divine promise their fathers had received; the promise of a land flowing with milk and honey. Accordingly, they assembled under Joshua their leader, took their weapons, crossed the River Jordan, and conquered the land of Canaan. Mighty cities such as Jericho were besieged and entered. Canaanite armies that boasted horse-drawn chariots driven by warriors in armor were overthrown in surprise attacks by the Israelite infantry. Victory was consistently attributed to Yahweh. A typical passage reports that "Yahweh said to Joshua, Do not fear them, for I have given them into your hands. Not a man shall stand before you" (Josh. 10:8). Though Joshua's warriors are described as fearless, Yahweh's miraculous intervention was instrumental in securing victory. The mighty walls of Jericho "fell down flat" after the Israelite priests had carried the sacred ark, their war palladium, seven times around the city. According to another battle report, we are told that Yahweh showered "large stones from heaven" onto the enemy, so that "there were more who died because of the hailstones than the Israelites killed with the sword" (Josh. 10:11).

Thus the ideology associated with the national god seems to be straightforward: whenever the army sets out on a campaign, Yahweh fights on its side, leading the army to victory and triumph. As is to be expected, this ideology was not infrequently contradicted by the facts, because Israel's army was defeated as often as it triumphed. In these instances, pragmatic explanations could easily be offered: the other side had many more soldiers and was better equipped, for instance, with the chariots that the early Israelite army did not have. But the pragmatic approach was not taken to be satisfactory. The issue of divine participation in war had to be considered. In the biblical text, we find several ideological explanations of defeat in warfare: (1) the priests failed to perform the proper rituals, or performed them insufficiently; (2) someone in the army broke a rule and thereby prevented divine assistance from being effectual; (3) the Israelite war god did not have any authority outside the territory owned by him; (4) when the Israelite people were in a state of sin, Yahweh was angry with them, withdrew his favor, and, to punish them, gave them into the hands of their enemies.

The Bible discusses the first two items of this casuistry in the form of precedents, short accounts of specific cases that serve to explain the relevant rule: during the lifetime of Moses, victory in war depended upon Moses himself, because only as long as he was able to lift his hands in prayer could his army prevail against the enemy (Exod. 17:11–12). When the warriors of Joshua failed to conquer the city of Ai, drawing lots revealed the culprit, and so Achan, who had stolen some of the booty that was meant to belong to Yahweh, had to be stoned to death (Josh. 7).

The third idea, Yahweh's lack of power beyond his own territorial boundary, is suggested by Israel's enemies after their defeat: "A god of the mountains is their [the Israelites'] god. For this reason, they have beaten us. But if we should perhaps fight with them on the plain, it is likely that we would beat them" (1 Kings 20:23). Israel's god, so they argue, cannot help his army if the battle takes place in the plain. This is clearly wrong, argues the biblical narrator, and goes on to describe how Yahweh gave the enemy, assembled in the plains, into the hands of the Israelite army. This apologetic account may sound convincing for a moment, but it evades a real problem of early Israelite warfare: Israelite warriors were more successful in the hill country than in the plains of Palestine.[1] The book of Judges explains this by reference to the poor equipment available to the Israelites: they lacked the most advanced war technology characteristic of Egypt and its Palestinian allies, namely, "chariots of iron" (Judg. 1:19). Theologically, this meant that Israel's god was a god of the hills, and not of the plains, and, as a consequence, unable to support his warriors in the plains. This fact is not expressed unambiguously, but implied in one paradoxical aside in the book of Judges: "And Yahweh was with Judah, and he [Judah] seized the hill country; but he did not seize the inhabitants of the valley because they had iron chariots" (Judg. 1:19; my translation). The paradox consists in the juxtaposition of Yahweh's solidarity with his people, and that people's inability to conquer the valleys.

The fourth and final idea, that Israel's warriors are defeated because of Israel's general sinfulness, is well known from the Old Testament. It is frequently invoked, and not only to account for the final overthrow of the states of Israel and Judah by the Assyrians and Babylonians. Israel's first (and legendary) military leader, Joshua, warns the Israelites not to forsake their god by worshiping other deities beside him, for this sin would have severe consequences: it would provoke divine anger. Israel's god "is a warlike god [*el qannô*]. He will not forgive your transgressions or your sins" (Josh. 24:19). Divine punishment for sin would take the form of military defeat.

The Universal God

Yahweh is portrayed as being fundamentally associated with one people, the Israelites or descendants of the patriarch Israel, but this is not the case with the creator god, to whom we now turn.

Israel, in its early period, had a high god called Elyon, or the Most High. In order to understand the Most High, we have to distinguish between this deity's original activity in creation and his later activities.

In the books of Genesis and Proverbs, the Most High is the Creator who by his word and through his wisdom gave the universe its form, stability, and regularity. Genesis 1, in prose, and Proverbs 8, in poetry, show him in full control of the universe. When he pronounces the words "Let there be light," light appears, suddenly and miraculously. He "made the Pleiades and Orion" (Amos 5:8). He "builds his upper chambers in the heavens, and founds his vault upon the earth" (Amos 9:6). To him, ancient mythology attributed the organization of people into national groups, apparently seventy in number. Each was assigned to the care of one of his sons, who served as patron deities (Deut. 32:8–9).

After the structure of the universe has been founded and the peoples are organized, the world is meant to function by itself, as it were, automatically, without further input and guidance. Accordingly, the Creator can retire, or semiretire. From now on, his activity is "restrained and occasional" (Handy 1994, 66). Thus he might, on occasion, assign new places to peoples, for instance by bringing Israel out of Egypt, the Philistines from Caphtor to Palestine, and the Syrians from Kir to Damascus (Amos 9:7). His intervention generally consists in the reestablishment of the original moral order when that order has been disturbed. This moral order, though not codified, is the same for all people: the Israelites of the northern kingdom, the Judeans of the southern kingdom, the Arameans of Damascus, and the Assyrians. We may speak here of "natural law" (Gonçalves 2010, 599–613). God's moral order, and how god reacts to its infringement, can be studied in two prophetic books, both dating from the eighth century BCE, which is Israel's archaic period: the book of Amos and the book of Isaiah (or, more precisely, Isaiah 1–39). If you violate god's natural law, as the Assyrians did by changing the borders, deporting entire populations, imposing heavy tribute, then you commit a serious crime, which can lead only to serious punishment. Israel and Judah, to be sure, are not innocent of crimes. They too will be punished, as the two prophets insist, for example, the wealthy elite of the Israelites for their exploitation of their peasant dependents. It is impossible, says the god of Amos, to escape from my punishment: "If they dig into Sheol, from there shall my hand take them; if they climb up to heaven, from there I will bring them down . . . and if they hide from my sight at the bottom of the sea, there I will command the serpent, and it shall bite them" (Amos 9:2–3). Yahweh, as understood by Amos and Isaiah, is a ubiquitous god. He is none other than Elyon, the Most High.

The world over which the Most High presides and which he owns is not stable. The order is often threatened by evil, destructive forces, sometimes identified as flood or drought, famine or plague, but sometimes also as hostile peoples or tyrannical conquerors. Then the Most High either reestablishes order himself or he appoints kings to do so on his behalf. But the reestablished order never remains untroubled, and must continually be restored. Accordingly, the cosmic battle between good and evil, cosmos and chaos, must continue in perpetuity. The administration of justice, the performance of religious rituals, and the waging of war are all aimed to combat evil and restore the proper order. When the ancient Israelites celebrated the festival of New Year, the horn of warfare was blown, most likely a survival of a dramatic representation of warfare ending in divine victory over evil powers.[2] Similarly, on the Day of Atonement (which actually forms part of the New Year rites) the sins of the people were symbolically placed upon the head of the scapegoat that was subsequently sent into the desert—a ritual of cleansing by elimination. The annual performance of such rituals echoes the cyclical nature of time and reveals the fragility of the divine order. Divine victory over evil remains temporary, a fact that reflects human experience.

Interim Reflections

Of the three aspects of god—personal god, national god, and universal god—only the first was recognized by all Hebrews, not the national god and the universal deity. According to a standard interpretation, which I find convincing, the later divine portraits correlate with the national organization of the Hebrews into two kingdoms: Israel in the northern and Judah in the southern part of Palestine. Yahweh, the national god, represented the leading theology of the North, but the Judeans of the South, with Jerusalem as its capital, focused in their religion on Elyon, the Most High, creator of the universe.

It makes sense to seek the background of Yahweh and Elyon in different social and political contexts. The context for Yahweh, the god of Moses, was tribal. He was the deity of a mountain in the desert, the patron deity of a tribe, or group of tribes, of peasants and nomads. The people who worshiped Yahweh lived in villages or small settlements, some of which may have been temporary rather than permanent, because their residents were wandering pastoralists. By contrast, Elyon, the universal god of Jerusalem, was an urban deity, related through the urban setting to the world of advanced civilizations, such as those of Egypt and Mesopotamia. The god of Jerusalem was the god of merchants, kings, and scribes, a cosmopolitan deity of people in contact with, and with an interest in, the wider world.

As long as all was well with the individual and wider society, the ancients felt that they were supported by a most powerful deity. This conviction even survived its most serious challenge: the destruction of Jerusalem and its temple by the Babylonians in 586 BCE. As we might expect, prayers pronounced in the aftermath of this event asked Yahweh to intervene and change the fate of the people. These prayers also included appeals to divine power. But rather than simply saying that Yahweh has the power to do what he is asked to do, the psalmist reminds god

Roles	*Tasks Performed*	*Gifts Bestowed on Humans*
Personal god	Sustaining care for the individual (animal and human)	Health, fertility, and wealth
Tribal or national god	Exercise of supreme power in the interest of the social group—people, state, and king	Victory in war
Universal god	Supreme cosmic authority, creator and owner of the universe	Wisdom and the sense of justice

Table 10.1. *Roles, tasks, and gifts of the biblical god.* In the literature of the Bible, three distinct portraits of God can be recognized. Originally, they might have been designated by different names: Shaddai, Yahweh, and Elyon. Interestingly, these correspond to the three types of gods—fertility gods, gods of war, and gods of supreme authority—that Georges Dumézil finds in some ancient mythological systems (Lang 2002, 3–11; Lang 2008, 22–26).

of how he acted in the past—by crushing the heads of the sea monsters when he created the ordered universe and guiding his people through the sea as they fled from slavery in Egypt (Pss. 74 and 77). God can still act in the same powerful manner. This, at least, is the psalmist's conviction.

In the cases referred to, the psalmist conceives of his god as the national deity or the cosmic, universal deity, as discussed above (and as summarized in table 10.1). But after the destruction of Jerusalem, some Israelites felt they had to rethink their notion of god, and of divine power. This rethinking led to what we may consider the birth of the notion of divine almightiness.

The Invention of the God Almighty

A new image of god appears toward the end of the biblical period. The main sources that depict his image, the book of Daniel and the book of Revelation, date from the second century BCE and the first century CE, respectively. The best way to understand the new god is to see him as the synthesis of two earlier notions of god: the universal and the tribal deity. As we have seen, the Most High was the universal god of some of the prophets. He was also worshiped as the god of Jerusalem. This universal god came to be closely associated with, and indeed identified with, Israel's tribal god, who, in this capacity, was worshiped mainly in northern Israel.

The book of Daniel and the book of Revelation combine the two notions of the deity to form a ditheistic system, a pantheon consisting of just two deities: the Most High and the young warrior.

At the top level, we find the Most High, a god portrayed as an older figure. As a wisdom god, he is the judge. He echoes the "Creator" of the old system, and keeps his name. He charges a younger figure, a warrior, with the task of fighting. This younger figure echoes Yahweh, the tribal or national deity, but he is given a new name. In the book of Daniel, the younger figure is called the "Son of Man," a designation of uncertain meaning; apparently it meant the angel Michael, to whom the Most High gives "dominion and glory and kingdom, that all peoples, nations, and languages should serve him" (Dan. 7:14), and from later chapters in Daniel it is clear that Michael must fight against God's enemies to establish his dominion. In the book of Revelation, the younger figure is Jesus Christ; as the rider of a white horse, he makes war, "clothed in a robe dipped in blood" (Rev. 19:13). The idea that Christ's power has been delegated appears elsewhere in the New Testament, notably when Jesus says, "All authority in heaven and on earth has been given to me" (Matt. 28:18), a statement that, upon reflection, implies that there is a still higher authority. Thus there are always two levels of authority, and we may invoke the dictum *le roi règne, mais il ne gouverne pas* (the king rules, but does not govern). The Most High does the universal ruling, and the second deity the governing, with specific interest in just one group: the Israelites (in the book of Daniel) or the church (in the book of Revelation). Both groups find themselves in a precarious position and in need of divine intervention, the Israelites or Jews as a stateless minority scattered throughout the Mediterranean region, and the church as a minority whose founder was killed and some of whose disciples were persecuted and their lives threatened by the Jewish religious authorities. Although there seems to be no realistic hope, the apocalyptic writings announce an imminent dramatic change: soon, god will intervene, and soon, all will be well again.

Unlike the universal god or the Most High, the apocalyptic god, helped by his warlike subsidiary, does not simply assert his superiority periodically over the forces of evil, but he aims, instead, at suppressing them throughout the world, eliminating them universally once and for all. Both Daniel and the book of Revelation anticipate a great turning point in history, precipitated by dramatic divine intervention and culminating in the total annihilation of evil. Israel's enemies will be punished and extinguished. They will no longer be remembered. The book of Daniel depicts the antagonistic forces in terms of ancient cosmic-battle mythology, as huge monsters that emerge from the sea: a winged lion, a winged leopard with several heads, a bear, and an elephant (Dan. 7:3–7). Unlike Job's monsters, which are only restrained and not killed, Daniel's beasts are "consumed and totally destroyed" (Dan. 7:26). God's victory is not confined here to a specific time, place, or country, nor is it partial, permitting further demonic assaults. Rather, it is both final and universal. The new kingdom that God will establish will supersede dominion by monsters for all eternity. The new ruler will no longer be endangered, for "all peoples, nations, and languages must serve him. His dominion is an everlasting dominion that shall not pass away, and his kingship is one that shall never be destroyed" (Dan. 7:14).

As the Norwegian scholar Nils Messel reminds us, early Jewish apocalyptic eschatology was completely and exclusively nationalistic in outlook (Messel 1915; see also Buchanan 1993, 58–89). The apocalyptic language of the books of Daniel and Revelation is bold and daring, but we must not misunderstand its actual meaning. It does not mean what it seems to be saying. Whenever universalistic and forever unalterable features seem to be attributed to the expected state of affairs, they should be taken as mere poetic embellishments and exaggerations. Thus when "eternal life" is promised, this means life without evil experiences, but death in old age is not excluded. Or when we hear of a change in the cosmic frame of the world, the authors actually are speaking about a new political situation in the land of Israel. Although brought about by miraculous events, the new state of affairs promised by Daniel and Revelation is not meant to last forever. It would continue for a long period of time, possibly for a thousand years—the millennium (Rev. 20:2, 7)—but eventually, the forces of evil would regain their power and renew their attack on the faithful. Accordingly, God must again intervene on behalf of his people, as he has done before, and so the cycle of history continues, forever repeating its pattern with but little variation.

Can this cycle be stopped, can history come to a final halt? Truly universalistic, cosmic eschatology that would imply the end of history as we know it was devised only in the postbiblical, post–New Testament period by Christian theologians who adopted and adapted apocalyptic scenarios to form their own, new worldview. Beginning with Church Father Irenaeus in the second century (Campenhausen 1979, 20–62), Christian thinkers began to construct what moderns call the "history of salvation" (*Heilsgeschichte*), a doctrine that received its standard form in Augustine's *City of God*, written in the early 400s. Augustine rejected the worldview of the ancients with its undulating, cyclical, never-ending sequence of good and bad periods, and replaced it with a novel, linear view of time and history (*City of God* 12.14, 12.20–21). Augustine saw history no longer as just *one* single, disorderly, never-ending sequence of conflicts and struggles interrupted by relatively conflict-free times of happiness and prosperity, but as a sequence of *three* distinct eras: (1) the primordial time in which the Creator established earth, sky, and humankind, all perfect and good; (2) the historical period in which we live. This period began when the state of perfection was lost and the evil powers gained the upper hand. This historical era is approaching its end, and the third era comes into view: (3) the era in which perfection will be restored. Now the Creator can renovate the world, and perfect happiness will ensue and endure forever, though of course only for God's elect. All evil powers will be eliminated by a cataclysmic event called the Last Judgment. This Augustinian view became the standard one in Christianity.

What is the source of Augustine's notion of salvation history, and, specifically, of his idea of the complete elimination of evil? Augustine certainly read the Bible with due consideration, but there must have been something that inspired or led him to read it in his own particular way. My feeling is this: he must have had a fundamental conviction about the radical dualism of good and evil. The more radical the contrast that is felt between the two, the more probable is the conviction that goodness must have the ultimate victory, because a compromise is impossible. Augustine's dualistic worldview no doubt reflects the teaching of the Manicheans, a Gnostic church of which he was a member before he finally joined the church catholic. Their doctrine of the absolute ending of world history through the final banishment of evil to hell is Manichean in character (Sundermann 1998), and is perhaps ultimately indebted to the Zoroastrian religion (Cohn 2001; Lincoln 2007, 62–65). Thus we may regard Augustine's linear concept of history and his salvation-historical teachings as the legacy of Manichaeism.

Accordingly, as the dreams of poets and visionaries hardened into doctrine, it was decided that Christians should believe in "the resurrection of the dead and the life of the world to come" (Neuner and Dupuis 1983, 9), gifts that could only be granted by an exceedingly powerful deity intent on completely renovating the world. All creeds that repeat the promise of "the resurrection of the dead and the life of the world to come," first included in the book of Daniel and the book of Revelation, begin with some form of the following assertion: "We believe in one God, the Father almighty, maker of heaven and earth, of all things visible and invisible." The price to be paid for the translation of apocalyptic fiction into eschatological dogma should not be overlooked: unlike the apocalyptic authors, most early-Christian thinkers no longer believed in imminent intervention by the almighty God. For them, the end of all evil and the dawning of the new world became an event of the distant, remote future.

ALMIGHTY: THE STORY OF A DIVINE ATTRIBUTE

Words such as *pantokrátôr* and *omnipotens,* the Greek and the Latin words for "almighty," reflect the in-built tendency of the Greek and Latin languages to form compound words. They also respond to the rhetorical need to use epithets and words that serve as alternatives to proper names for the sake of linguistic variation. Unlike Greek and Latin, biblical Hebrew does not have many adjectives, and epithets regularly attached to names do not figure in the repertoire of ancient Hebrew style. As a result, the move from the Hebrew linguistic world to that of Greek and Latin necessitated the creation or adoption of new words and new ways of saying things, including that the Jewish god rules over the entire universe. The words chosen were *pantokrátôr* and *omnipotens*. But how exactly was the new terminology adopted by Greek-speaking Jews and followers of Jesus and, eventually, by Latin-speaking Christians?[3]

To find the answer, we must consider two prominent pagan deities: Isis and Jupiter. In the Hellenistic and Roman worlds, their devotees were convinced of their almightiness in the sense of their ability to rule over everything, the entire universe. Their almightiness can be defined as "hierarchical supremacy and unbridled dominance wielded by one god over the whole universe" (Versnel 2011, 393).

From Almighty Isis to God Almighty

The ancient Egyptians venerated Isis as a nurturing mother goddess. She was associated with birth and rebirth. In the Hellenistic and Roman periods, her portrait was thoroughly revised. She had become the queen of the universe: a supreme goddess credited with rule over all countries, all individuals, all peoples, the entire universe. Throughout the Hellenistic world, she was venerated as a benefactor of humankind, a personal helper of men and women, yet a ruler of the universe. In the second century CE, Apuleius calls her *regina caeli* (queen of heaven), *elementorum omnium domina* (mistress of all elements), and *summa numinum* (highest of the divinities) (cf. Apuleius, *Metamorphoses* 11.2, 5).

In the first century BCE, the Egyptian poet Isidorus celebrated her deeds and virtues in several hymns that have survived as inscriptions. The first of these addresses the goddess as follows: "O wealth-giver, queen of the gods, Hermouthis, lady omnipotent [*pantokráteira*], Agathe Tychy, greatly renowned Isis, Deo, highest discoverer of all life" (Isidorus, *Hymn* 1; Totti 1985, 76). From a second-century CE Egyptian litany found on a papyrus we learn that the goddess was called "in Naucratis fatherless, joy, savior, almighty [*pantokráteiran*], most great" (Oxyrhynchus *Papyrus* 11.1380; see Totti 1985, 64). The claim made by her devotees that Isis was a goddess of sovereign power does not come as a surprise to those who are familiar with the Latin epithets that Apuleius heaps upon Isis, or with the Greek hymns that celebrate her mighty deeds and spread her cult and fame in the Hellenistic and Roman worlds (Streete 2000; Haase 2005).

Apparently, the noun *pantokrátôr*, "the almighty," or, more properly, "the ruler of everything," belonged to the honorific vocabulary used, or perhaps even invented, by Greek-speaking Egyptian priests and poets who sought to extol Isis and spread her cult. And they were listened to, especially, by the young generation of Hellenized Jews. In the Hellenistic period, Egypt boasted a substantial Jewish population. As a matter of fact, Egypt was the center of Greek-speaking Judaism. In the book of the

Wisdom of Solomon (written in Greek in the first century BCE or first century CE), we have a document that reflects Jewish interest in the goddess Isis. The author, an Alexandrian Jew, boldly assimilates the personified Wisdom of his native Jewish tradition to the goddess Isis. The Jewish Sophia (as the author calls Wisdom, in Greek) is portrayed as strongly resembling Isis. In making Sophia look like Isis, the Jewish author contends that Jews in Roman Alexandria are just normal people whose mythology is just a variety of then-fashionable beliefs (Kloppenborg 1982). "Yes, we do worship Isis," young, well-educated Jews would answer when asked about the matter. Jews, they might add with a disarming smile, are not as strange as they might appear to some of their contemporaries; the same contemporaries that are credited with the invention of the earliest form of anti-Semitism.

According to the author of the book of Wisdom, Sophia, very much like Isis, "can do all things" (Wisd. 7:27). She is credited with much power, but not with sovereignty. Accordingly, the author refrains from calling her "almighty" in the sense of "ruling over everything." The reason for doing so has to do with monotheism. In monotheistic Judaism, Sophia may be extolled, but actual sovereign power must remain in the hands of the Jewish god.

Jewish writers other than the author of the book of Wisdom found the word *pantokrátôr* eminently useful, for it seemed appropriate as a description of their own deity, the monotheistic Jewish god. Their Hebrew scriptures did not include a word they could with confidence render as *pantokrátôr*, but they nevertheless introduced the word into the sacred books that they translated from the original Hebrew into Greek. In the Greek Bible of Alexandria, known as the Septuagint, *pantokrátôr* is frequently used for the mysterious divine names Shaddai, Yahweh, and Zebaoth. Jewish books directly composed in Greek also use the word occasionally. An example can be found in the legendary account of a meeting between Jewish sages and Ptolemy II, king of Egypt, where one of the king's guests addresses his host as follows: "May god the almighty [*ho pantokrátôr theós*] fill you, o king, with all the blessings he has created" (*Letter of Aristeas* 185; ca. 170 BCE). As is to be expected, Philo of Alexandria (ca. 20 BCE–50 CE), the most famous of all Egyptian Jews, also uses the word (*De sacrificiis Abelis et Caini* 63; *De gigantibus* 64; *De somniis* 2.172).

Early Christians knew and used *ho pantokrátôr*, as the Jews did, for the monotheistic god. But only two of the twenty-seven books in the New Testament use it: one of the letters of Paul uses it once (2 Cor. 6:18), and the book of Revelation several times as part of an extended form of the divine name—*kyrios ho theós ho pantokrátôr*, "the Lord, the God, the Almighty" (Rev. 4:8, 11:17, and frequently). It was only much later, in the third and fourth centuries, that *pantokrátôr* became a stock word of Christian discourse. The most authoritative use of the word is in the creed formula endorsed by the bishops who met in the city of Nicaea in Asia Minor in 325 CE: *pisteúomen eis héna theón, patéra pantokrátôra* (we believe in one god, the father, the almighty). This text, known as the "Symbol of Nicaea," was subsequently considered a standard Christian creed, and many similar creedal statements were derived from it, including some that are used today.

From Jupiter *pater omnipotens* to the Christian *pater omnipotens*

For the Latin-speaking Romans, the god Jupiter was the unrivaled head of the pantheon. His well-known description as "father of the gods and king of humankind" (*divom pater atque hominum rex*; Virgil, *Aeneid* 10.2) not only echoes a Homeric expression: Zeus is "father of gods and men" (*Odyssey* 1.28). It also made sense in a society in which the father, undisputed head of a patriarchal *familia*, was notoriously strong, to the point of being able, by virtue of his *patria potestas*, to kill his slaves and even his children. Furthermore, the notion of Jupiter as the king of humankind resonated with a Mediterranean world controlled by the Roman Republic and eventually dominated by the Roman Empire.

The earliest substantial works of Roman literature date to around 200 BCE. In this first body of literature, the word *omnipotens*, applied to the gods, appears, thus constituting an essential word of religious discourse. "Immortal, almighty gods," exclaims the young Carthaginian Agorastocles, one of the characters in the comedy *Poenulus* (Plautus, *Poenulus* 275). It comes quite naturally to the lips of Agorastocles and, we must assume, to the lips of Plautus. Is this then an established expression? It is with Ennius, rather than Plautus, that *omnipotens* begins to develop into a regular epithet of just one god: Jupiter.

Ennius (239–169 BCE), one of the originators of Latin poetry, wrote a hexameter epic called *Annales* in which he told the story of the Roman state from its origins to his own time. Ennius aspired to be the Homer of the Romans: the author who gave his people a national epic that promoted Roman identity and inspired civic pride. Although this epic is now lost, certain fragments did survive and help us understand the impression it made on its first readers and subsequent generations of the intellectual elite. One such fragment gives us a glimpse of how Ennius understood Jupiter: as *Iuppiter omnipotens*, "almighty Jupiter" (Ennius, *Annals*, fr. 450/51; Warmington 1935, 168). Ennius apparently was the first to apply the adjective *omnipotens* to Jupiter, and his reasons can be discerned. First, Ennius emulated his Greek model in seeking equivalents of Homeric expressions, such as "all-seeing Jove," "blue-eyed Athena," and "rosy-fingered Eos." Second, Jupiter, the Roman equivalent of Zeus, is considered the unrivaled leader of the pantheon, hence the appropriateness of the epithet for the supreme deity. Third, the word has unique prosodic qualities. It fits well into the hexameter, the metric pattern introduced by Ennius into Latin epic literature. The expression *Iuppiter omnipotens* stuck; as late as the fourth century CE, five centuries after the epic was written, the Latin grammarian Servius knew and remembered it (Warmington 1935, 168).

Soon after Ennius, *Iuppiter omnipotens* reappears in literature, now generally in the form *pater omnipotens*, "father almighty," used as a description of Jupiter in poetic discourse. Lucretius, Virgil, and Ovid use it as a stock expression.[4] The shift from *Iuppiter omnipotens* to *pater omnipotens* can be explained. The very name "Jupiter" includes the word *pater*. The Latin speakers understood well, as modern linguists do, that the god's name means "divine father." We may call *pater omnipotens* an etymologizing paraphrase of the name Jupiter. Virgil (70–19 BCE), in particular, uses the expression frequently in the *Aeneid*, another national epic of the Romans, and one that was to supplant the *Annales* of Ennius. Virgil applies *omnipotens* frequently to Jupiter (*Aeneid* 10.1, 10.100, 10.668, etc.), and more rarely to the goddess Juno (*Aeneid* 4.693, 7.428). Juno is Jupiter's sister and, at the same time, Jupiter's spouse. Something of Jupiter's almightiness rubs off on her. Virgil, however, does not seem to imply that she is the coruler who, at the side of Jupiter, governs the universe. The couple Jupiter and Juno echoes another divine couple: Zeus and Hera. Neither Hera nor Juno governs the universe as Zeus and Jupiter do.

Virgil's *Aeneid* came to belong to the canon of Latin literature that had to be studied by everyone who in the first four centuries of the Christian era acquired a higher education. Thus we know from Augustine that he read it to the point of weeping at one of its emotional episodes, the self-inflicted death of Queen Dido, whom Aeneas left behind in Carthage in order to complete his political mission (Augustine, *Confessiones* 1.21). From Virgil, Christians adopted the expression *pater omnipotens* (Dölger 1940). The Christian creed in use in Augustine's church in Hippo, in North Africa, in the fourth century clearly echoes pagan language: *Fides ergo haec et salutis regula, credere nos in deum patrem omnipotentem* (this, then, is the faith and the rule of salvation, that we believe in God the father almighty) (Augustine, *Sermo* 215.2; Migne 1865, 1072). Augustine (*Confessions* 1:4) never tired of celebrating God Almighty even at the expense of linguistic correctness, addressing him as "you—the most highest, most good, most potent, most omnipotent" (*summe, optime, potentissime, omnipotentissime*). Once a description of the Roman god Jupiter, *pater omnipotens* was now the Christian god. This transfer of the expression from pagan Jupiter to the Christian god also implied a subtle shift of meaning: a merely descriptive and honorific term came to designate what was, in the eyes of Augustine and his Christian contemporaries, a metaphysical reality.

SUMMARY

We began by considering three roles of God in the older parts of the Bible. First, God understood as the personal god, that is, the protector of the individual. Second, God understood as the tribal deity or national god, a deity closely and indeed exclusively associated with one people, the Israelites. Third, God understood as the universal god who created heaven and earth, who owns and supervises the entire universe. A quick review of some biblical texts associated with each of these three roles does not reveal any notion that the biblical God's power is limited. However, nor does there seem to be an explicit concept of divine almightiness. Turning to the last stage of biblical religion, we considered the emergence, or invention, of a new notion of God. This notion originated when two earlier roles of God were combined: the universal deity originally venerated in the southern kingdom of Judah, and the tribal deity that served as the national god of the northern kingdom of Israel. These two merged toward the end of the biblical period, when the two kingdoms no longer existed.

The combination of the two divine images provided the basis for a third image of the deity: the apocalyptic god. Though at first devised only as a strong statement of the hope for the restoration of Israel as a state, Christian authors such as Augustine integrated apocalyptic notions into their linear, rather than cyclical, view of human history. God was expected to change the course of human history dramatically, even to end that history through the elimination of all evil. The very act of this elimination requires not just power, but supreme power. Thus the apocalyptic deity of the late biblical period became the God Almighty.

The term "almighty," though, is a different matter. It was introduced into the Jewish and Christian traditions in Hellenistic and Roman times, not so much as a consequence of the new, apocalyptic notion of a most powerful deity, but as a consequence of the encounter with certain pagan deities for whom their devotees claimed almightiness in the sense of wielding authority over the entire universe. In Hellenistic, Greek-speaking Egypt, where priests called the goddess Isis *pantokráteira*, "the almighty lady," Jewish intellectuals began to adopt the word, calling the biblical god *ho pantokrátôr*, "the almighty." A similar development occurred in the Latin-speaking world, where Virgil's cliché *pater omnipotens*, applied to Jupiter, was adopted by the Christians for their monotheistic deity.

Today, in the Western, monotheistic religious tradition, the attribute of "almighty" comes immediately to mind when we think of the one god, creator and ruler of the universe. It is said to sum up all other possible attributes. All early Christian creeds start with the same assertion: "I believe in God, the Father Almighty," or "We believe in one God, the Father Almighty, the maker of all things visible and invisible" (Neuner and Dupuis 1983, 3–11). In his *Church Dogmatics*, Karl Barth explains: "As is well known, the oldest confessions of faith were content to attribute this *one* characteristic to God"; "they apparently saw in this attribute the sum of all others, as it were the compendium of them" (Barth [7]1987, 2/1:587; my translation).

NOTES

1. The distinction between hilly country and level country is not only a major feature of the Palestinian landscape, but it also accounts for much of its history. The mountain land represents Israel's proper and longest-standing possessions, while all

relatively level land, such as valleys and plateaus, was, without superior chariots, more difficult to defend against invading armies (Smith 1966, 58–59).

2. The best-documented ancient new-year festival, Babylonia's Akitu festival, was celebrated in the spring. The priests recited and presumably performed the mythological drama of the god Marduk's victory over the powers of evil.

3. There is substantial scholarly literature on the relevant terminology: on *pantokrátôr*, see Feldmeier [2]1999; Kate 2001; Bachmann 2002, 113–95; Batut 2009; on *omnipotens*, see Dölger 1940; Schubert 1984; Kate 2001, 155–75; Batut 2009, 492–95. I have considered these valuable studies but offer my own interpretation.

4. Lucretius, *De rerum natura* 5.399 (ca. 53 BCE); Virgil, *Aeneid* 10.100 (ca. 19 BCE); Ovid, *Metamorphoses* 2.304 (ca. 8 CE).

REFERENCES

Bachmann, Michael. 2002. *Göttliche Allmacht und theologische Vorsicht: Zu Rezeption, Funktion und Konnotationen des biblisch-frühchristlichen Gottesepithetons pantokrator*. Stuttgart: Katholisches Bibelwerk.

Barth, Karl. [7]1987. *Die kirchliche Dogmatik. Vol. 2/1, Die Lehre von Gott*. Zurich: Theologischer Verlag Zürich.

Batut, Jean-Pierre. 2009. *Pantocrator: "Dieu le Père tout-puissant" dans la théologie prénicéenne*. Paris: Institut d'Études Augustiniennes.

Berkhof, Henrikus. 1986. *Christian Faith: An Introduction to the Study of the Faith*. Grand Rapids, MI: Eerdmans.

Buchanan, George Wesley. 1993. *New Testament Eschatology*. Lewiston, NY: Edwin Mellen.

Campenhausen, Hans von. 1979. "Die Entstehung der Heilsgeschichte: Der Aufbau des christlichen Geschichtsdenkens in der Theologie des ersten und zweiten Jahrhunderts." In *Urchristliches und Altchristliches*, 20–62. Tübingen: Mohr.

Cohn, Norman. [2]2001. *Cosmos, Chaos and the World to Come: The Ancient Roots of Apocalyptic Faith*. New Haven, CT: Yale University Press.

Dölger, Franz Joseph. 1940. "Juppiter omnipotens (Ad Scap. 4)." In *Antike und Christentum*, 6:70–71. Münster: Aschendorff.

Feldmeier, Reinhard. [2]1999. "Almighty." In *Dictionary of Deities and Demons in the Bible*, edited by Karel van der Toorn et al., 20–23. Leiden: Brill.

Frye, Northrop. 1983. *The Great Code: The Bible and Literature*. New York: Harcourt Brace Jovanovic.

Gonçalves, Francolino J. 2010. "Fondements du message social des prophètes." In *Congress Volume Ljubljana 2007*, edited by André Lemaire, 597–620. Leiden: Brill.

Haase, Mareile. 2005. "Epithets of Isis." In *Brill's New Pauly*, vol. 6, edited by Hubert Cancik and Helmuth Schneider, 967–68. Leiden: Brill.

Handy, Lowell K. 1994. *Among the Host of Heaven: The Syro-Palestinian Pantheon as Bureaucracy.* Winona Lake, IN: Eisenbrauns.

Kate, Albertus A. S. ten. 2001. "Avec dévouement total: La lutte de Dieu contre toute puissance. Origine et évolution de la toute-puissance." PhD diss., University of Amsterdam.

Kloppenborg, John S. 1982. "Isis and Sophia in the Book of Wisdom." *Harvard Theological Review* 75 (1): 57–84.

Lang, Bernhard. 2002. *The Hebrew God: Portrait of an Ancient Deity.* New Haven, CT: Yale University Press.

———. 2008. "The Hebrew God: A Very Short Summary." In Bernhard Lang, *Hebrew Life and Literature*, 13–28. Farnham: Ashgate.

———. 2011. "Le Dieu de l'Ancien Testament est-il un dieu jaloux? Essai de réponse." In *Jalousie des dieux, jalousie des hommes*, edited by Hedwige Rouillard-Bonraisin, 159–71. Turnhout: Brepols.

Lincoln, Bruce. 2007. *Religion, Empire and Torture: The Case of Achaemenian Persia.* Chicago: University of Chicago Press.

Messel, Nils. 1915. *Die Einheitlichkeit der jüdischen Eschatologie.* Gießen: Töpelmann.

Migne, Jacques-Paul, ed. 1865. *Patrologiae cursus completus, Series latina.* Vol. 38. Paris: Migne.

Neuner, Joseph, and Jacques Dupuis, eds. 1983. *The Christian Faith in the Doctrinal Documents of the Catholic Church.* Revised ed. London: Collins.

Schubert, Werner. 1984. "Der Begriff *omnipotens* in der lateinischen Literatur." *Gymnasium* 91 (5): 369–78.

Silverman, Jason M. 2012. *Persepolis and Jerusalem: Iranian Influence on the Apocalyptic Hermeneutic.* New York: T & T Clark International.

Smith, George Adam. 1966. *The Historical Geography of the Holy Land.* London: Collins.

Streete, Gail Corrington. 2000. "An Isis Aretalogy from Kyme in Asia Minor, First Century B.C.E." In *Religions of Late Antiquity in Practice*, edited by Richard Valantasis, 369–83. Princeton, NJ: Princeton University Press.

Sundermann, Werner. 1998. "Manichaean Eschatology." In *Encyclopaedia Iranica*, edited by Ihsan Yarsatir, 8:569–75. New York: Center for Iranian Studies.

Totti, Maria. 1985. *Ausgewählte Texte der Isis- und Sarapis-Religion.* Hildesheim: Olms.

Versnel, Hendrik S. 2011. "God: The Question of Divine Omnipotence." In *Coping with the Gods: Wayward Readings in Greek Theology*, 379–438. Leiden: Brill.

Warmington, Eric H. 1935. *Remains of Old Latin I: Ennius and Caecilius.* Loeb Classical Library. Cambridge, MA: Harvard University Press.

Whedbee, William J. 1998. *The Bible and the Cosmic Vision.* Cambridge: Cambridge University Press.

CHAPTER 11

God's Omnipotence

RICHARD SWINBURNE

God, the doctrinal definitions of the Church often affirm, is "omnipotent" (*omnipotens* translating παντοκράτωρ), which is naturally interpreted as meaning "has power over everything" or "can do anything," and the latter in turn is naturally understood as "can do any intentional action." However, there are plenty of things that seem like "actions" no one can do, such as making me exist and not exist at the same time, or causing an uncaused state, or (perhaps) causing a state of affairs already past. And it seems that for each particular being there are further actions only that particular being, and so not God, can do—for example, no one except Jones can bring it about that Jones freely mows the lawn. (I use "freely" in a libertarian sense such that for S to act freely entails that there is no sufficient cause of the event of S-doing-that-act.) And it seems that there are actions a being with the other supposed divine properties of the Judeo-Christian God cannot do, for example, a bad act. Hence the attempts of many modern philosophers, following Aquinas and other earlier philosophers, to give a plausible explication of the concept of omnipotence.[1] In the belief that none of them have reached an entirely satisfactory explication, I shall try to do better. My aim here is to reach a coherent definition of "omnipotent" that is consonant with normal theistic thinking in the Judeo-Christian tradition about what God is like. To do so will involve taking stances on some contentious philosophical and theological issues.

I begin with some definitions of terms I shall use in candidate definitions of "omnipotent." I shall understand "brings about" as equivalent to "causes." I shall understand by an "event" either a continuing or timeless state of affairs or a change of state, and I shall assume that this consists in the instantiation of a property in a substance(s) (or in properties or events), or the coming into existence or the ceasing to exist of a substance, at a time (or a timeless moment). All actions consist in bringing about events distinct from the action of bringing them about, apart from actions of trying to bring about some event. But since anyone, however weak, who tries to bring about some event will always succeed in trying, we need not concern ourselves further with acts of trying. What distinguishes powerful beings from others is that they can bring about more events of the normal kind distinct from the action of bringing them about. I understand by "metaphysical necessity" the strongest kind of necessity, by "metaphysical impossibility" the strongest kind of impossibility, and so by "metaphysical possibility" the weakest kind of possibility. A metaphysically necessary event is one that necessarily occurs; a metaphysically impossible event is one that necessarily does not occur, and so on. An event is necessary (or whatever) iff the proposition that reports its occurrence is necessary (or whatever). A central case of a metaphysically impossible proposition is one that entails a contradiction, and a central case of a metaphysically necessary proposition is one whose negation entails a contradiction. My subsequent discussion of what is necessary (or whatever) concerns metaphysical necessity (or whatever).

On the assumption (which I will later abandon) that the omnipotent being exists in time, the above considerations suggest the following simple definition:

> [α] S is omnipotent at a time *t* iff S is able at *t* to bring about any event M which it is metaphysically possible that he could bring about at *t*.

I understand "is able to" as entailing "knows how to." So [α] requires that, to be omnipotent, S has both the power and the knowledge of how to cause M. Erik Wielenberg (2000) has defined a being as "omnipotent" in such a way that any being who is sufficiently powerful is omnipotent, and he explicitly denies that a being could lack omnipotence simply because he lacked the knowledge of how to bring about some effect. But, although etymology might suggest that being "omnipotent" is merely a matter of having enough power, I think that hardly anyone in a monotheistic tradition would count a being as omnipotent if he did not know how to bring about some effect that it was within his power to bring about—for example, if he did not know how to talk to Adam because he did not know where Adam was.[2]

Definition [α] has the consequence that, to be omnipotent, S is not required to bring about either a metaphysically impossible event (such as me existing and not existing at the same time) or a metaphysically possible event that it would be metaphysically impossible for S to bring about (such as some agent other than S freely mowing the lawn). Unfortunately, it runs into the "McEar" objection. McEar is a being who essentially (that is, necessarily) can do no action except scratching his left ear; so long as he can scratch his left ear, he comes out as omnipotent on this definition—which is absurd.[3] And in different ways several definitions have fallen foul of the McEar objection.[4] So instead of "metaphysically possible that he bring about," some have substituted "metaphysically possible that some being bring about." This suggests:

> [β] S is omnipotent at *t* iff S is able at *t* to bring about any event M which it is metaphysically possible that some being bring about at *t*.

Definition [β] is, however, like definition [α], unsatisfactory in not elucidating which actions of bringing about events are metaphysically possible for a being to do at some time. But in order to provide such a definition, we need to take a stance on two issues about time. The first issue concerns the relation between events occurring at instants of time and events occurring over periods of time. One theory I shall call the "I-theory," which seems to be adopted unthinkingly in many definitions and so incorporated by me into [α] and [β], is that an event can take place at an instant of time, independently of what happens over any period of time. The rival theory, the one I hold and I shall call the "P-theory," is that all talk about an event happening at some instant can always be analyzed more fundamentally in terms of what happens over a period or periods of time.

The normal sense of "instant," made precise in mathematics, is as designating a zero-dimensional boundary to a one-dimensional period of time. Most ordinary talk about an event beginning or ending at an instant is analyzable as the event lasting for a period of which that instant is the initial or terminal boundary. Talk about an event happening at an instant is analyzable as the event happening for a period divisible into two periods, each bounded (terminally or initially) by that instant, and which in a loose sense may therefore be said to "include" that instant. Talk about an object of a certain volume traveling from A to B occupying a certain region of the same volume as the object at a certain instant can be analyzed as it occupying smaller and smaller regions, including that volume for smaller and smaller periods of time, each divisible into two periods bounded (terminally or initially) by that instant. Talk about objects having some rate of change, such as velocity at an instant at some point, is analyzable in terms of that velocity (or whatever) being the limit of the distance (or whatever) covered over time as we take smaller and smaller distances (or whatever) from that point (on one side or the other of the point—and of course the "instantaneous velocity" may differ in the two cases—which couldn't happen if it was what it was independently of how it was measured). No mundane phenomenon requires us to postulate that there is a sense of "instant" in which an event can happen at an instant, which is not analyzable more fundamentally in terms of what happens over periods. And it seems difficult to understand what it would mean to say that an event happened at an instant without that entailing anything about the period that "includes" or begins or ends with that instant. It seems an ungrounded assumption to postulate that there is a sense of "instant" in which this does have a meaning, and I cannot see why talk about God needs to postulate this. Hence I shall in future assume the P-theory, although my subsequent results can be reexpressed in terms of the I-theory.

The second issue about time is whether either simultaneous or backward causation is metaphysically possible, that is, whether the exercise of causation over a period can cause an event at exactly the same period or at an earlier period. Definition [β] allows both possibilities. Whether these kinds of causation are possible depends on how we understand causation and temporal succession. I am going to make the highly contentious assumptions,[5] which I develop in the next three paragraphs, that both causation and temporal succession are unanalyzable relations, which we come to understand by acquaintance with paradigm instances of them.

Exerting causal influence just is the kind of process of which we are aware when we try to perform a basic action, such as moving a hand (or when we feel something else exerting causal influence on us, for example, by pushing or pulling us), and actually causing just is the kind of process of which we are aware when we succeed in performing the basic action (or succumb to the effects of an action on us). We are substances and we cause events intentionally by intentionally exercising our causal powers. We then come to recognize the same kind of process at work in the nonintentional regularities of the natural world. "Laws of nature" are summaries of the regularities in the possession by substances of powers to exercise causal influence and their (nonintentional) liabilities to do so under certain conditions. Many of the paradigm instances of causation, as when we perform a basic action, are ones in which the succession cause/effect is experienced as a single event. And we recognize this relation as transitive: if a substance A causes a substance B to have a property ϕ, and B is such as to have a liability when having ϕ to exercise its power to cause κ, then A causes κ. If I light the gunpowder, and gunpowder is such as to explode when alight, then I cause the explosion.

It is important to distinguish between the cause, the substance (either a person or an inanimate object) exercising its causal power, and it succeeding in producing the effect. That distinction is clear in the case of intentional causation where the agent (a person) having the intention ("trying") is the exercise of causal power, and it having its intended effect is it causing that effect. What the causing substance does is to exert a certain amount of causal influence, which—if the surrounding circumstances are of a certain kind—will produce a certain effect. Where nothing turns on the distinction between "causing" and "exercising causal influence," I shall often—in order to make the exposition easier to follow—ignore it, and just speak of "causing" when "exercising causal influence" would have been more accurate. Also, although I have just been describing the operation of causation in what I have argued as the correct way as causation by a substance, where nothing turns on it, I shall sometimes speak of events causing effects, and ask the reader to bear in mind that causation by an event is really causation by a substance exercising some power.

Temporal succession—one event being succeeded by another—just is the relation of which we are aware when it holds between our experiences in the specious present—for example, the experience of a flash being succeeded by the experience of a bang. In talking about events in the public world succeeding one another, we are attributing to them the same relation as we experience when we experience one experience being followed by another one. And we also are aware in the specious present of instances of one event being succeeded by a second event, and this by a third event, and since they are also instances of the first event being succeeded by the third event, we come to understand the concept of temporal succession, like the relation of causation, as transitive.

With these understandings of causation and temporal succession, and of the times at which events occur, I come back to the issue of whether backward or simultaneous causation is metaphysically possible. Given the P-theory, simultaneous causation is to be understood as a substance exercising its causal influence over exactly the same period of time as the effect it causes, and backward causation is to be understood as a substance beginning to exercise its causal influence after the beginning of the effect it causes. However, in order to make my exposition simpler to follow, I shall sometimes describe these alleged possibilities in terms of an event (a substance exercising causal power) causing another simultaneous or earlier event.

I now introduce three connected concepts. By an event occurring at a (period of) time *T* being "*fixed*" at a time *T**. I mean that it is such that of metaphysical necessity no agent, however powerful and however hard he tried at *T**, could cause any part of that event not to occur at *T*. As I noted earlier, one may cause an event E by causing another event that in turn causes E. A cause that is the most immediate (or direct) cause of an event, and actually brings it (the whole event) into existence (when but for the act of causing, E would not occur) may be said to "fix" it. By an event being "fixable" at a time *T*, I shall mean that the event is such that it is metaphysically possible that a strong enough agent could at *T* have fixed either it or its nonoccurrence, whichever he chose, whatever else metaphysically possible might be the case at any other time or timeless moment.

One could only bring any event E into existence at a time by beginning to act at a time when E does not exist. The most immediate act of exerting causal influence—what I call "direct" causation—moves E from nonexistence to existence, and so must begin to be exerted when

E does not exist. So E cannot be fixed by a direct cause acting at exactly the same time as E. Hence some part of E is fixed at any period during E's occurrence. For this simple reason, I suggest, simultaneous direct causation is impossible. That is compatible with some direct cause acting during the first part of the occurrence of some event so as to bring about the second part of that event (or indeed, barring my arguments below, a direct cause acting during the second part of the occurrence of some event so as to bring about the first part of that event). But what, I argue, is impossible is that a direct cause could cause an effect by acting over exactly the same period of time as the effect.

And not merely must the direct cause C of some effect E exert causal influence at some time other than when E occurs, but it must continue to do so over the whole time while E occurs, if C is to fix E. For if C exerted causal influence so as to cause E for only a part of the time when E occurred, the substance involved in E itself or some other substance could fix the other part of E. So, I claim, a substance that fixes an event must exert its influence for a period of time that includes both a time when the event does not occur and the whole time when the event does occur. We can see this principle at work in mundane examples. When a moving billiard ball A hits a stationary billiard ball B and causes B to move, A must be moving as it touches B, and so—given P-theory—for some period ending after the instant at which it touches B in order to transmit its motion, and it will only be the direct cause of B's subsequent motion for the period while it is still in contact with B. After contact is broken, the direct cause of B's subsequent motion is B itself.

I turn next to the issue of whether (direct or indirect) backward causation—a cause causing an earlier effect—is possible. Among the paradigm examples of a substance being the direct cause of an event is a person having an intention to cause some bodily movement being followed immediately by a brain event that causes the movement ("followed immediately" in the sense that the effect begins after the beginning of the cause and ends at the same instant as it). In these cases it follows from the argument of the previous paragraph that some part of the first event (e.g., a person having a certain intention) is fixed at a time at which no part of the second event (the brain event) is fixed. And although most of the successions of temporally overlapping events ending at the same instant of which we are aware are not ones in which, we believe, the first event causes the second event, we nevertheless derive from the paradigm examples the understanding that a time at which persons can cause events directly is any time immediately following the beginning of their exercise of causal influence and ending at the same instant as it. It follows that if at the time of the occurrence of the first event we had formed an intention to stop the second event occurring and/or to cause a different event instead and were strong enough, we would have succeeded—for this would be a succession just like the actual ones in which our intentions cause events. It follows that all successions of temporally overlapping events that end at the same instant, not merely the ones of which we are aware, are ones where some part of the event that begins earlier is fixed at a time when no part of the later event is fixed but the whole later event is fixable.

Yet no (whole) event E at a time T_2 is fixable at any time T_0 ending earlier than the end of T_2. This is because, however strong an agent was at T_0, it remains possible that he changes his mind at a later time T_1, beginning at the end of T_0 and ending at the end of T_2, and so stops (some part of) E occurring. Yet since an agent at any earlier time T_0 could (unless he changed his mind, or was prevented by the act of a stronger agent before the end of T_1) cause (though not necessarily fix) what happens at T_1, and thereby what happens at T_2, what happens at T_2 cannot be fixed at T_0. Hence every event is unfixed at every time ending before it ends.

Further, no event E at T_2 can be caused by any agent acting at any subsequent time. For, as we have seen, the direct cause C of an event E must last for the whole time while E occurs, and also for some further time when E does not occur. In the case of a supposed subsequent cause D, that further time T_3 would be immediately subsequent to E. But only what is fixed can fix some other event; until D's action is fixed, that action cannot fix E. For, while an event is not fixed, an agent F of sufficient strength acting at an immediately earlier time could always prevent that event from happening. So D's action at T_3 could be caused not to occur by an agent of sufficient strength acting at T_2, and then its effect (E) would not happen. So, if it could be fixed by a later cause, E would not be fixed before the end of T_3. Now consider the event consisting of E and D's supposed action of fixing E. By previous arguments, the first part of this total event E is fixed at the end of T_2, when the event of D's action is still fixable. So D's action could not fix E. Yet if an event cannot have a later direct cause, it cannot have any later cause. For such a later cause would be connected to the direct cause by a chain of causes, each of which was the direct cause of the next one, and some of these direct causings would be of an earlier event by a later one, and so ruled out by the previous pattern of argument. So every event is fixed at all times after it has occurred. In summary and very loosely, since forward causation is always metaphysically possible, backward causation is never metaphysically possible.

It follows that not merely direct simultaneous causation but indirect simultaneous causation is impossible. For if C caused a simultaneous event E by directly causing F, which directly caused E, F would have to occur either (1) simultaneously with C, or (2) after C, or (3) before E. (1) is impossible because of the impossibility of direct simultaneous causation; (2) involves backward causation in the respect that F causes E; and (3) involves backward causation in the respect that C causes F. And the same problems arise for any postulated longer chain of causes by which C causes E. Hence the metaphysical impossibility, not merely of backward causation, but of all simultaneous causation.[6]

Before we can apply this principle, we need an account of what it is for an event to occur at a certain time. Despite the principles stated above, there are certainly propositions reporting some event purportedly occurring at some time that can be made true or false (wholly or partially) by some later event, where it seems fairly evident that the proposition reporting the purportedly earlier event does not report merely what happened at that earlier time. And it is for this reason that the event reported by the former proposition is not caused by a later event. For example, suppose that John met Mary in 1988 and they married two years later; then by marrying Mary in 1990 John made it true that "John met his future wife in 1988." But it is evident that "John met his future wife in 1988" is only partly about a meeting in 1988; it is also about what happened at some unspecified later time. While it may seem in general obvious whether or not some proposition purportedly about some time is really totally about that time and so made true or false solely by events at that time, it is desirable to have a definition that makes a clear distinction between such propositions and other propositions that are purportedly about one time but are really (at least partly) about another time, and so made true or false solely or partially by events at that other time. Various philosophers have sought to make such a distinction in terms of "hard" and "soft" facts. In my view, the definitions they have offered[7] have not been satisfactory, and I shall now provide what I believe to be a more satisfactory one.

I define an event E as a hard event at *T* iff neither the occurrence nor the nonoccurrence of times before or after *T* (or a timeless state—if there are any such states) is metaphysically necessary or metaphysically sufficient for the occurrence of E, given the occurrence of the other events that occurred at *T*, for the occurrence of which neither the occurrence nor the nonoccurrence of times before or after *T* (or a timeless state) is metaphysically necessary or metaphysically sufficient. (To avoid too clumsy a definition, I am counting the nonoccurrence of an event of some kind at *T* as an event at *T*.) An event at *T* that is not a hard event at *T* is a soft event at *T*.

I now illustrate how this definition works. Me-having-a-pain-at-midday is a hard event at midday, because whether it occurred is metaphysically independent of the occurrence of any time before or after midday (or any timeless state), given everything else that happened at midday for the occurrence of which neither the occurrence nor the nonoccurrence of times before or after *T* (or a timeless state) is metaphysically necessary or metaphysically sufficient. Whether or not there were such times, I could still have had or not have had a pain at midday. Likewise, a-world-war-beginning-in-1914 is a hard event at 1914. Its occurrence is independent of there being any time before or after 1914 (or any timeless state). Whether or not there was a year 1800 and whether or not there are timeless truths is not metaphysically necessary or sufficient for the occurrence of the world war, given everything else that happened in 1914, for the occurrence of which neither the occurrence or the nonoccurrence of an earlier or later year (or a timeless state) is either metaphysically necessary or metaphysically sufficient. But John-being-thirty-years-old-in-1988 is a soft event in 1988, since it is metaphysically necessary for its occurrence that there were thirty years before 1988. Necessarily, if there were no times before 1988, that event could not have occurred. Intuitively John-meeting-his-future-wife-in-1988 is a soft event in 1988. But neither the occurrence nor the nonoccurrence of times before or after 1988 (or a timeless state) is metaphysically sufficient or metaphysically necessary for its occurrence, for John could have married Mary later in 1988. To get the result that this is a soft event in 1988 we need the clause: "given the occurrence of the other events that occurred at *T*, for the occurrence of which neither the occurrence nor the nonoccurrence of times before or after *T* (or a timeless state) is metaphysically necessary or metaphysically sufficient." For, given what else happened in 1988, including the event of John-not-marrying-Mary-in-1988, the occurrence of times later than 1988 are necessary for the occurrence of John-meeting-his-future-wife-in-1988, and the occurrence of other times apart from 1988 (or a timeless state) is not metaphysically necessary or sufficient for the event of John-not-marrying-Mary-in-1988. The-beginning-of-the-First-World-War in 1914 is a soft event in 1914 because the nonoccurrence of any years before 1914 is metaphysically sufficient for its occurrence, given the occurrence of a-world-war-beginning-in-1914 (for the occurrence or nonoccurrence of which, as noted above, neither the occurrence nor the nonoccurrence of times before or after 1914 [or a timeless state] is metaphysically necessary or metaphysically sufficient).

Given the impossibility of both simultaneous and backward causation, we now have the core of a theory of which actions of causing metaphysically possible events are themselves metaphysically possible.[8] But before providing that definition, we need to take into account two further complications. First, we must insist that to be omnipotent a being does not need to be able to cause a metaphysically necessary event, since the latter will occur, whatever an omnipotent being does. Second, we need to take into account that although some event may be a hard event at *T*, in the sense that it is the event it is, quite independently of what happened at earlier or later times, it may be not merely causally but metaphysically necessary that it occur, given what has in fact happened earlier. For example, if there is in fact a being P who has essentially (that is, of metaphysical necessity) the power at some time to cause an event E to occur after an interval of time, he might at some earlier period of time *T* have caused some event E to occur after a certain interval of time at a period of time *T**. Then, if P did so act, no person (including P) could cause E not to occur by any action between these two periods (that is, it would be metaphysically impossible that if P existed and had so acted, any actual person could be strong enough to do so). All these considerations together now allow us to replace definition [β] by a more satisfactory definition:

> [γ] S is omnipotent during some period of time *T* iff S is able to bring about by an act beginning at any instant *t* during that period any [hard] event M beginning at any instant t_1 later than *t*, which is metaphysically contingent given the state of the world before t_1.

This definition deals with a problem to which I have drawn attention but so far have not provided a solution for. We must not require S to be able to cause during *T* any free action beginning at t_1 of any other agent P, which—being a "free" action on my definition—is uncaused. (The action consists in P causing some effect. The action being "free" consists in there being no cause of P-causing-that-effect.) [γ] does not require S to be able to do this, since—given the impossibility of backward or simultaneous causation—neither an uncaused event nor one not caused by S at a period beginning at t_1 is a hard event at that period, since necessarily if there were no times before that period every event and so every action would be uncaused. The nonexistence of every such time would entail that any event beginning at t_1 is uncaused. S can cause P-causing-that-effect at a period beginning at t_1, since that is a hard event at that period. But what he is not required to be able to cause in order to be omnipotent at *t* on [γ] is the event of P-freely-causing-that-effect at a period beginning at t_1, because that would not be a hard event at that period.

The requirement that an omnipotent being need at any time only to be able to bring about hard events enables us also to cope with the difficulty that certain kinds of action are only (metaphysically) possible for certain kinds of agent, and being an agent of one kind entails not being an agent of another kind. For example, an unmarried being S cannot get divorced (immediately, that is). For S to get divorced he would have first to get married, that is, cause himself to be a married person. A married being cannot get remarried (immediately, that is); he or she would have first to get divorced or become an unmarried widow(er). But no being can be both married and divorced-or-an-unmarried-widow(er) at the same time. So there will always be one of these two actions S cannot do. But in both cases the action consists in bringing about a soft event, S-getting-divorced or S-getting-remarried, events that entail the occurrence of previous events. Hence if S is unable immediately to bring it about that he is divorced or that he is remarried, that does not count against S being omnipotent. In future I shall mean by an "event at *T*" a "hard event at *T*."

However, [γ] allows the possibility of an essentially omnipotent being (as God is supposed to be) bringing about the whole future both of himself and of the universe by an act beginning at *t*, which was the time of the beginning of the universe. In that case, since (for all t_1) no subsequent events would then be "metaphysically contingent given the state of the world before t_1," the essentially omnipotent being would, after that act at t, be unable to affect the subsequent development of the universe or of himself in any way. Even those theologians who believe that God has predestined the whole future of the human race would surely be unhappy with supposing that God is able to predestine his own whole future. And if God had predestined the whole future of the universe, while still essentially omnipotent (on definition [γ]), he would have become in effect no longer theistic (in charge of the universe at all times), but instead deistic (a God who caused the universe to begin to exist and then left it alone); he would have made the universe no longer depend on him for its continuing existence. If this were possible, it would follow that the universe is not in its nature such that it does not have the causal power to exist without God's sustaining action, which again seems contrary to the normal theistic view. It seems better consonant with normal theism to maintain that the universe does not have this causal power, and so to have an understanding of omnipotence compatible with that. These considerations lead to:

> [δ] S is omnipotent during some period of time T iff S is able to cause by an act beginning at any instant t and ending at any instant t_2, both during that period, any metaphysically contingent event M beginning at any instant t_1 later than t and ending at t_2.

By omitting the "given the state of the world before t_1," this definition ensures that nothing earlier in history, such as the act of S or some other being omnipotent in sense [γ], could cause any event during T unless during T S allows it to do so. On [δ], while S exists, he can bring about any metaphysically contingent event by an act beginning before that event and lasting until the event has fully occurred. Given [δ], S continuously sustains the universe during T with its "laws of nature," or permits other substances, animate or inanimate, to do so; so everything that happens during that period happens because he causes it or permits someone or something else to do so. This entails nothing about what must be the case at any other period, and so nothing about whether—if there were no God then—there could be a universe, but it is metaphysically necessary that if there is a being S who is essentially omnipotent (in sense [δ]) and essentially everlasting, no universe could ever have the causal power to exist independently of his sustaining action.

On one natural understanding of "is able," a being may be omnipotent even if there are actions he cannot choose to do—even if he is totally caused by some other being or some internal irrational inclination to choose to do as he does. But it fits theism much better if we think of God as "omnipotent" in the sense that not merely has he the ability to do any particular action, but he is not caused or in any way influenced to do any particular action. We should think of him as having a choice of which of the actions he is able to do (as this has been understood so far), he will do, and that entails that he knows the range of that choice, and so knows what he cannot do (e.g., which events are metaphysically impossible). But—given a plausible assumption from moral philosophy—it is not possible for God, if he has moral beliefs, to have a choice of actions within the range so far delimited. The assumption is moral internalism,[9] which I construe as the theory that to believe an action to be morally good involves having some natural inclination to do it (rather than not to do it), to believe an action to be the best (or equal best) available is to have a stronger natural inclination to do it than to do any incompatible less good action, and to believe an action to be morally bad involves having some natural inclination not to do it (rather than to do it). The only way in which an agent can do an action other than one he believes the best or equal best, or otherwise what he believes a good action, is if he allows himself to yield to irrational influences. It would be very odd to suppose that God does not have moral beliefs, or that he is subject to irrational influences.

A being subject to no irrational influences is "perfectly free" in one natural sense of that expression. It is of course a paradoxical sense because it entails that certain choices are not open to a perfectly free being. But the obvious alternative sense of "perfectly free," that a perfectly free being is one for whom all choices, whether he believes them good or bad, are open, entails—given moral internalism—that a perfectly free being who has moral beliefs is one subject to irrational causal influences; so that is also a paradoxical sense of "perfectly free." So, given that omnipotence entails "perfect freedom" in the former sense (the sense in which I will use that expression in future), it follows that an omnipotent being believing an action to be the best to do will do it; if there is—he believes—no best action but two or more equal (incompatible) best actions, he will do one of them; or if there is—he believes—no best or equal best action, but an infinite number of (incompatible) good actions, each less good than some other action, he will do one of them, but he will not do any action that he believes bad.

But "God can do no evil" requires a further qualification to a definition of omnipotence.[10] I suggest that the moral quality, as perfectly good or whatever, of an agent at a certain time *T* depends only on whether she does the best in the light of what is knowable by that agent at that time. Thus an agent is only morally culpable for doing an action that has bad consequences if those consequences were foreseeable by her at that time. So "God can do no evil" at time *T* requires not merely that God does at time *T* only what he believes to be the best (or whatever), but also that he has all true beliefs (which I shall construe as knowledge) about which actions are the best, good, or bad, which he is able to acquire. Whether an action believed good (bad, or whatever) is good (or whatever) depends on the necessary moral principles and on the contingent nonmoral truths that determine their application (see Swinburne 2016, 204–9). Among the actions available to an omnipotent being are—for all *p*—the actions of telling humans that *p*. In order to know whether it is good to tell humans that *p*, he needs to know whether or not *p*, and any necessary truth about whether it is always bad to give false information—I shall assume that it is always bad. (If someone thinks that sometimes it is not bad, my main point can easily be stated in a more complicated way, so as to take account of this.) So in order to know whether it would be good to tell humans that Adam did not sin, or that $2 + 2 = 5$, he needs to know that it is always bad to give false information, and he needs to know that Adam sinned or that $2 + 2$ does not equal 5. It is not metaphysically impossible that a being who is omnipotent at some time in sense [δ] should know everything that is already fixed at that time, and so all necessary moral principles and other necessary truths, and all contingent truths about the past. So if God is to be omnipotent and yet unable to do evil, we must incorporate into the definition of omnipotence that an omnipotent being is omniscient in the sense that he knows all necessary truths and all contingent truths about the past, and whatever they entail. These truths will determine what it is best, good, or bad to bring about in future.

Also, among the actions that a being omnipotent in sense [δ] can do at a time is to promise to bring about some future event M. So, given that it is always a best action to fulfil one's promises (and, again, if someone thinks otherwise, my main point can easily be stated in a more complicated way), it follows that an omnipotent being who has made such a promise will subsequently believe it to be a best act to bring about M, and so will do it. But it would seem to follow—despite very influential contrary views in the Christian theological tradition—that an omnipotent being can know infallibly about the contingent future only to the extent to which he knows what would be the best action to do in any situation, which may sometimes be because he has made some action the best action. Plausibly there will often at any time *T* not be a unique best action for such a being to do, and so he will often have a wide choice of actions, between which to choose. It may sometimes be that such a person will choose to allow other rational agents (who may or may not be perfectly free) to make free choices, and in that case—it would seem—he will not know how they will choose. So if an omnipotent being has a free choice of which actions to do (including which actions to allow other beings to do freely), omnipotence ought not to be defined so as to require knowledge of all future events. For, having the freedom to choose between alternatives is incompatible with knowing infallibly how you will choose.

The knowledge requirement on a perfectly free omnipotent being that he knows at a time *T* everything knowable by him at *T* is therefore to be understood as the requirement that he knows all necessary truths (including necessary moral truths), all events already fixed at *T*, and which actions he will do himself in all possible future circumstances because they are the best to do, including which actions he will allow others to do, because it would be the best action to allow them to do those actions. The considerations of the last few paragraphs lead to this final definition of the omnipotence of a being existing in time:

> [ε] S is omnipotent during some period of time *T* iff S knows all metaphysically necessary propositions and all metaphysically contingent true propositions about every event at any time earlier than the beginning of his action and all propositions which those propositions entail; he is not influenced by any nonrational influences, and is able to cause by an act beginning at any instant *t* and ending at any instant t_2, both during that period, any metaphysically contingent event M beginning at any instant t_1 later than *t* and ending at t_2, which does not require him to be influenced by nonrational influences in order to do that act.

The clause "and all propositions which those proportions entail" captures the requirement that S needs to know which actions he will do himself in all possible future circumstances because they are the best to do, including which actions he will allow others to do, because it would be the best action to allow them to do those actions. That these actions are the best to do may be entailed by metaphysically contingent truths about the past although they are not themselves metaphysically contingent truths about the past—as when some past action determines (via necessary moral principles) which future actions would be obligatory.

It might seem that a being could be omnipotent until a time *T*, but then cease to be omnipotent. That could only happen if that being permitted it to happen—for, being omnipotent, he could always bring about his own continued existence (including doing so by preventing any other being bringing about his nonexistence). But unless it would be a nonbest action to do so, he could deprive himself (or allow some other being to deprive him) of his omnipotence or even his existence. He could commit suicide or make a stone too heavy for himself to lift. If every such action would be bad, then any omnipotent being would be necessarily omnipotent and forwardly (but not necessarily backwardly) everlasting. If, however, some such acts (e.g., abandoning his omnipotence while making some other being omnipotent instead) are not nonbest acts, then there are actions a contingently omnipotent being could do, which an essentially omnipotent being could not do. For if S is an essentially omnipotent being, as God is normally supposed to be, abandoning his omnipotence would involve bringing about a metaphysically impossible event (e.g., S continuing to exist while no longer omnipotent). And if S is essentially everlasting and also essentially omnipotent, there are further actions S would be unable to do, that is, any actions that entail his ceasing to exist.

All the other traditional divine properties, I suggest, follow from omnipotence in sense [ε]—with two exceptions to be discussed shortly. An omnipotent being could not be tied down to acting through a body, and so will not be essentially embodied. He will be omnipresent in the sense that he can cause effects at any place directly as a basic action (i.e., not via doing something somewhere else) and that he knows what has happened at every place without that knowledge depending on anything else. He will be the creator and sustainer of any physical universe there may be, in the sense that either he brings it about or he allows some other agent to bring it about. As we have seen, he will be perfectly free (in my sense) and perfectly good, but omniscient only in the weak sense of knowing everything compatible with being perfectly free. If S is essentially omnipotent, then he will have these properties that follow from omnipotence, essentially.

If—as I am still assuming—the omnipotent being is in time, one traditional divine property that is not entailed by omnipotence at any one time, or even by essential omnipotence at any one time, is everlastingness. It follows, as we have seen, that if he exists at one time, he will exist at all future times—given that he would regard it as bad to eliminate himself, or even to abdicate and permit some other being to eliminate him. But essential omnipotence at a time does not entail always having existed. So whether or not he would regard it as bad to abandon his omnipotence in order to get a normal kind of theism, essential everlastingness needs to be added.

The definitions I have offered are suitable only for an omnipotent being whose acts of causing are temporally related to their effects, and so one who exists in time; it cannot apply to a timeless being whose causal action has no temporal relation to its effects. For a timeless being the definition can be simplified to:

> [ζ] S is omnipotent iff S knows all true propositions, and is able to bring about any event at any period of time *T* or any timeless event M, which is metaphysically contingent, the occurrence of which does not require him to be influenced by nonrational influences in order to do so.

Note that, since by "any event" I mean "any hard event," this definition deals in the same way as earlier definitions with the problem that we must not require an omnipotent being to be able to cause any uncaused event (including any free action of some other agent). An "uncaused event at *T*" is not a hard event at *T* or at any other time, since the nonoccurrence of any other time before *T* (and of any timeless state) is metaphysically sufficient for any event at *T* to be uncaused. Omniscience in a strong sense follows from omnipotence in this timeless sense without any temporal restrictions on such knowledge. The other divine properties that follow from [ε] follow also from [ζ]—apart of course from everlastingness.

But if a timeless being exists at all, he exists at every moment (and there is only one moment) of his timeless existence. There are in my view very substantial objections to there existing any timeless being who satisfies [ζ],[11] but I provide this definition as the natural development of the definition of omnipotence I have provided for a being in time.

The other traditional divine property (apart from everlastingness) not entailed by omnipotence, or even by essential omnipotence, is aseity. It does not follow from S being essentially omnipotent and essentially everlasting or timeless that he does not depend for his existence on some other being Q (without Q also depending on S). For Q may be such that it would be a bad act for S to eliminate Q. So the normal theistic view that God does not and could not depend for his existence on some other independent being requires us to add aseity as a separate essential property of God.

It does not, however, in my view, follow that a being essentially omnipotent in any of the senses analyzed is a metaphysically necessary being, that is, that "an essentially omnipotent being exists" is a metaphysically necessary proposition in the strong sense in which "no squares are round" or "water is H_2O" are necessarily true propositions, as many contemporary philosophers of religion think that it is.[12]

NOTES

I am most grateful to Brian Leftow for criticisms of two previous versions of this chapter, to members of an Oxford philosophy of religion group for criticisms of the later of these versions, to the members of the Bochum conference for criticisms of the penultimate version, and to Dean Zimmerman, Tim Pawl, and, especially, Alexander Pruss for criticisms of some crucial ideas contained herein. This chapter is a shorter version of chapter 9 of Swinburne 2016, and some of it is repeated in an article published in *Topoi*, "Causation, Time, and God's Omnipotence." The account of a "hard event" given here is a corrected version of that contained in Swinburne 2014.

1. For a brief history of patristic and medieval accounts of omnipotence, see Leftow 2009.

2. Leftow's (2009) own definition of a being S being omnipotent at a time *t* has a clause "(i) it is not the case that at *t* that there is some event S is unable to bring about at least partially through lack of power." His reason for adding this clause is that otherwise S might be able to cause any event, but some events only with difficulty, and an omnipotent being needs to be able to cause any event without difficulty. But if S cannot bring about M without difficulty, that—I suggest—entails that he

cannot bring it about immediately—and so not at *t*. But if my suggestion is mistaken, the phrase "without difficulty" needs to be added after "is able" in [α] and all subsequent definitions.

3. "McEar" was introduced into modern discussion by Plantinga (1967, 170). Flint and Freddoso (1983, 109n4) refer to prior medieval awareness of a McEar-type objection.

4. Those of Swinburne (1973) and Wierenga (1989, 25). Flint and Freddoso (1983) and Wielenberg (2000) have all been accused—in my view, correctly—of being open to a variant of the McEar objection. For this criticism of myself, see Flint and Freddoso 1983, 112n23; for this criticism of Wierenga, and of Flint and Freddoso, see Wielenberg 2000, 31 and 36; and for this criticism of Wielenberg, see Leftow 2009, 185.

5. For brief argument in defense of these assumptions, see Swinburne 2016, chap. 9.

6. Most philosophers seem to allow the possibility of simultaneous causation. Kant ([1787] 1933, B 248) claimed that "the great majority of efficient causes are simultaneous with their effects." He justified this claim by means of examples of purported simultaneous causation, without giving any argument to show that they were not examples only of (at most) very fast forward causation; his least plausible example was "the stove, as cause, is simultaneous with its effect, the heat of the room." Hume ([1739] 1888, bk. 1, pt. 2, sec. 2) claimed that "the priority of time in the cause before the effect" is a relation "essential to causes and effects," but he admitted that this principle is "not universally acknowledged" because some thinkers allowed the possibility of simultaneous causation.

7. This definition is based on the account of a "hard fact" given in Hasker 1989 (81–90), as (in effect) one that has truth conditions independent of whether there is any future time subsequent to the mentioned time. I have narrowed this by requiring that the truth conditions be also independent of whether there is any past time precedent to the mentioned time.

8. Leftow (2009) sees omnipotence as having the "intrinsic" power to cause events, and distinguishes this from having the "opportunity" to do so. It is only because of a lack of opportunity that, in Leftow's view, an omnipotent being cannot cause a past event. But most theists would only count a being as "omnipotent" if any inability to act due to lack of opportunity was due to a lack of opportunity that it was not metaphysically possible for any being to have. I am seeking a definition that elucidates which deficiencies of opportunity are metaphysically necessary (186–87).

9. For analysis and discussion of moral internalism, see, for example, Smith 1994, 60–91.

10. Oppy (2005, 81–83) presses the point that the Judeo-Christian God is not pictured as maximally powerful but as limited to doing what is good. I emphasize that a being could only be maximally powerful in Oppy's sense if it was subject to irrational influences.

11. See Swinburne 2016, chap. 12.

12. Alvin Plantinga (1974, chap. 10) is among the many recent Anglo-American philosophers who think that there can be a metaphysically necessary being, and that God is such a being. For a defense of my view that it is not metaphysically possible that there should be such a being, see Swinburne 2016, 262–71. Many of those who think that there can be a metaphysically necessary being think that either by his free action or in virtue of his nature one such being, God, can bring it about that propositions have the modal status that they do. See, for example, Morris and Menzel 1986. For a very thorough development of the view that it is because of God that all necessary propositions are necessary, see Leftow 2012. See Swinburne 2016, 262–71, for my reasons for rejecting that view.

REFERENCES

Flint, T. P., and A. J. Freddoso. 1983. "Maximal Power." In *The Existence and Nature of God*, edited by A. J. Freddoso, 81–113. Notre Dame, IN: University of Notre Dame Press.

Hasker, William. 1989. *God, Time, and Knowledge*. Ithaca, NY: Cornell University Press.

Hume, David. (1739) 1888. *A Treatise of Human Nature*. Edited by L. A. Selby-Bigge. Oxford: Clarendon.

Kant, Immanuel. (1787) 1933. *The Critique of Pure Reason*. 2nd ed. Translated by N. Kemp-Smith. London: Macmillan and Co.

Leftow, Brian. 2009. "Omnipotence." In *The Oxford Handbook of Philosophical Theology*, edited by T. P. Flint and M. C. Rea, 167–98. Oxford: Oxford University Press.

———. 2012. *God and Necessity*. Oxford: Oxford University Press.

Morris, T., and C. Menzel. 1986 "Absolute Creation." *American Philosophical Quarterly* 23:353–62.

Oppy, Graham. 2005. "Omnipotence." *Philosophy and Phenomenological Research* 71:58–83.

Plantinga, Alvin. 1967. *God and Other Minds*. Ithaca, NY: Cornell University Press.

———. 1974. *The Nature of Necessity*. Oxford: Oxford University Press.

Smith, Michael. 1994. *The Moral Problem*. Oxford: Blackwell.

Swinburne, Richard. 1973. "Omnipotence." *American Philosophical Quarterly* 10:231–37.

———. 2014. "Time and Causation." *American Philosophical Quarterly* 51:233–45.

———. 2016. *The Coherence of Theism*. 2nd ed. Oxford: Oxford University Press.

Wielenberg, Erik. 2000. "Omnipotence Again." *Faith and Philosophy* 17:26–47.

Wierenga, Edward. 1989. *The Nature of God*. Ithaca, NY: Cornell University Press.

CHAPTER 12

Infinite Power and Finite Powers

KENNETH L. PEARCE

The etymological meaning of "omnipotent" is "all-powerful." Being all-powerful is usually understood as having all of the powers. The problem of analyzing omnipotence is then seen as the problem of determining over what class the quantifier "all" in this assertion ranges. An omnipotent being would be one that has all of the powers—all of *what* powers?

Attempts to answer this question have run into serious difficulties. Responding to these difficulties, Alexander Pruss and I have proposed a way of analyzing omnipotence that makes no reference to the notion of power at all (Pearce and Pruss 2012). However, this raises an obvious worry: if our analysis is not related to the notion of power, then how can it count as an analysis of *omnipotence*, the property of being *all-powerful*, at all?

In this chapter, I propose to answer this question by appeal to a particular (nonmathematical) notion of infinity found in the tradition of philosophical theology. On the view I have in mind, the relation of the infinite to the finite is the relation of the *universal* or *general* to the *particular*. Thus, to be infinitely powerful would be to have power in general, rather than having particular powers. Proponents of this understanding of infinity see the infinite as conceptually simpler than the finite: infinite concepts differ from finite concepts in that the latter have conceptual ingredients the former lack, namely, *limitations*. My aim here is to use this notion of the infinite to understand the conceptual relationship between the Infinite Power of God and the finite powers of creatures.

I begin by reviewing some of the difficulties involved in analyzing omnipotence. After this, I explain in more detail the notion of infinity with which I am working and the understanding of Infinite Power I wish to defend. The bulk of the chapter is then taken up by a discussion of the various ways in which finite powers are limited, as compared to the Infinite Power of God.

My hope is that this investigation will be of use to theologians and philosophers of religion by clarifying some of the puzzles about divine omnipotence and the nature of the infinite gulf between creature and creator. I also hope that it will be of interest to metaphysicians and philosophers of action, even those who are not theists, by showing how an idealized notion of Infinite Power, or of power *simpliciter*, can have considerable theoretical utility, regardless of whether one believes in an infinitely powerful being.

HAVING ALL OF THE POWERS

According to Thomas Aquinas, "All confess that God is omnipotent; but it seems difficult to explain in what His omnipotence precisely consists: for there may be doubt as to the precise meaning of the word 'all' when we say God can do all things" (*ST* I, q. 25, a. 3).[1] Aquinas himself holds that God has the power to perform any consistently describable action. This, however, cannot be correct. *Making a stone one cannot lift*, *coming to know that one has never been omnipotent*, and *failing* are all self-consistent descriptions of actions, yet it is impossible that any of these actions should be performed by an omnipotent being (Pearce and Pruss 2012, 403–4).[2]

Most recent treatments of omnipotence in analytic philosophy have rejected *act theories* like Aquinas's, which say that God has the power to perform any action, in favor of *result theories*, which say that God has the power to bring about any state of affairs (Pearce 2011). Again, the quantifier must be restricted. Thus most recent analyses of omnipotence have the following form:

> x is omnipotent = df. for any state of affairs s, if s satisfies condition C, then x has the power to bring about s.

Different analyses differ in what condition they substitute for C in this schema.

The very first thing to consider in judging the success of a proposed analysis is extensional adequacy: Does it rule in everything that ought to be ruled in, and rule out everything that ought to be ruled out? In the case of omnipotence, this means, at a minimum, that a successful analysis must rule *in* the Anselmian God, and rule *out* any being who is not impressive with respect to power. Richard La Croix has argued that no consistent analysis of this general form can satisfy even this minimal condition (La Croix 1977): some definitions are too strong, requiring, for instance, that an omnipotent being have the power to do evil. This is problematic insofar as it is traditionally thought to be impossible that God should do evil.[3] Others are too weak, and count unimpressive beings as omnipotent.

It is often thought that the problem of the impossibility of God's doing evil can be solved by saying that for God to be omnipotent requires only that God have the power to do anything consistent with God's nature (see, e.g., Wierenga 1983). However, if omnipotence does not require the power to act contrary to one's own nature, then it seems that McEar, whose "nature" will not permit him to do anything other than scratch his ear, is omnipotent provided he can scratch his ear (La Croix 1977, 183, 187).[4] Other weakenings run into similar problems.

There have been a variety of ingenious attempts to escape La Croix's argument. Since my aims here are primarily constructive rather than destructive, I will not pause to criticize these attempts individually; this has already been done by others (Wielenberg 2000; Oppy 2005). Suffice it to say that every attempt so far to analyze omnipotence by saying *which powers an omnipotent being must have* has either required of an omnipotent being that it have the power to do things which it is impossible that God should do, or else has classified unimpressive beings as omnipotent. Furthermore, even supposing this problem could be solved, a theory of this sort would have an important limitation: it would depend on a prior notion of power or ability, a notion whose analysis has proven quite difficult.

OMNIPOTENCE AS INFINITE POWER

In light of these difficulties, I propose that we understand powers in terms of omnipotence, rather than attempt to understand omnipotence in terms of powers. I begin with two ingredients: a notion of divine infinity, which I borrow from Nicolas Malebranche, and an analysis of omnipotence recently proposed by Alexander Pruss and me.

According to Malebranche, God is an "infinite being" in the sense that "God is all being, since He is infinite and comprehends everything; but He is no being in particular" (Malebranche [1674–1675] 1997, 231).[5] The idea of God is, for Malebranche, the idea of infinite being, or being in general, and the ideas of particular beings are got by adding *restrictions* or *determinations* to this idea (232).

I will be defending a thesis about the *power* of God analogous to Malebranche's view about the *being* of God. Thus in claiming that God is all-powerful (omnipotent) I shall mean that God possesses Infinite Power, which "comprehends everything," but God, strictly speaking, has no particular powers.[6] It is misleading (if not outright false) either to affirm or to deny that God has the power to perform a particular action or bring about a particular state of affairs; strictly speaking, God simply has *power*, full stop.

If we wish to understand what finite, particular powers are, I will argue, we should *begin* with this notion of Infinite Power and place various restrictions or limitations on it. These include restrictions of scope and efficacy, among others. It is because these limitations do not apply to God that God does not, strictly speaking, possess any particular powers. On the other hand, it should not be thought that God therefore *lacks* these powers, if by God's lacking these powers one means that God is less powerful than God might otherwise be.

Instead of analyzing omnipotence in terms of our ordinary notion of power, omnipotence should be analyzed as follows (Pearce and Pruss 2012, 405):

(1) x is omnipotent = df. x has perfect freedom of will and x has perfect efficacy of will.

This is to be combined with the following definition of perfect efficacy of will (Pearce and Pruss 2012, 407):

(2) x has perfect efficacy of will = df. $(\forall p)\Box((x \text{ wills that } p) \mathrel{\Box\!\!\rightarrow} (x \text{ intentionally brings about } p))$.

In this definition, "$\Box\!\!\rightarrow$" is a subjunctive conditional, and it is stipulated that, in order for a being to be omnipotent, all substitutions for *p*, including those that result in counterpossible conditionals, must be nontrivially true.[7] Thus, in order for God to be omnipotent, it must be nontrivially true that

(3) (God wills that $2 + 2 = 3$) $\Box\!\!\rightarrow$ (God intentionally brings it about that $2 + 2 = 3$).

It might be thought that this suggestion implies that it is possible that God does the impossible, and is therefore incoherent. However, this is incorrect. On standard logics for subjunctives,[8] $p \Box\!\!\rightarrow q$ can be true even if *q* is impossible, provided that *p* is also impossible. Thus the correct conclusion to draw is that it is impossible that God should will the impossible (Pearce and Pruss 2012, 410).

To complete the analysis, an account of perfect freedom of will should be provided. However, we can make a good deal of progress without completing this difficult task. It is now widely recognized by both libertarians and compatibilists that there are cases in which the impossibility of willing in a certain way, or making a certain choice, does not count against an agent's freedom. Philosophers have arrived at this position by consideration of so-called Luther cases (Frankfurt 1982, 263–66; Williams 1993, 61; Kane 1996, 39–40, 77–78). It is believed that Martin Luther may have identified some genuine limitation on his willing when he said, "I can do no other," and may nonetheless have been free and responsible in his refusal to recant because of the way in which this limitation on his willing arose from his character and/or choices. The conclusion, then, is that a being can be omnipotent only if the laws of logic are deeply embedded in that being's character, or perhaps freely endorsed by that being,[9] in such a way that the impossibility of willing illogically is not prejudicial to that being's perfect freedom. A similar line of thought can be used in defense of God's necessary moral perfection (Pearce and Pruss 2012, 411–12).

Now that we have our two basic ingredients—the Malebranchean notion of infinity, and the Pearce–Pruss theory of omnipotence—I shall proceed to make some observations about how these two ingredients fit together.

First, I am proposing that the notion of Infinite Power be taken as prior to the notions of particular, finite powers. Thus by making the analysis of omnipotence do the work I want it to do here, I am taking a controversial stand regarding the analysis of free will: since particular finite powers are to be understood in terms of Infinite Power, and Infinite Power is in turn analyzed in terms of perfect freedom of will, I am committed to the claim that perfect freedom of will is *not* to be analyzed in terms of either Infinite Power or finite powers. In particular, this means that it is no part of the *analysis* of perfect freedom of will that a perfectly free being should have the power to will or do otherwise. This might be thought to be a significant cost of my view. However, I think the following considerations show that the price is not so high as it might at first appear. First, this approach does not preclude freedom (either perfect or imperfect) from *entailing* the power to will or do otherwise. Indeed, I do not even need to challenge the claim that one has free will *if and only if* one has the ability to will or do otherwise. Furthermore, my approach is consistent with analyzing freedom of *action* in terms of the power or ability to do otherwise. Finally, this approach does not preclude a completed analysis of perfect freedom from somehow involving the *possibility* of willing or doing otherwise, although we should note that there are, on this account, some actions which it is impossible that God should will or do; thus perfect freedom must not require the possibility of willing or doing just anything.

Understood thus weakly, I find this restriction unproblematic. Power is about reality's conforming to an agent's choices. Freedom of will is about how those choices come about. Thus it is perfectly natural that freedom of will should be conceptually prior to power: genuine, robust power involves being free to choose how reality will be.

I shall not provide here an analysis of freedom of will, either perfect or imperfect. I shall work instead with our intuitions about what sorts of things *impair* freedom.[10] We will find the following intuitive gloss helpful: when I choose (will) freely, I *own* the choice (volition) and any action that follows from it. That is, it is genuinely *my* choice and *my* action. This sort of ownership clearly comes in degrees. This is recognized, for instance, in discussions of mitigating factors in law. Suppose Smith kills Jones in a fit of rage. The choice to murder Jones is Smith's choice, and Smith is the one who commits the murder, so in this sense Smith owns both the choice and the action. However, at least in the case where Smith's possession of a violent temper is due to his genetics or environment rather than his choices, we can say that Smith would have owned the choice and the action *to a greater degree* if his decision to kill Jones had been preceded by calm deliberation. Because Smith still owns the choice and the action, he is still a murderer and is to be punished, but because he owns it to a lesser degree, he receives a lesser punishment. In saying that God has perfect freedom of will, I mean that God always owns God's choices and actions in the strongest possible sense.

A second thing to note about my approach is that rather than answering the question posed by the Stone Paradox, this approach rejects that question as ill posed. The Stone Paradox asks whether an omnipotent being would have a particular power: the power to create a stone one lacks the power to lift. This question, however, hinges on the way of thinking about omnipotence I rejected above: it is a question about which of the many finite powers God has. As a result, any straightforward answer to it is bound to mislead.[11] Avoiding use of "can," "ability," "power," and related vocabulary, let p_1 be the proposition that there is a stone such that it is impossible that God should lift that stone. To say that God could not create the stone would suggest that God would fail to create such a stone if God tried, and this is false, for if God willed p_1 God would intentionally bring about p_1. On the other hand, to say that God could create the stone would suggest that p_1 is possible. However, since God is essentially omnipotent, p_1 is not possible. Similar remarks apply to tasks that are impossible *simpliciter*, or impossible for an omnipotent being, or impossible for God due, for example, to God's necessary moral perfection. Thus it is better to refrain from attributing particular powers to God at all.

ABSOLUTE POWER OVER A PROPOSITION

My thesis is that omnipotence (Infinite Power) is a relatively simple, unified property, whereas particular finite powers are complex properties that can be analyzed by specifying the ways in which they fall short of omnipotence. This approach is offered as an alternative to the more conventional approach, which analyzes omnipotence as "having all of the powers." However, it might be thought that, even on the analysis of omnipotence

I favor, omnipotence really *does* amount to "having all of the powers" of a certain sort, namely, *having power over every proposition.*

For someone to have power over a proposition is for it to be *up to* that person whether that proposition is true.[12] Now, we humans sometimes have powers of this sort. For instance, it is up to me whether I eat breakfast on a particular day. But these powers are not *absolute*. The fact that it is possible that I should be hindered in the exercise of my choice does not mean that it is not up to me, but it does mean that it is not *absolutely* up to me. If someone had *absolute* power over a proposition, that person would not require any external cooperation in order effectively to choose whether that proposition would be true.

The suggestion, then, is that God's omnipotence consists in God's having absolute power over every proposition, that is, in its being absolutely up to God whether any proposition is true. We might define "absolute power over a proposition" as follows:

> (4) x has absolute power over p = df. x is perfectly free with respect to willing p and with respect to willing $\neg p$ and x has perfect efficacy of will with respect to p and with respect to $\neg p$

It would then be claimed that omnipotence, on the analysis given, is just the possession of all of these powers.

The claim I am defending is that one needs to *start* with omnipotence and work from there to an account of finite powers by applying various restrictions or limitations. This commits me to the denial of the claim that omnipotence should be *analyzed* as having all of the powers, but it does not commit me to the claim that omnipotence does not *entail* having all of the powers. I have denied that our ordinary notions of powers are strictly applicable to God, but (as we will see below) the notion we are dealing with here is far removed from the notions we use in attributing powers to humans. Thus I need not deny that an omnipotent being has absolute power over every proposition, in the sense defined. Instead, I need only argue that it is more fruitful to begin with omnipotence, as the simpler concept, and regard the possession of absolute power over particular propositions as the possession of, as it were, *pieces* of omnipotence.

This is the case with both the freedom condition and the efficacy condition. In both instances, the infinite case is the simpler one, and the difficulty involved in understanding or analyzing the finite case is precisely the difficulty of applying bounds or limits to the infinite concept.

Take freedom first. It is widely agreed that there are such things as psychological impairments to freedom. These include psychological disorders, such as phobias, compulsions, or delusions, and more ordinary forms of impairment, such as *akrasia*, emotional disturbances,[13] being under the influence of a drug, being unaware of (or having false beliefs about) the consequences of one's actions, or being unaware of (or having false beliefs about) the alternatives available to one. Libertarians hold that in addition to these psychological impairments, there can be *metaphysical* impairments to freedom. That is, according to libertarianism, even if none of these psychological factors is present, if the agent's choice is deterministically caused by something outside the agent, then the agent is not free in that choice. Indeed, the libertarian may wish to claim that the psychological impairments are impairments of freedom precisely because they stem from external causes. Given that psychological impairment comes in degrees, the libertarian might also want to admit degrees of metaphysical impairment of freedom. These might take the form of problematic external *influences* that do not actually *necessitate* the particular action in question.

Some degree of impairment is certainly compatible with the sort of freedom we ordinarily attribute to ourselves and other human beings, and the sort of freedom necessary for moral responsibility. However, *no* degree of impairment is consistent with perfect freedom. A perfectly free being would know all of the alternatives available to it, and the outcomes of any of these actions.[14] Furthermore, a perfectly free being would suffer no emotional disturbances,[15] and could not be phobic, or even akratic, *in any degree*. If libertarianism is true, then a perfectly free being must also be "a prime mover unmoved" (cf. Chisholm 1966, 23), that is, none of its choices or actions can have a source outside itself.

The absence of impairments is clearly only a necessary, and not a sufficient, condition for perfect freedom. Before one's freedom can be impaired, one must *have* freedom, and I have not said what this involves. However, from what we have seen so far, it should be apparent that perfect freedom involves a holistic property of an agent's psychology: nothing can interfere with the perfectly free agent's deliberation. This is among the reasons that a perfectly free agent (such as God) owns its choices and actions in the strongest possible sense: these choices are made without any interference, whether from inside or outside the agent.

If definition (4) is the correct analysis of absolute power over a proposition, and omnipotence is absolute power over every proposition, then the perfect freedom of will required for omnipotence must be freedom with respect to willing any proposition. Adopting this approach would commit us to the claim that God is free with respect to willing propositions that it is impossible that God should will. It is far from clear that this is the best approach; it might be better to say that one can be perfectly free without being free with respect to willing, for example, that $2 + 2 = 3$. Let us, however, waive this objection and instead focus on trying to understand what it would mean for a being who was not perfectly free to be perfectly free with respect to a particular choice.

It seems that this would mean, at least, that the agent experienced no impairments *in the course of making that choice*. But what does this amount to? When one makes a choice by rational deliberation, one employs all sorts of relevant beliefs, some of which have usually been arrived at by deliberation. In order to meet the freedom condition of (4), the agent needs to *know* what alternatives she has. These are the following: willing p, willing $\neg p$, and willing neither. She also needs knowledge of the outcomes of any of these choices, which will include all the contingent propositions entailed by p and all the contingent propositions entailed by $\neg p$. It will also include the causal consequences of p and of $\neg p$. Many of the things that impair freedom can also impair theoretical deliberation in such a way as to prevent knowledge, so we will need to assume that the beliefs at issue here were formed in a way that did not involve, for instance, irrational responses to emotion.

This, however, is still not enough. Suppose that the agent knows that q will be a consequence of her willing p, and she has previously decided to attempt to bring about q. This will quite properly figure into her deliberations about whether to will p. However, if she was not perfectly free in deciding to attempt to bring about q, then it appears that the influence of this prior decision on her deliberation about whether to will p will render her less than perfectly free. For an extreme instance, imagine that the agent has an overwhelming psychological compulsion to bring about q by any means necessary. Surely the fact that she is deliberating quite properly in noticing that p is a means to one of her ends, and considering this as a reason in favor of bringing about *p*, does not guarantee that her choice to bring about *p* is perfectly free if the end in question was chosen by a radically unfree process.

Being perfectly free with respect to a particular choice is thus a far more complicated matter than being perfectly free *simpliciter*. In order to understand what it is for an agent to be perfectly free with respect to a proposition, we have to manage somehow to cordon off from the rest of her deliberations that part of her deliberation that is to be utterly unimpaired. However, because of the close relation between one course of deliberation and another, it is not clear that this is even possible. If it is possible, it is surely far more complicated than perfect freedom *simpliciter*.

We turn now to efficacy. It is possible to possess perfect efficacy over only one individual proposition or some narrow class of propositions. To see this, suppose the proposition that *x* wills *p* is logically independent of the proposition that *x* wills *q*, and that it is possible that *x* should will either of these things. Then there is a possible world, *w*, at which *x* does not will *p*, but *x* does will *q*. The claim that *x* has perfect efficacy with respect to *p* has no consequences whatsoever for what happens at *w*, but the claim that *x* has perfect efficacy with respect to *q* entails that, at *w*, *x* intentionally brings about *q*. Therefore, the claim that *x* has perfect efficacy with respect to *q* entails something not entailed by the claim that *x* has perfect efficacy with respect to *p*, and so (by the transitivity of entailment) the claim that *x* has perfect efficacy with respect to *p* cannot entail that *x* has perfect efficacy with respect to *q*.[16] Thus, quite generally, unless there is a logical connection between the agent's *willing* one proposition and the agent's *willing* another, there will be no logical connection between the agent's having perfect efficacy with respect to one proposition and perfect efficacy with respect to another. Now, although there may be possible agents, or kinds of agents, who have these sorts of logical connections between their willings, it does not seem that there are any propositions such that, necessarily, *any agent at all* who wills one must will the other, nor do there seem to be any propositions such that, necessarily, no agent, no matter how confused or irrational, can will both of them. An agent who wills two contradictory propositions must be to some degree confused and/or irrational, and if the contradiction is *obvious*, then the agent must be *very* confused and/or irrational, but this is perfectly possible.

Note, however, that the argument here for the independence of efficacy with respect to one proposition and efficacy with respect to another depends on the possibility of the agent's willing irrationally. As a result, this does not tend to show that absolute power over one proposition is independent of absolute power over another, since absolute power requires perfect freedom, and hence rationality. Thus, for instance, since being perfectly free with respect to willing a proposition requires being able to deliberate properly about it, and knowing all of its consequences, having absolute power over a proposition will entail having absolute power over every proposition logically equivalent to it.

Things will become rather more complicated if we consider propositions that it is impossible that the agent will. In the absence of an adequate logic for counterpossibles, we cannot come to any firm conclusions on whether there might be entailment relations between perfect efficacy with respect to one unwillable proposition and another.

Although it is not difficult to see what it would amount to for an agent to have perfect efficacy with respect to some propositions and not others, there is nonetheless something about this situation that cries out for explanation in a way that the notion of an omnipotent being does not. The world, in its entirety, necessarily conforms to the will of an omnipotent being, but the being we are imagining is one to whose will the world necessarily conforms *in part*, but not in its entirety. What accounts for this distinction, between those aspects of the world that necessarily conform and those that do not? If nothing accounts for this distinction, then we have a puzzling brute necessity on our hands. If something does account for this distinction, then we are likely to run into demarcation problems. Hence it seems that, with respect to the efficacy condition also, omnipotence is the simpler concept.

If we want to say that a particular agent has absolute power over certain propositions, the thing to do is to start with the concept of omnipotence, and then *limit* that concept to certain propositions. This limiting is a messy business, and it gives rise to the messier, more difficult concept of a finite, particular power. However, as we shall see, despite the messiness involved here, this is the simplest sort of finite, particular power. The powers possessed by more limited agents, such as ourselves, are much more complicated than this.

ABSOLUTE POWER TO PERFORM AN ACTION

So far we have been speaking of power over propositions, that is, of its being up to one whether a proposition is true. It seems likely that God's power is, fundamentally, of this sort: God has merely to utter (will) "Let there be light," and light is. Human powers are, however, not like this. Although it is often up to us whether a particular proposition is true, we never have this kind of direct, or fundamental, control over the truth values of propositions.[17] Rather, we have powers to perform actions. In particular, our most fundamental powers are powers to perform certain *basic actions*. Whatever other powers we may have can be exercised only by exercising these basic powers (Danto 1963, 1965).

Power to perform actions cannot be straightforwardly reduced to power over propositions. My power to raise my arm is not power over the proposition that my arm goes up, or even power over the proposition that I raise my arm. This can be seen from the fact that it is possible for someone to have power over the proposition that my arm goes up without that person having the power to raise my arm. For instance, someone might have planted a posthypnotic suggestion whereby I raise my arm every time I hear a certain word. Any person who knows the word has power over the proposition that my arm goes up, and indeed over the proposition that I raise my arm, but only I have the power to raise my arm. This is because the person who says the word is not the one who raises my arm: *I* am the one who raises my arm, although I do so unfreely.

What this suggests is that, to get closer to the human case, we need to replace our original efficacy condition with a new one:

(5) x has perfect efficacy with respect to action A = df. $\Box((x \text{ wills to do } A) \mathrel{\Box\!\!\rightarrow} (x \text{ intentionally does } A))$.

The freedom condition does not need any revision to handle this case, even though it inherits all of the complexities and difficulties about having perfect freedom with respect to some things and not others. Thus:

(6) x has absolute power to perform A = df. x has perfect efficacy with respect to A and x has perfect freedom with respect to A.

Again, the fact that the power is supposed to be *absolute*, that it requires no cooperation from the external world, greatly simplifies matters. The fact that we require this sort of cooperation from the external world in the exercise of our powers is one of the reasons for the failure of simple conditional analyses of the ordinary attribution of powers (abilities) to humans.

Note that, even if performing one action necessarily involves performing another, it does not follow that having perfect efficacy with respect to the first implies having perfect efficacy with respect to the second. To claim that one has perfect efficacy with respect to an action is to claim that reality would respond in a certain way to one's willing a certain action. Nothing follows from this about what would happen if one willed a different action. In other words, it may be that if you willed *A* you would do *A* and this would necessarily *involve* doing *B* but, nevertheless, nothing would happen if you willed *B* alone.

Again, however, as in the case of absolute power over propositions, once we put efficacy and freedom together, it does seem that one power may entail another. Recall that the perfect freedom condition requires that one have full knowledge of the consequences of one's action, and that nothing disrupts one's deliberation with respect to that action. Now this consequence relation is transitive, so if one has full knowledge of the consequences of an action, then one has full knowledge of the consequences of its consequences, and so on. Furthermore, since deliberating with perfect freedom about an action involves deliberating about its consequences, it seems that the latter deliberation must be perfectly free if the former is to be. So it would seem that, although perfect efficacy with respect to an action does not, on its own, imply perfect efficacy with respect to other actions one could perform *by* performing that action, absolute power to perform an action does imply absolute power to perform any action that can be performed *by* performing the first one.

A concrete example will make this clearer. Suppose Alice has absolute power to remove Bob's tumor, and, necessarily, if she *does* remove Bob's tumor, she will thereby save Bob's life. Now, Alice's absolute power to remove Bob's tumor implies knowledge of the consequences of this action, one of which is that Bob's life will be saved. Alice's deliberation about whether to remove the tumor will involve deliberation about whether the consequences of removal are desirable. Hence part of her deliberation about whether to remove the tumor will be deliberation about whether to save Bob's life, and the latter deliberation must be perfectly free if the former is to be. Furthermore (we are imagining), if Alice removes the tumor she will necessarily thereby save Bob's life, thus she cannot fail to save Bob's life if she tries to. So it seems that Alice has both perfect freedom and perfect efficacy with respect to saving Bob's life.

This argument has a hidden assumption, namely, that removing the tumor is the only means available to Alice whereby Bob's life might be saved. If Alice is to be perfectly free in her decision to remove the tumor, it seems that she must know about any other possible means to the end of saving Bob's life, and their chances of success. However, if there are, in addition to the *certain* means of saving Bob's life, also *chancy* means available to Alice, then there will be possible worlds in which she wills to save Bob's life but fails to do so, namely, worlds in which (for some reason) she chooses the chancy means instead of the certain means. We can say, at least, then, that if the *only* way for an agent to perform some action *A* is by performing some action *B*, then absolute power to perform *B* entails absolute power to perform *A*, and hence that powers of this sort are not neatly separable from one another.

RESISTIBLE POWER

So far we have been speaking of absolute powers, but, as I have mentioned several times, none of *our* powers are absolute, and it is this feature of our powers that adds the greatest difficulty to their analysis. The first way in which our powers are nonabsolute is that they are *resistible*. For instance, I have the power to open doors, but if someone stronger than I is holding the door from the other side, my attempt to exercise this power will fail. This does not speak against my having the power to open doors, but we might be inclined to say that I lack the power to open *this* door, under *these* circumstances (see Honoré 1964; Vihvelin 2013, 11–13).

This limitation stems from the finitude, or particularity, of our powers in a direct way: the finite powers are distributed among us in such a way that in exercising our powers we sometimes come into conflict. Furthermore, we are limited not only in that our power can be overcome by resistance from other agents, but also in that our exercise of power may fail simply due to an uncooperative world. This may be the case with Austin's golfer (Austin 1961, 166n1), who *can* make a putt from this distance, but nevertheless fails to do so when he tries. Austin says that the golfer could have made the putt given "conditions as they precisely were," but presumably he means precisely as they were *immediately before the putt*. Suppose, then, that a sudden gust of wind, immediately after the ball is struck, causes the ball to miss the hole. The golfer had the power to make the putt, but the fact that he isn't the only causal factor present means that his power may be *resisted* by an uncooperative world.

What this suggests is that we need to add a *ceteris paribus* clause to our efficacy condition:

> (7) x has (at least) *resistible efficacy* with respect to A = df. $\Box$((x wills to A and no unusual interfering conditions occur and no unusual helping conditions occur) $\Box\!\!\rightarrow$ (x intentionally does A)).

The *ceteris paribus* clause needs to exclude unusual helping conditions along with unusual interfering conditions to deal with cases like the following.[18] Suppose that, because of local weather conditions and geography, it is true that *if the wind hadn't blown the ball away from the hole, it would have blown the ball into the hole*. Then it will be true that if the golfer had willed to make the putt and the interfering condition had been absent, he would have made the putt. However, if it would, in this case, have been only due to assistance from the wind that the putt went in, then the golfer will not count as having the power to make the putt.

This formulation inherits all of the well-known difficulties with *ceteris paribus* clauses and their inherent vagueness (Cartwright 1980; Hempel 1988; Pruss 2006, 107–12). Thus when dealing with resistible powers we must, in addition to specifying their *scope*, specify their *failure conditions*.[19] This is not to say that we need to know exactly what the failure conditions are in order to ascribe resistible power to an agent, any more than we need to know exactly what conditions must be satisfied to avoid exceptions to a *ceteris paribus* law in science before we can know it to be a law. But, whether the details are known to us or not, these failure conditions exist "in the background" of all of our power ascriptions to human beings. This is one of the reasons for the great difficulty in analyzing such power ascriptions, and the ready availability of counterexamples to proposed analyses: we do not know, with sufficient precision, exactly what powers the people in question have, for we are not in a position to specify the failure conditions. This is also a reason why our ordinary talk about powers is not suitable for talk about God: the powers that you and I have, which we ordinarily designate by phrases like "power (ability) to open doors," are all powers with built-in failure conditions.

It is unclear whether powers to perform basic actions can be resistible. According to some philosophers (e.g., McCann 1974), the only basic actions are acts of the will (volitions). It is not even clear whether it is coherent to speak of efficacy here, since this would presumably involve talk of willing to will (cf. Locke [1690] 1975, §2.21.25). Indeed, if there is such a thing as power to will, it seems that "power" must be used here equivocally (cf. Reid [1788] 2010, 201). Thus it may well be that only the freedom condition is relevant to basic actions. If, on the other hand, there are no such things as acts of will (or if acts of will are not genuine actions),[20] then actions of more familiar sorts, such as bodily movements, may be taken as basic (Danto 1963, 1965). In this case, it seems that our basic actions *are* resistible, both by unexpected external impediments, and by nervous system malfunctions.

Resistibility is a limitation on one's efficacy. It does not affect freedom of will (though of course it diminishes freedom of *action*). Thus it is possible to have perfect freedom of will with respect to an action even though one's efficacy with respect to that action is limited. An agent like this would have to have perfect knowledge of the failure conditions for exercises of her power in order to deliberate in full freedom about whether to try to exercise it.

MANIPULABLE POWER

In addition to *resisting* our power, foreign agents may *manipulate* our exercise of power. This comes in more and less subtle forms. Thus, for instance, one advises regarding a mischievous child, "Don't give him any ideas." One can prevent the child from acting in a certain way just by preventing it from occurring to him to act that way. Or consider Gary

Watson's case of "Cozy Clara" (Watson 2004, 90–93). Making Clara *very* comfortable in her bed, Watson suggests, may be just as effective a means of preventing her from getting to work as chaining her to the bed-post. Humans can also be manipulated by providing them with false, or misleadingly incomplete, information. Finally, there are extremely un-subtle forms of manipulation, such as drugs, hypnosis, physical manipulation of the brain, threats, or torture. All of these are ways not of preventing people from carrying out their will, but of preventing them from willing in the first place.

As Cozy Clara makes clear, not every case of manipulation eliminates ordinary (imperfect) freedom or responsibility. Even if a malicious agent is trying to prevent Clara from getting to work, if all he does is make her comfortable, or supply her with a DVD collection of her favorite television series, she is culpable for failing to come to work. Nevertheless, the fact that a foreign agent can manipulate her in this way shows that she is not *perfectly* free.[21]

Like resistibility, manipulability is not merely a matter of susceptibility to the actions of other agents. Sometimes we are "manipulated," not intentionally by others, but by our environment. In some cases, we even seem to be "manipulated," so to speak, by our own genetics. In fact, there are forms of "manipulation" that seem to be universal among humans, such as the disruptive effects that our bodily appetites can have on our deliberations, which prevent us from deliberating with perfect freedom. Similarly, our incomplete knowledge can be just as great an impairment to our freedom of will as if some malicious agent had intentionally provided us with misinformation.

What this shows is that we humans always operate with impaired freedom of will. Nevertheless, we succeed in drawing a contrast between free and unfree choices. What has not always been adequately recognized is that this contrast is something that comes in degrees. Our choices can be more or less free, more or less our own.

What degree of freedom, in one's deliberations about an action, is necessary for one to have the power to perform that action? How strong must a compulsion or phobia be before it takes away that power? These questions have not been adequately answered, and they may have no answers, for it may be that the very notion of finite power is vague. Perhaps the best we can do is this:

(8) x has the power to A = df. x has (at least) resistible efficacy with respect to A and x is (at least) somewhat free with respect to A.

CONTINGENT POWER

An especially tricky problem for conditional analyses of dispositions and abilities is that it is possible to gain and lose dispositions and abilities (Martin 1994). This wreaks havoc with simple conditional analyses because it raises the possibility that an ability or disposition may be "finkish" (Lewis 1997). Suppose, for instance, that God has resolved that if I should ever attempt to read a Sanskrit text, God would perform a miracle whereby I would instantaneously learn Sanskrit. God's resolution is sufficient to make the conditional—if I tried to read Sanskrit, I would succeed—true, but since God has not performed any such miracle (and I have not learned Sanskrit the old fashioned way), I do not have the power to read Sanskrit. One can also construct cases going in the other direction, where one *has* a power, but if one chose to exercise it one would lose it.

We have been implicitly assuming that the agents we are discussing have their powers *essentially*. Once this assumption is dropped, the problem of these "finkish" powers arises. However, we can easily adapt the solution suggested by David Lewis and others (with respect to dispositions), that of positing *a categorical basis* for the power in question (see Vihvelin 2004; Vihvelin 2013, §6.3). That is, we suppose that there is some intrinsic, categorical (i.e., nondispositional) property possessed by the agent that is a nomologically sufficient condition for the agent to have (resistible) efficacy with respect to the action in question. Only the efficacy condition involves a conditional, so it is the one we need to worry about. This gives us the following:

(9) x has (at least) *contingent, resistible efficacy* with respect to A = df. $(\exists B)$(x has B and $\Box_n$ ((x wills to A and x retains B and no unusual interfering conditions occur and no unusual helping conditions occur) $\Box\!\!\rightarrow$ (x intentionally does A))).

Another issue that arises at this point is the interpretation of the necessity operator. So far, we have been supposing that the connection between the agent's will and the effect is a matter of logical or metaphysical necessity (this, at least, would seem to be the case with God). Here, however, the connection is a matter only of nomological necessity. I indicate this with the subscript "n." It seems that we are prepared to ascribe power (of some kind, in some degree) to an agent, provided that the agent has contingent, resistible efficacy, and is somewhat free. However, the more easily the agent could lose the categorical basis *B*, and thereby lose the power, the less inclined we are to ascribe the power to the agent in the first place. This is especially so if the agent's tendency to lose *B* is somehow connected to the agent's willing the action. The stability of the categorical basis of the agent's efficacy, and the question of what the categorical basis *is*, is another source of complexity and vagueness in the notions of finite powers.

CHANCY POWER

One final complexity remains. I assumed earlier that Austin's golfer missed the putt because of a gust of wind, that is, an unusual interfering condition. But it seems there might also be powers that are intrinsically chancy, quite independently of cooperation from the environment. In fact, all of our physical powers are probably like this, for there is a tiny, but non-zero, probability that a quantum fluctuation could prevent the signal from my brain from reaching the muscle that needs to contract for the action to be performed. In this case, different environmental conditions might not increase my success rate, and certainly there are no conditions so ideal as to bring my probability of success to one. Thus:

> **(10)** *x* has (at least) *chancy, contingent, resistible efficacy* with respect to *A* = df. (∃*B*)(*x* has *B* and $\Box_n$((*x* wills to *A* and *x* retains *B* and no unusual interfering conditions occur and no unusual helping conditions occur) □→ (probably, *x* intentionally does *A*))).

The "probably" in the consequent refers to the objective probability immediately after the agent's willing (see Lewis 1987, 176–77). This of course introduces a new dimension of vagueness. How probable is success? How probable does it have to be before we are prepared to ascribe the power to the agent? This is another factor that must be described if we are to identify a particular, finite power fully.

CONCLUSION

God possesses Infinite Power. That is, God possesses general or universal power, of which finite powers are *particularizations*, or *determinations*. Our finite powers can be understood by adding a series of limitations to God's Infinite Power. Each limitation introduces complexity and vagueness. In particular, if we wish to identify a finite, particular power fully, we must answer the following questions:

1. What is the *scope* of the power?
2. What *unusual interfering conditions* would prevent the successful exercise of the power?
3. What *impairments of freedom* might prevent the agent from attempting to exercise the power?
4. What is the *categorical basis* of the agent's efficacy within the scope of the power?
5. How easily could the categorical basis be lost? Are the conditions for the loss of the categorical basis connected with the conditions in which the agent would try to exercise the power?
6. If no unusual interfering conditions occur, and the categorical basis is retained, how likely is it that the agent's attempt to exercise the power *might fail anyway*?

The narrower the scope, the more likely the interfering conditions, the more impaired the agent's freedom, the less stable the categorical basis, and the more likelihood of failure under ideal conditions there is, the farther removed the particular finite power is from the Infinite Power of God. As we get farther and farther removed, we become more and more hesitant to ascribe the power to the agent at all.

Plain English locutions like "the power to lift stones" carry with them implicit answers to the above questions. Because these answers are vague, no precise answer can be given to the question of what it is to possess such a power. It is because "power to lift stones" implies some (vague) degree of limitation that this phrase is not (univocally) applicable to God. God does not possess such particular powers; rather, God possesses that ideal of power to which every particular power is a mere, limited approximation.

NOTES

I thank Kenneth Silver, Chad Marxen, Alexander Pruss, Lindsay Cleveland, Kadri Vihvelin, and Gary Watson for helpful comments on previous drafts. This chapter also benefited significantly from discussions with participants at the Bochum infinity conference, especially Brian Leftow.

1. Translations are from the text of the *Summa* online: http://newadvent.org/summa/.

2. These sorts of problems were not unknown to the medieval philosopher-theologians, and, of course, the medievals were deeply concerned with the problem of God's inability to sin (see Aquinas, *ST* I, q. 25, a. 3, obj. 2). However, the solutions proposed by them are now widely regarded as unsuccessful. See Ross 1969, 196–210.

3. It is sometimes thought that God might have the power to do evil even if it is impossible that God should do evil (Talbott 1988; Wielenberg 2000). In Pearce 2012, I defend the attribution of this view to Thomas Reid. However, divorcing power from possibility in this way renders the already puzzling notion of power all the more obscure. See Hasker 1989, 89–90n18; Rowe 2004, 147–49. On my view, necessarily unexercised powers are possible if, but only if, it is possible to be free to choose an alternative which it is impossible that one should choose. See section below titled "Absolute Power over a Proposition."

4. Wierenga objects, with considerable plausibility, that McEar is impossible (Wierenga 1983, 374–75). However, more plausible examples can easily be constructed (Wielenberg 2000, 29–31; Pearce and Pruss 2012, 405–6).

5. For the historical antecedents of this view, see the essays by Krainer (chapter 2) and Carroll (chapter 4) in this volume.

6. The terminology here is being drawn from the contrast, in metaphysics, between universals and particulars. It is unrelated to the distinction between general and particular powers introduced by Honoré (1964).

7. On the nontrivial truth of counterpossibles, see Pearce and Pruss 2012, 406–8; Pearce 2016; Pearce 2017.

8. Introducing nontrivial counterpossibles involves rejecting the standard (Lewis–Stalnaker) *semantics* for subjunctives, but this does not affect this logical point.

9. Since perfect efficacy of will can only be an essential property (Pearce and Pruss 2012, 405), it must be *de re* impossible, of any being with perfect efficacy of will, that that being should will contrary of the laws of logic. As a result, anyone who wishes to say that the impossibility of God's willing contrary to the laws of logic is due to God's freely endorsing those laws must hold that freedom is compatible with necessity, at least in certain cases. See Rowe 2004, 70–71, 142–50.

10. On impaired freedom, see Watson 2004, but note that my notion of unimpaired freedom is far more demanding than Watson's, since Watson is drawing a contrast among human choices, while I am drawing a contrast between human and divine choices.

11. Aquinas may be expressing a similar idea when he says, of contradictory tasks, that "it is better to say that such things cannot be done, than that God cannot do them" (Aquinas, *ST* I, q. 25, a. 3): despite the fact that it is impossible that God should draw a round square, there is something misleading about saying "God cannot draw a round square."

12. Van Inwagen (1983) makes extensive use of the notion of its being "up to" a person whether a proposition is true. "Relative possibility" analyses of "can," such as those found in Lewis 1976 and Kratzer 1977, are also naturally interpreted by means of this idiom.

13. By an emotional *disturbance* I mean an instance in which emotions disrupt proper deliberation. I do not mean to claim that all emotion is inimical to proper deliberation.

14. This appears to imply that a perfectly free being would have comprehensive counterfactual knowledge, including knowledge of counterfactuals of creaturely freedom, but this is not the place to wade into that thorny issue.

15. But this does not necessarily mean that it would have no emotions. See note 13, above.

16. In this argument, I rely on a notion of entailment on which necessary truths (and falsehoods) may differ in what they entail. Giving an account of such a notion is a difficult task, but I think we have an intuitive grasp of it. Thus, for instance, it seems intuitively correct to say, "That water is H_2O has no consequences for (entails nothing about) mathematics."

17. As Gary Watson pointed out to me, there may be a few counterexamples to this claim, such as my power over the proposition *that I am willing something*. These cases can safely be set to one side.

18. I thank Alexander Pruss for this point. The example is my own.

19. The clause about "helping conditions" does not require additional specification; it should be seen, instead, as saying that the failure conditions must not include those conditions we consider "normal."

20. Although I have been assuming that it makes sense to talk of "willing *p*" and "willing to *A*," this need not involve the assumption that "willing to *A*" involves a mental action of a special sort.

21. Perhaps, as Strawson (1994) argues, we sometimes confusedly employ an extremely robust notion of responsibility that does require perfect freedom. My view, according to which our notions of freedom and power are notions of approximation to an ideal, and hence admit of degrees and of vagueness, neatly explains how this mistake might be made: we simply slide, unnoticed, up the scale. However, it seems clear that the rejection of this confusion should have little influence on our ordinary practice of holding people accountable. Whether (as Strawson suggests) the rejection of this confusion causes problems for belief in eternal reward and punishment in an afterlife is a more difficult question, and cannot be addressed here.

REFERENCES

Aquinas, Thomas. 1920. *The Summa Theologica of St. Thomas Aquinas*. Translated by Fathers of the English Dominican Province. London: Burns Oates & Washbourne.

Austin, John L. 1961. "Ifs and Cans." In *Philosophical Papers*, edited by James O. Urmson and Geoffrey J. Warnock, 153–80. Oxford: Clarendon.

Cartwright, Nancy. 1980. "The Truth Doesn't Explain Much." *American Philosophical Quarterly* 17 (2): 159–63.

Chisholm, Roderick M. 1966. "Freedom and Action." In *Freedom and Determinism*, edited by Keith Lehrer, 11–44. New York: Random House.

Danto, Arthur C. 1963. "What We Can Do." *Journal of Philosophy* 60 (15): 435–45.

———. 1965. "Basic Actions." *American Philosophical Quarterly* 2 (2): 141–48.

Frankfurt, Harry. 1982. "The Importance of What We Care About." *Synthese* 53 (2): 257–72.

Hasker, William. 1989. *God, Time, and Knowledge*. Ithaca, NY: Cornell University Press.

Hempel, Carl G. 1988. "Provisoes: A Problem Concerning the Inferential Function of Scientific Theories." *Erkenntnis* 28 (2): 147–64.

Honoré, Anthony M. 1964. "Can and Can't." *Mind* 73 (292): 463–79.

Kane, Robert H. 1996. *The Significance of Free Will*. Oxford: Oxford University Press.

Kratzer, Angelika. 1977. "What 'Must' and 'Can' Must and Can Mean." *Linguistics and Philosophy* 1 (3): 337–55.

La Croix, Richard R. 1977. "The Impossibility of Defining 'Omnipotence.'" *Philosophical Studies* 32 (2): 181–90.

Lewis, David. 1976. "The Paradoxes of Time Travel." *American Philosophical Quarterly* 13 (2): 145–52.

———. 1987. "Postscripts to 'Causation.'" In *Philosophical Papers*, 2:172–214. Oxford: Oxford University Press.

———. 1997. "Finkish Dispositions." *Philosophical Quarterly* 47 (187): 143–58.

Locke, John. (1690) 1975. *An Essay Concerning Human Understanding*. Edited by Peter H. Nidditch. Oxford: Oxford University Press.

Malebranche, Nicolas. (1674–1675) 1997. *The Search after Truth*. Edited and translated by Thomas M. Lennon and Paul J. Olscamp. Cambridge: Cambridge University Press.

Martin, Charles B. 1994. "Dispositions and Conditionals." *Philosophical Quarterly* 44 (174): 1–8.

McCann, Hugh J. 1974. "Volition and Basic Action." *Philosophical Review* 83 (4): 451–73.

Oppy, Graham. 2005. "Omnipotence." *Philosophy and Phenomenological Research* 71 (1): 58–84.

Pearce, Kenneth L. 2011. "Omnipotence." In *The Internet Encyclopedia of Philosophy*, edited by James Fieser and Bradley Dowden. http://www.iep.utm.edu/omnipote/.

———. 2012. "Thomas Reid on Character and Freedom." *History of Philosophy Quarterly* 29 (2): 159–76.

———. 2016. "Counteressential Conditionals." *Thought: A Journal of Philosophy* 5 (1): 73–81.

———. 2017. "Counterpossible Dependence and the Efficacy of the Divine Will." *Faith and Philosophy* 34 (1): 3–16.

Pearce, Kenneth L., and Alexander R. Pruss. 2012. "Understanding Omnipotence." *Religious Studies* 48 (3): 403–14.

Pruss, Alexander R. 2006. *The Principle of Sufficient Reason: A Reassessment.* Cambridge Studies in Philosophy. Cambridge: Cambridge University Press.

Reid, Thomas. (1788) 2010. *Essays on the Active Powers of Man.* Edited by Knud Haakonssen and James A. Harris. Edinburgh: Edinburgh University Press.

Ross, James F. 1969. *Philosophical Theology.* Indianapolis: Bobbs-Merrill.

Rowe, William L. 2004. *Can God Be Free?* Oxford: Clarendon.

Strawson, Galen. 1994. "The Impossibility of Moral Responsibility." *Philosophical Studies* 75 (1/2): 5–24.

Talbott, Thomas B. 1988. "On the Divine Nature and the Nature of Divine Freedom." *Faith and Philosophy* 5 (1): 3–24.

Van Inwagen, Peter. 1983. *An Essay on Free Will.* Oxford: Clarendon.

Vihvelin, Kadri. 2004. "Free Will Demystified: A Dispositional Account." *Philosophical Topics* 32 (1/2): 427–50.

———. 2013. *Causes, Laws, and Free Will: Why Determinism Doesn't Matter.* Oxford: Oxford University Press.

Watson, Gary. 2004. "Volitional Necessities." In *Agency and Answerability: Selected Essays*, 88–122. Oxford: Clarendon.

Wielenberg, Erik J. 2000. "Omnipotence Again." *Faith and Philosophy* 17 (1): 26–47.

Wierenga, Edward R. 1983. "Omnipotence Defined." *Philosophy and Phenomenological Research* 43 (3): 363–75.

Williams, Bernard. 1993. "Moral Incapacity." *Proceedings of the Aristotelian Society* 93:59–70.

CHAPTER 13

Infinite God, Open Future

WILLIAM HASKER

According to the *Westminster Shorter Catechism*, "God is a spirit, infinite, eternal, and unchangeable in his being, wisdom, power, holiness, justice, goodness, and truth." After "spirit," which functions as the substance term, the first thing mentioned is God's infinity. Similarly, H. P. Owen, writing in the *Encyclopedia of Philosophy*, states that "Christian theologians . . . if they are orthodox, regard infinity as the first among God's attributes" (Owen 1967, 191). Indeed, that God is infinite is very much taken as a given by the theological tradition; anyone speaking of a finite deity reveals immediately an intention to deviate markedly from that tradition. Yet it is beyond question that, among the divine attributes, infinity has received comparatively little attention in recent philosophy and theology—far less than omnipotence and omniscience, or even such attributes as eternity and immutability. Hence, I take it, the present volume. This chapter therefore is of an exploratory nature; I hope to point us in the direction of some of the issues that need to be addressed concerning divine infinity. And I shall be especially concerned, near the end, with issues that arise specifically in the context of the view of God's nature that I myself embrace, the view known as "open theism," or "the openness of God."

Anyone considering divine infinity at the present time must inevitably give some attention to the developments, in the nineteenth and twentieth centuries, in the mathematical study of infinity, developments pioneered especially by Cantor. How should philosophical theology respond to these developments? To begin with, I propose that we should accept the mathematical theory of infinity as a legitimate part of mathematics. It is true that there are significant objections to this theory that are entertained by some mathematicians and some philosophers.[1] But at least for those of us who are not experts in this area, it seems wisest to go with what seems to be decidedly the majority view, and to accept "Cantor's heaven" as a mathematical reality. But given the theory of infinity, what can we make of it for theology? Both Cantor and Bolzano desired to employ the mathematical theory of infinity in order to elucidate the concept of God,[2] and it would be gratifying to be able to fulfill their wishes. I suspect, however, that we shall be able to do this only very partially. The theory of infinity is first of all a theory of infinite *numbers*, and we shall be hard pressed to find very much about the divine nature to which we can usefully attach a number.[3] True, there is exactly one God, and exactly three divine Persons. But the issues to which those numbers point us do not seem to require, or indeed to allow, an appeal to a mathematical theory of infinity for their solution.[4] I suggest, therefore, that the mathematical theory of infinity has its most evident theological application in addressing the issues of space and, especially, time—topics certainly not unrelated to philosophical theology. We shall return to this subject later.

The classical theologians, however, did not have the modern mathematical theory of infinity available to them, so in order to understand what they made of infinity we need to consult the history of the topic. For purposes of my study here, only a few high points in that history can be mentioned. In early Greek philosophy, the infinite, or *apeiron*, was seen as that which lacks boundaries or definition; it was the indefinite, which required a limit (*peiras*) in order to assume any determinate character. H. P. Owen states that "in classical Greek philosophy infinity represents a substratum which is formless, characterless, indeterminate. It is a pejorative word. An entity is good to the extent that it is limited by form" (Owen 1967, 190). Because of this, "Plato could not envisage God (or the divine) as infinite. If God is perfect, he must represent the principle of limit" (191). This characterization of the infinite as undefined and featureless has not persisted in its original form, yet I will suggest later on that some vestiges of the idea remain with us today.

A crucial turning point in ancient thought is found in Plotinus. On the one hand, he reflects earlier thinking by holding that matter is infinite and therefore "evil because it trends intrinsically to formlessness." One the other hand, the divine Mind, the first of the emanations from the One, is "infinite because of its endless power, its complete unity, and its self-sufficiency" (Owen 1967, 190). Viewed in this light, infinity seems eminently suited to be an attribute of God, and in the hands of Christian theologians that is exactly what it became.

Another line of thought concerning infinity, one that had momentous consequences, stems from Aristotle.[5] Aristotle was perhaps the first to grasp clearly the mathematical idea of infinity, conceived as that which goes on and on without limit. He argued, for a variety of reasons, that there cannot be an infinite totality existing all at once. However, he was unable to dismiss entirely the idea of infinity. It seems that the process of dividing something is one that can, in principle, go on indefinitely without ever reaching an end point. Also, space itself seems to go on without end; whatever point one has reached, there is always more beyond. Aristotle's solution was found in a distinction between the "actual infinite," an infinite collection of items actually coexisting, which he regarded as impossible, and what he termed the "potential infinite," which is something that can go on indefinitely without ever reaching a stopping point. An example of the latter can be found in future time. If some person were to be literally immortal, it would seem correct to say that she has an infinity of life before her. Yet there will never come a time when she has lived for an infinite number of years.[6] The idea can be expressed in a pair of logical formulas: where x and y represent quantities of something that is potentially infinite, we have

> $\forall x \exists y (y > x)$, which says that for any such quantity there is a greater quantity, but not $\exists y \forall x (y \geq x)$, which says that there is a quantity which is greater than any (other) quantity. (Moore 2001, 42)

In Aristotle's own words: "In general, the infinite exists through one thing being taken after another, what is taken being always finite, but ever other and other" (cited in Moore 2001, 34). This distinction between actual and potential infinity, with actual infinity taken to be impossible, has exercised a long-lasting influence.

In particular, Aristotle's doctrine prevailed generally among the philosophers and theologians of the Middle Ages. And this led to the recognition of a problem for Aristotle's own view that the world was eternal. If this is so, then all the infinitely many past years must actually have occurred: Is this not, then, an "actual infinity" of the sort Aristotle had ruled out? The most common response was to accept the argument, and to regard it as a proof that the world could only be finitely old; this was the key contention of what has come to be called the *kalam* cosmological argument. Thomas Aquinas, however, demurred, claiming that it is only the simultaneous existence of an infinite multitude that is ruled out. Aquinas, of course, also held that the world has a finite age, but he accepted this on the basis of revelation and not because of a philosophical argument. Later in the Middle Ages, however, the objections to an actual infinity began to be challenged; this led eventually to the "infinite world" championed by Nicholas of Cusa and Giordano Bruno.

Infinity as an attribute of God was also present in medieval thought. This conception, however, was kept somewhat distinct from the ideas we already discussed concerning actual and potential infinities, because number and quantity were held to be inapplicable to God. And in spite of Owen's remark quoted earlier, infinity did not always play an especially important role in the list of divine attributes. Aquinas, for example, affirms that God is infinite; he derives infinity primarily from divine simplicity, by considering various ways in which a being might be limited, and showing that none of them apply to God (Aquinas, *ST* I, q.7). One might say, then, that for Aquinas infinity is a negative relational property; it consists in God's not being related to other entities in certain ways.[7] Infinity is not, however, a central element in his philosophical theology.

One medieval thinker for whom divine infinity did play an important role is John Duns Scotus. Infinity figures in his treatment of a number of topics, but for our present purposes the most important is the relationship between infinity, the divine attributes, and theological predication.[8] It is well known that Scotus, unlike Aquinas and some other medievals, insisted that there must be a univocal element in our talk about God. Following Anselm, we must ascribe to God all possible perfections—"whatever it is better to be than not to be." These perfections, however, are known first in creatures, but then if we are to ascribe those perfections to God they must be *the very same properties* that were ascribed to the creatures; otherwise the Anselmian project would have gone off the rails. This must not, to be sure, mean that we are denying or overlooking the differences between God and, for instance, human beings. The perfection of knowledge, for example, is present in human beings as limited—that is to say, in a finite mode. We first abstract the "pure perfection" from the finite mode in which it is attributed to human beings, and then we ascribe that perfection to God in the "infinite mode." Considered in this light, infinity is not so much a distinct attribute all by itself, but rather is the modality in which other attributes—the "pure perfections"—are ascribed to God. The result, as stated by Thomas Williams, is that infinity is both "the key component of our best available concept of God and a guarantor of the success of theological language" (Williams 2001, sec. 2.3). This contrasts with the situation in Aquinas, where the key role is played by simplicity, which "acts as an ontological spoilsport for theological semantics" (sec. 2.3), ultimately making it unclear whether we are able to say anything positive about God that is both true and intelligible to us.

But how is this conception of infinity to be understood? According to Scotus, we should begin by thinking in terms of the "potentially infinite in quantity"—some magnitude that can be increased indefinitely, without a limit ever being reached. Then, he asks us to imagine that all of the parts of this infinity are in existence simultaneously. Williams describes the next step as follows: "Scotus then asks us to shift from thinking about an actual quantitative infinity to thinking about an actual *qualitative* infinity. Think of some quality (say, goodness) as existing infinitely; so that there is, as it were, no more goodness that you could add to that goodness to make it any greater" (Williams 2001, sec. 2.3).

Note, however, that infinite qualitative goodness is not made of up of parts, as is a quantitative goodness; instead, Scotus describes it as "a measure of intrinsic excellence that is not finite." Scotus insists, furthermore, that infinity must be an intrinsic property of God, not a relational property, as was held by Aquinas.

One final point to notice is that this understanding of infinity and theological predication leads Scotus to adopt a significantly weakened doctrine of divine simplicity. In a divine attribute such as goodness the "pure perfection" is formally distinct from the "infinite mode" in which the perfection is possessed; additionally, the divine attributes are formally distinct from each other and from the divine essence. Scotus holds that there cannot be in God parts that are *really* distinct from each other—that is, such that one part can actually exist without the others—but this is a far weaker claim of simplicity than the one made by Aquinas. Some may find this weakened doctrine of simplicity to be a serious weakness in Scotus's theology. It might, on the other hand, be viewed as an additional benefit that flows from his position.[9]

There is one more episode that needs to be included in this all-too-brief survey of the history of the infinite. In the nineteenth century, Hegel distinguished between the "false infinite" and the "true infinite," reasoning as follows: The infinite is conceived as that which stands opposed to the finite. But whatever stands in opposition to something else is defined, and therefore limited, by that other thing. So if the infinite stands in relation to the finite, it is defined and limited by the finite, and is therefore itself finite—a spurious infinite, rather than a true infinite. The true infinite, then, must transcend the difference between the spurious infinite and the finite, and must include both of them as moments. Thus we reach the conclusion that the true infinite must be a single, all-inclusive reality; this is monism or, in theological terms, pantheism.

The pantheistic conclusion of this argument would not be in any way disturbing for Hegel himself. William Craig (2012) has pointed out, however, that several contemporary theologians, who by no means intend to be pantheists, have adopted this argument as their own. These theologians include Wolfhart Pannenberg, LeRon Shults, and Philip Clayton. Craig articulates the argument as developed by Pannenberg as follows:

1. The finite is that which is defined by its distinction from something else.
2. The infinite is defined as that which is not finite.
3. Therefore, the infinite is defined by its distinction from something else.
4. Therefore, the infinite is finite.

Since (4) is a logical contradiction, either (1) or (2) must be false. Pannenberg, following Hegel, seems to reject (2) in favor of 2′ "The infinite is that which includes the finite" (Craig 2012, 5).

Similar, though not identical, arguments are presented and endorsed by both Shults and Clayton.

But now we are confronted with a problem. The conclusion of these arguments seems to amount to a metaphysical monism or pantheism, which may have been congenial to Hegel, but which none of these theologians is eager to embrace. (Their actual views are better described as panentheist—all things are *in* God, but God is not identical with the totality of things.) How then can they avoid pantheism? The answers given by each of them are strikingly similar. The antithesis between God and the world, which emerges when we consider the infinite as that which is not finite, is overcome by emphasizing the various relationships God establishes with the world. Pannenberg states:

> The thought of the holiness of God and the understanding of the essence of God as Spirit bring us closer to a resolving of the contradiction. They express the fact that the transcendent God himself is characterized by a vital movement which causes him to invade what is different from himself and to give it a share in his own life. . . . God gives existence to the finite as that which is different from himself, so that his holiness does not mean the abolition of the distinction between the finite and the infinite. (Pannenberg 1991, 400)

Along with divine holiness, divine love also establishes a relatedness that overcomes the antithesis between God and the world: "By the unity of reconciliation by love which embraces the world and bridges the gulf between God and the world, the unity of God himself is realized in relation to the world. . . . By the love which manifests itself in his revelatory action God's unity is constituted the unity of the true Infinite which transcends the antithesis to what is distinct from it" (Pannenberg 1991, 446).

Similar points are made with regard to several other divine attributes. Moves of the same general sort are made by both Clayton and Shults.

Our reaction to these moves may be a combination of appreciation and puzzlement. On the one hand, the emphasis on the relationship between God and the world is fundamentally biblical, and serves to calm our apprehension that these theologians may be covertly endorsing pantheism. On the other hand, however, it may occur to us that this emphasis really does not succeed in countering the implications of the Hegelian argument they have embraced. As Craig points out, Pannenberg avoids "the contradiction that the infinite is finite by rejecting

2. The infinite is defined as that which is not finite

in favor of

2′. The infinite is that which includes the finite."

Craig goes on to say that, according to Pannenberg, "The perceived antithesis of God to the world is a sort of antagonism which is removed by God's being related to the world in affirming ways." Craig notes, however, that "the ontological distinctness between God and the world is actually presupposed by, rather than undermined by, such relatedness. But if this is all that is meant by God's (or the truly infinite's) inclusion of the world (or the finite), then (2′) is just impotent to resolve the original contradiction" (Craig 2012, 14).

Similar moves made by Shults and Clayton have similarly unsuccessful results. These theologians have embraced arguments that have as their conclusion a metaphysical monism or pantheism, and the strategy by which they seek to avoid that conclusion is simply inadequate. Craig states: "So long as they continue to endorse the premises of those arguments, they will be stuck with pantheism in spite of themselves" (Craig 2012, 19). In a more irenic vein, we might observe that these theologians, who by no means are or intend to be pantheists, have got themselves into this difficulty by endorsing a defective conception of infinity.[10]

But is this concept of infinity really defective? We certainly may define infinity in this way if we choose to do so. But there is no necessity to do this, since other conceptions of the infinite are readily available. For instance, in the mathematical theory of infinity the sequence of even numbers is said to be infinite, even though it is distinguished from the series of odd numbers, which is also infinite. And there are other options, as we have seen in our discussion of Duns Scotus. (More on this below.) I suggest, in fact, that this conception of the infinite is really an echo, at a considerable historical distance, of the ancient Greek idea of the *apeiron*, the indefinite, which, since it has no *peiras* (limit), cannot be anything in particular because it cannot be distinguished from anything else. We have little to gain, certainly not in theology, by returning to this long-abandoned conception in the twenty-first century.[11]

My more particular topic, however, is the relationship between divine infinity and open theism. To begin with, what exactly is open theism? The term "open theism" came into use subsequent to the publication in 1994 of *The Openness of God: A Biblical Challenge to the Traditional Understanding of God* by Clark Pinnock, Richard Rice, John Sanders, William Hasker, and David Basinger. It should be noted that the view thus articulated had already been advocated in its essentials by a number of other philosophers and theologians, not least by Richard Swinburne in his groundbreaking work *The Coherence of Theism.* Our present task, however, is not to trace the history but to give a concise overview of the openness viewpoint.

A central concern of open theists is to maintain a robust realism concerning the character and activities of God attested in the Bible, and to present such a view in a framework that is clear, readily intelligible, and philosophically and theologically defensible. We do not advocate a wooden literalism; clearly much that is said about God in scripture is figurative and requires a nonliteral interpretation. However, we seek to avoid the situation found in some other approaches, where an acceptance of the biblical witness is substantially undermined by an extremely abstract characterization of God that can only be reconciled with that witness by heroic intellectual gymnastics. My purpose here is expository rather than polemical, but a few illustrations will help to make the general point clear. The traditional view that God is pure actuality, lacking any element of potentiality, is difficult to reconcile with the biblically based assertion that God has done many things that he need not have done, and is capable of doing a great many other things that in fact he never does. The view of God as timelessly eternal apparently has the unfortunate consequence that at any given present moment God does not know what is happening *right now*, or what *has happened* in the past, or what *may still happen* in the future. A very strange consequence of this is that God does not know the truth that is enshrined in the liturgy: "Christ *has* died, Christ *has* risen, Christ *will* come again."[12] Finally, it has often been noted that the strong doctrine of divine simplicity, as affirmed, for example, by Aquinas, makes it very difficult to make sense of the distinction between Father, Son, and Holy Spirit that is required for the doctrine of the Trinity. Now, it is acknowledged that in each case there are elaborate attempts to answer the apparent difficulty, and to show that the philosophical and theological doctrines are not after all in conflict with what the Bible tells us about God. We cannot take the time here and now to adjudicate these issues; I merely make the point that it is a prima facie advantage for a theological viewpoint if it does not generate such conundrums in the first place.

What is it, stated in positive terms, that open theists say about God? We hold, following Anselm, that God is absolutely perfect in power, wisdom, and love; in general, God is "whatever it is better to be than not to be." Our differences concern not the truth that God is perfect, but rather the question of what that perfection consists in. God, we say, is eternal, not in the sense of being timeless, but in the sense of everlastingness: he always has existed, and always will exist. (This is, by the way, the *only* conception of divine eternity that is clearly present in the scriptures.) God is unchangeable in his nature, but undergoes changes of state, including God's affective states; he genuinely rejoices when we flourish, and sorrows when we suffer, especially when we suffer due to our self-imposed alienation from him. In the words of Clark Pinnock, God is not the Unmoved Mover but rather the "Most Moved Mover" (Pinnock 2001). The *relationality* of God is extremely important; God is "really related" to us his creatures, and we are really related to him. I add an additional point, though one that may not be especially central to our considerations in this volume.[13] Open theists are strongly inclined toward a "social" conception of the Trinity: if we attach such great importance to the relationality of God to creatures, we are disposed also to find such relationality within the Godhead itself.

Much of this, it is readily acknowledged, is held in common with other Christian theists who find such a dynamic, relational conception of God congenial. The most distinctive tenet of open theism, and the one that has generated much of the controversy about the view, is open theism's conception of "dynamic omniscience"—the term is from John Sanders—in particular of God's knowledge of the future. We hold that much (not all) of the future is known by God as what *may happen*, and as what *will probably happen*, but not as what *will definitely* take place. It is primarily this feature that has led to the denigration of open theism as a theology of God's "lesser glory," and even to the accusation that we are proclaiming a finite god. We shall return to these complaints presently.

But how does open theism appropriate the notion of divine infinity? To begin with, we thoroughly repudiate the pantheistic notion that, since God is "unbounded," God cannot be distinguished from any other thing, or other things from him. That simply is not the way infinity needs to function in Christian theology. How then shall we understand infinity as a characteristic of God as conceived by open theism? Here I propose that we take a page from Duns Scotus and consider infinity as the *mode* or *manner* in which the various divine attributes are possessed. God is wise, powerful, and good; indeed, he is *infinitely* wise, powerful, and good. (Thus the *Westminster Catechism* shows genuine insight: God is infinite *in* his various attributes; infinity is the *manner* or *mode* in which he exhibits those attributes.[14])

This in turn presupposes, again with Scotus, that there is indeed a "univocal core" of meaning in the attributes we ascribe both to God and to creatures. If this is not so, then the entire Anselmian project threatens to collapse. It is certainly true that wisdom, as we know it in some human beings, is a perfection, something it is better to have than not to have. But if divine wisdom has literally nothing in common with human wisdom, how are we to know that this divine "wisdom," whatever it may be, is really a perfection? Nor is it sufficient to invoke here the causal relation between the divine attributes and the corresponding human characteristics. If "God is love" means *only* that God has that characteristic, *whatever it may be*, that causes or enables humans to experience and exercise love, this tells us nothing whatsoever concerning the mind and heart of God himself. Indeed, I am strongly persuaded that those who deny any univocal common meaning are nevertheless covertly relying on such a meaning in their personal appropriation of the various divine attributes. Scotus did not deny that our language concerning God is analogical, but by insisting on a univocal core of common meaning he prevented analogy from degenerating into sheer equivocation. The dissimilarity in meaning, on the other hand, is in large part captured by the assertion that humans and other creatures possess these perfections in a finite mode, whereas God possesses them in the infinite mode.[15]

But how shall divine infinity be understood, given that we cannot attach numbers to the divine being? (Infinite divine power cannot mean that God has an infinite number of *kratons*, or units of power.) Many of the divine attributes, however, admit of *degrees*: we readily understand the idea that one person has more of power, or knowledge, or wisdom than another.[16] But what is meant by saying that God is infinite in some attribute, say, in power? Two things at least, I should say, though it may be that more could be added. First of all, that God's power is *inexhaustible*; no matter how much God has done, or might ever do, his power will always suffice for doing still more. And this clearly distinguishes God from the finite agents, however powerful, with which we are acquainted. But also, and importantly, that it is *incoherent to suppose* that some entity should have more of the attribute in question than God: in Anselm's terms, "nothing can be conceived" that is greater than God in any respect whatsoever.

But what of the mathematical theory of infinity? I believe this theory is of most use to theology when applied to the question of the possible infinity of space and time. One service the theory performs for us is that it mitigates, if it cannot completely eliminate, the impression of paradox that surrounds our talk of infinity. Take, for example, Hilbert's Hotel, a hotel that contains an infinite number of rooms, each of which is occupied by a guest. Disruption threatens when a tour bus pulls up at the front door, bringing an infinite number of additional travelers, each of whom desires a room in the hotel. The manager, however, has read his Cantor, and immediately comes up with the solution: each of the existing guests is assigned to one of the infinitely many odd-numbered rooms of the hotel, leaving the infinitely many even-numbered rooms for the new arrivals. To be sure, not everyone is satisfied with this solution of the manager's problem. One might argue that in the present physical universe the relocation of the guests could not be accomplished in any finite period of time, in view of the finite speed of physical processes, including the speed of light.[17] Another interesting difficulty has been raised (in discussion) by Richard Swinburne. Swinburne points out that, if each of the infinite rooms in the hotel is already occupied, there is no way that additional guests can be accommodated *in those very same rooms*. Rather, in order to accommodate the new arrivals, the manager will have to order the construction of new rooms. However, the number of rooms in the enlarged hotel will be the same as the number of rooms prior to the enlargement, namely, aleph-null, the transfinite number that is the cardinal number of the set of all natural numbers. So the manager does not need a *greater number of rooms* to accommodate the new guests, but he does need *additional rooms*, rooms that were not present in the hotel previously.

The point made by this little story in any of its versions is that an infinite set, unlike its finite counterparts, can be placed into a one-to-one correspondence with one of its proper subsets. In the Swinburne variation, for instance, the rooms in the expanded hotel, to which infinitely many new rooms have been added, are equal in number to the rooms in the already-full hotel prior to the additions. Admittedly this has an air of paradox; our practical experience, which exclusively involves finite sets, does not prepare us for this sort of thing. But there is no "hard" paradox here, no contradiction that would invalidate the theory. That an infinite set can be placed in a one-to-one correspondence with one of its proper subsets is simply a consequence of the theory of infinity, something we just have to accept even if we are unable to arrive at an imaginative grasp of this situation.[18]

Assuming we do accept this, what are the implications for theology? I can think of two: one more negative in its effect and one more positive. The negative implication is that it undermines those versions of the *kalam* cosmological argument[19] that rely on the alleged impossibility of an "actual infinity," in this case of a literally infinite past time.[20] If this is indeed impossible, then time must have had a beginning—and if a beginning, then most likely a creator. But maintaining this view now would seem to require refuting or invalidating the mathematical theory of infinity, not a task to be undertaken lightly. To be sure, one might hold that although a purely mathematical actual infinity is coherent, an infinity of concrete beings is nevertheless impossible. This point also, however, would need to be argued, and so far as I can see the prospects are not especially bright.[21] It should be noted that insofar as the *kalam* proponent is mounting an *argument* for the existence of a creator, the burden lies on the proponent to establish, by compelling reasons, that an actual infinity of past time is impossible. (It will not, then, suffice to challenge the objector to prove that an actual infinity is possible, thus shifting the burden of proof.) I judge that this task has been made substantially more difficult, once the mathematical theory of infinity has been accepted.

On the more positive side, the theory enables us to make free use of the possibility of a literally infinite past, not only in relation to the created universe (or multiverse), but also, if we are temporalists, in relation to the existence of God. And the question of the finite or infinite magnitude of the physical creation is left open, to be decided on whatever grounds we find compelling.

Finally, however, we come to open theism's "limitation" of divine knowledge—the limitation which, in the minds of its critics, shows that in the end open theism has only a finite God. The standard definition of omniscience holds that God knows all true propositions, but there are true propositions concerning the future that, according to open theism, God does not know. So, case closed: open theism denies an infinite God.

In fact, there is much less in this argument than meets the eye. It is widely acknowledged that the divine perfections must be *logically possible* perfections; the impossible ones we leave to the lovers of paradox, to do with as they will. We do claim to have shown, however, that divine knowledge of the contingent future is *not* logically possible. The basic idea is stunningly simple, though ingenious attempted replies have at times led to baroque complications. For a person's action to be free it must be genuinely possible for that person to act in any of two or more different ways. But this means that there does not exist, at the time of the action, anything that guarantees that the action will be taken in one particular way and not in any other. It has to be objectively uncertain, up until the action is taken, which course will be chosen. And on the other hand, divine knowledge, in the relevant sense, must be absolutely certain; mere probable belief will not in this context count as knowledge. But there is an evident contradiction if we suppose some knower to have absolutely certain knowledge of something that is objectively uncertain; that would be to know the future event as having a status that it does not in fact possess. Insofar as the future is objectively uncertain, a perfect Knower will know it precisely in that way—as objectively uncertain. The argument is simple, clear, and (I believe) intuitively compelling.

Even so, it may be replied, the argument leaves us with a situation in which divine knowledge is indeed limited, and in that sense finite, even if the limitation is said to be a necessary one. This is far from clear, however. That divine knowledge is infinite means, in this context, that the divine cognitive capability is *inexhaustible*; we will never reach a situation in which there is more to be known and God isn't up to knowing it. But in the envisaged situation, what more is there to be known? Presumably, the answer is this: the true propositions concerning future contingent events. Now, one response to this, one that has considerable merit and has attracted a good number of adherents, is to say that there simply are no such propositions: the propositions in question are neither true nor false until the ambiguity is resolved by the actual occurrence of the events in question. My own response, however, has been somewhat different. I acknowledge the fact that we do, in ordinary speech, regard as true or false at least some propositions concerning the contingent future, and I seek to make the best sense of this that I can. But if we are to acknowledge some such propositions as true, we are forced to conclude that their status is rather different than that of other true propositions that we deal with. These propositions, I believe, have to be considered as examples of what J. R. Lucas termed "valedictory truths." What I think he meant by this is that the truth of such propositions can only be recognized as we are looking backward, after the events in question have either occurred or failed to occur. After the fact, we consider predictions that have been fulfilled to have been true, and those that have not been fulfilled to have been false. But at the time the predictions were made, it *had not yet been decided* whether they would be true or false—the relevant truth-conditions did not yet exist. And given this, it is clear that they do not and cannot count as possible objects of knowledge, such that failure to know them would count as a limitation on the cognitive powers of God. With this in view, omniscience can be defined in a way that parallels a familiar definition of omnipotence: an omniscient being is one who knows everything that it is logically possible for a perfect being to know.[22]

So we have two ways of regarding propositions concerning the contingent future, but whichever one is chosen it will still be the case that divine knowledge is infinite and perfect. I may add that, for the purposes of open theism, I do not think it is crucially important which of the ways is chosen. Either way, we have an infinite God who has created a universe with an open future—and that is as it should be.

NOTES

1. For a discussion of these objections, see Oppy 2006 and Moore 2001.

2. For Cantor's views, see Newstead 2009; for Bolzano, see chapter 8 in this volume by Christian Tapp.

3. The studies in this volume by Benedikt Göcke (chapter 19) and Christian Tapp (chapter 8) both illustrate the limitations of interpreting divine attributes in terms of mathematical infinity.

4. None of the moves in Trinitarian theology discussed by Thomas Schärtl involves an application of the mathematical theory of infinity (see chapter 17 in this volume).

5. For Aristotle's views on infinity, see Moore 2001, chap. 2.

6. The idea is captured nicely in the final stanza of the hymn "Amazing Grace":

When we've been there ten thousand years
Bright shining as the sun,
We've no less days to sing God's praise
Than when we'd first begun.

7. "Simplicity, for Aquinas, entails infinity just because infinity is defined as the lack of a relation to a limiting entity" (Cross 1999, 40).

8. For Scotus's views on infinity, see Williams 2001 and Cross 1999, esp. chaps. 2 and 3.

9. At least, it may be viewed as an improvement over Aquinas's even stronger doctrine. Richard Cross gives reasons, based on Scotus's view concerning God's knowledge of contingent truths, to think that his doctrine of simplicity is still too strong (see Cross 1999, 51–55).

10. In correspondence, Philip Clayton pointed out that he and the other theologians mentioned did not begin by accepting Hegel's reasoning as a freestanding philosophical argument. Rather, they embraced it because it seemed a sympathetic philosophical representation of a theological idea—the idea of the "inclusion" of finite realities in God—which they were already disposed to affirm. Acknowledging this, I still believe Craig's criticism is correct, and that the counterarguments by Pannenberg and colleagues do not suffice to avoid the pantheistic implication of Hegel's argument. However, this by no means shows that these theologians are actually pantheists. Clayton stresses the difference between Hegel's view of "Absolute Spirit, unfolding itself by rational necessity" and his own view that "God, who created freely, might include His created beings 'within Himself.'"

11. Thomas Aquinas, in the thirteenth century, was already quite clear that this is not the way to go: "The fact that the being of God is self-subsisting, not received in any other, and is thus called infinite, *shows Him to be distinguished from all other beings, and all others to be apart from Him*" (Aquinas, *ST* I, q. 7, a. 1, reply to obj. 3; emphasis added).

12. Admittedly, these consequences can be avoided by adopting a four-dimensionalist theory of time, according to which all the instants of time coexist in the four-dimensional continuum and there is no privileged present moment. I believe four-dimensionalism is a bad option both philosophically and theologically, but there is no space to argue this here.

13. But see chapter 17 in this volume by Thomas Schärtl.

14. Open theists, however, will note with concern that *love* is not mentioned in the Westminster definition; the significance of this omission bears pondering.

15. For a more extensive defense of the view that it is possible to speak about God in language that is literal (with appropriate qualifications), see Alston 1989, pt. I.

16. But not, perhaps (*pace* Scotus), that one person has *more being* than another: "being," as ordinarily understood, is not a degreed attribute. Some special interpretation, then, will be needed to make sense of the idea that God is "infinite in being"—unless it simply means that God is a being who is infinite.

17. For discussion of this and some other objections to Hilbert's Hotel, see Oppy 2006, 51–53.

18. Interestingly, this point was anticipated by Gregory of Rimini in the fourteenth century. Moore explains his thought as follows: "The paradoxes of the infinitely big showed that in the infinite case it was possible for one collection to be contained within another *and* for their members to be susceptible of a pairing off. In such circumstances, the former collection would be smaller than the latter in one sense, but not in another (a distinction that could not arise in the finite case)" (Moore 2001, 53).

19. The *kalam* cosmological argument, which originated among Muslim philosophers in the Middle Ages, has recently come back onto prominence through the publication of Craig 1979. A recent statement of Craig's view may be found in Craig and Sinclair 2012.

20. Note that this objection does not affect versions of the *kalam* argument that rely on scientific evidence that the universe has a finite age. The majority of the Craig and Sinclair (2012) study is devoted to such physics-based arguments.

21. This is, however, Craig's position; the Craig and Sinclair (2012) study accepts the validity of the mathematical theory of infinity but relies heavily on the alleged absurdities generated by an actual infinity, with Hilbert's Hotel playing a starring role. In particular, the absurdities appear because admitting infinite sets of concrete objects requires us to deny:

> (ii) There are more things in M than there are in M′ if M′ is a proper submultitude of M.

A denial that is held to be absurd. It would seem, however, that if it is absurd to deny (ii) as applied to infinite sets of concrete objects, it is equally absurd to deny it as applied to infinite sets of numbers. But if that is absurd, the mathematical theory of infinity cannot stand.

22. Note that the claim that God knows all true propositions may be problematic for other reasons. It is widely held that there are first-person propositions, such as "I am now suffering from a toothache," which can be known only by the person to whom they refer. (To be sure, another person can very well know the third-person counterpart of such a proposition.)

REFERENCES

Alston, William P. 1989. *Divine Nature and Human Language*. Ithaca, NY: Cornell University Press.

Craig, William Lane. 1979. *The Kalam Cosmological Argument*. London: Macmillan.

———. 2012. "Pantheists in Spite of Themselves? Pannenberg, Clayton, and Shults on Divine Infinity." *American Theological Inquiry* 5 (1): 2–23.

Craig, William Lane, and James D. Sinclair. 2012. "The *Kalam* Cosmological Argument." In *The Blackwell Companion to Natural Theology*, edited by William Lane Craig and J. P. Moreland, 101–201. London: Blackwell.

Cross, Richard. 1999. *Duns Scotus*. Oxford: Oxford University Press.

Moore, Adrian William. 2001. *The Infinite*. London: Routledge.

Newstead, Ann. 2009. "Cantor on Infinity in Nature, Number, and the Divine Mind." *American Catholic Philosophical Quarterly* 83 (4): 533–53.

Oppy, Graham. 2006. *Philosophical Perspectives on Infinity*. Cambridge: Cambridge University Press.

Owen, Huw Parri. 1967. "Infinity in Theology and Metaphysics." In *The Encyclopedia of Philosophy*, vol. 4, edited by Paul Edwards, 190–93. New York: Macmillan.

Pannenberg, Wolfhart. 1991. *Systematic Theology*. Vol. 1. Translated by Geoffrey W. Bromiley. Grand Rapids, MI: Eerdmans.

Pinnock, Clark H. 2001. *Most Moved Mover: A Theology of God's Openness*. Grand Rapids, MI: Baker.

Pinnock, Clark H., et al. 1994. *The Openness of God: A Biblical Challenge to the Traditional Understanding of God*. Downers Grove, IL: InterVarsity.

Swinburne, Richard. 1977. *The Coherence of Theism*. Oxford: Oxford University Press.

Thomas Aquinas. 1981. *Summa Theologiae*. 5 vols. Edited by the Fathers of the English Dominican Province. New York: Benzinger.

Williams, Thomas. 2001. "John Duns Scotus." In *Stanford Encyclopedia of Philosophy* (substantive revision April 4, 2013). http://plato.stanford.edu/archives/sum2013/entries/duns-scotus/.

CHAPTER 14

Infinity and God's Atemporality

PAUL HELM

It is a strange fact that thinking of God's nature relative to the idea or ideas of infinity is a rather unfamiliar pastime. This strangeness is not lessened by the fact that relatively little attention is given to divine infinity in the tradition of discussion that we have all inherited. Katherin Rogers goes so far as to claim that the earlier medievals "steered clear" of using the term "infinite" of God. She suggests that this may be because of "a holdover from the ancient association of 'infinite' with 'incomplete,' 'unfinished,' and hence 'imperfect'" (2000, 11–12). Yet as far back as Augustine we find the language of infinity associated with God, as Augustine wrestles to free himself from Manicheism (Augustine 1992, 111; Rogers 2000, 12). And as Rogers points out, the use of "infinite" was more common by the time of Aquinas. That, I suppose, makes it safe for us to use it. Safe but not easy.

Traditionally, both in later medievalism and in the Reformed Orthodoxy that developed after the Reformation, affirming God's infinity has implications for his creation *ad extra*, as regarding space, in accounts of his omnipresence, and regarding time in accounts of divine atemporality. His infinity enables him to have immediate access to all places and at all times, and so in some sense to be "at" those places and times. In what follows I shall attempt to defend a version of this way of thinking against current objections to it, not all objections, but to some of the more prominent, and to try to clarify in what sense, on this view, God may be said to have immediate access to his creation. Before coming to concentrate on time, I shall make some remarks on space.

In doing this we need to observe the distinction between infinite as a hypothetical series that is indefinitely extendable and as an ontological term predicable of God, properly predicable of him and of no other. So Aquinas says: "The divine being, on which the notion of divine power is founded, is infinite existence, not limited to any kind of being, but holding within itself and anticipating the perfection of the whole of existence" (*Summa theologiae* [*ST*] I, q. 25, a. 3; translations are from Aquinas 1969).

INFINITY AND SPACE

It is tempting to think of infinity spatially. And then an infinite being would be one that occupies all space, and occupies it all equally densely. The physical universe, being spatial though continuously expanding, as we are told, is a part of the infinite. Such an infinite God could not be simple therefore. If God is infinite, then on such an understanding being infinite is being *the all*, and we are set fair for pantheism. But to be tempted to move in this direction, as many modern theologians have been tempted, according to William Lane Craig (2006, 135–56), is to forget that it is as God that God is infinite, and as God, at least the God of traditional theism, he is certainly not a container in which the universe is to be found, encompassed by his being, nor as a being who exists alongside the physical universe, in juxtaposition with it, for an infinite God has no sides or surfaces, and a fortiori no side that is alongside anything.

God is a distinct being, not in competition with other beings, in a category all by himself. To say that he is infinite as God has to do with the character of his essential properties, or his essential nature or essence. Any states of affairs that come about as a result of God's contemplation or action *ad extra* are not infinite; they are dependent on the goodness of their Creator, continuously so. So the created universe could not be infinite, or at least not infinite in the same sense as God is infinite.

To think of God's infinity in such spatial terms is to think of him in the manner of the Manichees. While under their influence, St. Augustine says, "I conceived even you, life of my life, permeating infinite space on every side, penetrating the entire mass of the world, and outside this extending in all directions for immense distances without end; so earth had you, heaven had you, everything had you, and in relation to you all was finite; but you not so" (Augustine 1992, 112).

So it is *as God* that the infinite God of classical theism is infinite and as such he is without limit. That does not mean that he is without physical limits or that he is physically limited, either. As God he is necessarily not a physical thing of any kind. Infinity may be a great-making property, but if so, not as wisdom, say, is a great-making property, but as an infinite being, possessing unbounded and unboundable wisdom, and love, and justice, and whatever else, is great.

So far, and briefly, we have thought about infinity in spatial terms, or on analogy with space, and tried to divest our minds of certain misconceptions.

Infinity is sometimes called a formal property. Craig suggests it is a mode of possessing properties rather than a distinct property in its own right. As Craig understands it, infinity is not a separate attribute or essential property of God, but qualifies all the essential attributes he has. It is a property of properties.

As Craig puts it, "Infinity serves as an umbrella term for capturing all the properties that serve to make God the greatest conceivable being. In saying that God is infinite, we mean that God is necessary, self-existent, omnipotent, omniscient, holy, eternal, and so forth" (Craig 2006, 155). Infinity, then, is a constituent of the greatness of Anselm's great-making properties, which are in turn constitutive of the greatest of all possible beings; wisdom is great-making, but unlimited or infinite wisdom is certainly more great-making. All things considered, it is better to be infinite than merely finite, however glorious and great-making one's properties otherwise might be.

INFINITY, TIME, AND ETERNITY

These opening comments have had in view infinity with respect to space. But what about infinity with respect to time? I say, the same applies. God, if he is infinite, is not in time, nor is he subject to time as his creatures are. So our first thought about God's infinity as it relates to time implies that he is not in time or bound by it in the way that his creatures are. He does not exist for a period at discrete dateable times, moments or years or eons. He does not have a memory. The pantheizing temptation is not so great with time as with space, perhaps, but then thinking of God's infinity with respect to time has temptations of its own.

In what follows, I shall use *possessing the ability immediately to access all moments of time* as a test of divine infinity, or, for lovers of the pure actuality of God, *immediately accessing all moments of time*. But I shall not claim that only this view of God's relation to time satisfies this test. So that not being subject to time God immediately creates or conserves or otherwise brings about, or permits, whatever occurs at each moment of time. Maybe a God who was very, very great in power, but not infinite, could be in such a state, or an atemporal God who is less than infinite in some other respect, but arguing in Richard Swinburne's vein at this point one could deploy an argument on the grounds of simplicity that it is reasonable to conclude that a God who has very great powers is infinite (Swinburne 1979, 52–57).

In the literature there are two conceptions of God's eternity—that of timeless eternity, which for clarity I shall call *atemporality*, according to which God is thought of as above or apart from or without time and its limitations; and that of God's *sempiternity*, his existence in time, and so in a temporal series that is forwardly and backwardly everlasting. I shall not make a general comparison between these conceptions, nor will I dwell on the strengths and weaknesses of each, except insofar as these arise from their relations with the infinity of God.

In the case of God's sempiternity, we have the view that God is part of the temporal series, as you and I are, and he is *in time*, as we might say. God's existence is conditioned by time's unidirectional topology, and by its evanescence. But in the case of God, but of none of his creatures, it is said that for any time, God is present at that time. Such sempiternity obviously has the advantages of neatness and simplicity, for we do not have to concern ourselves with God's relation to temporality from a standpoint outside it. But there is a converse problem of God's creation of time as a distinct metaphysical dimension or category, for it seems that on the sempiternalist account the existence of time is logically prior to God's own existence.

A-theorists, who treat the temporal series indexically, think that the reality of time is best understood from a standpoint within that series, and that the associated temporal indexical expressions are not simply necessary practical ways of referring to time as having a past, a present, and a future but are ways of referring to what time essentially is. Among clearer theorists there is more than one way of thinking about the reality of time. Some think that only the present, however measured, is real, the past and the future each being unreal. But other A-theorists think of the past being different in ontological status from the future.

There is certainly a marked difference between what has happened at a time in the past and a work of fiction that is set in the same time. For what is past is causally connected with what is present, connected by having helped to bring it about. So here there is a marked asymmetry between past and future; the past, as what has been, now has a shadowy actuality by comparison with the present. It cannot now not have been the case. It has no present actuality, though it has left its mark on the present. But unlike fiction, even historical fiction, the past is dignified by having been present. The future is "open."

People who talk of open futures at this point usually have in mind the occurrence of indeterminate events and actions. But an A-series view is not necessarily open, as not yet having a truth value, since if there are indeterminate events, these are contingently so. Sometimes this is qualified by this kind of A-theorist: the future is that segment of time that, when it is present, is effectible by free human choice and/or by divine choice and as future, and has no present actuality. According to a second kind of A-theorist, what we have just said about the past implies that it is currently nonexistent in a way that the future now is. It is important to note that determinists who are not fatalists don't think that coming to learn what later times are like is simply a matter of discovery, but rather the result of human and other agency.

I think we might say that divine sempiternity can be characterized as entailing access to all times, and that the mode of that access is conditioned by the serial character of time itself. God has had, has and will have, such access. He cannot now, in the temporal present, have access to the past in the way that he now has access to the present, however. He may know all about the past now, but he cannot now have access to the past as an agent. It is as closed to him now as it is to any creature existing in time now. And periods of the future are not accessible to him until their turn comes to be present.

So it might seem that the two A-series views of the series of time that we have briefly characterized fall sadly short of the requirement of immediate accessibility. For someone in the present, particularly one who possesses great powers of cognition and retention, would in each case now be able to access the past by the effortless exercise of "memory," by remembering the past in all the detail it possessed as a present occurrence, or set of such occurrences. So a sempiternal God has access to what is now past through present memory traces, traces in no sense cognitively defective, but ones ontologically distinct from what is present. So God is "bound" by time in that his own cognitive state has to accept the consequences of time, and what he knows changes to correspond with the "flow" of time, and more and more of his life is in the past. Even though there remains a limitless future, what is anticipated constantly changes as present events become past. Further, although a limitless future may remain, it may be unknowable in the present if there are indeterminacies as part of it, unknowable even to God.

Do these consequences, and particularly the consequence that given certain kinds of A-series, and the evanescence of time as in A-series thinking more generally, and given indeterminate futures, mean that God is not infinite? Well, these consequences certainly compromise the incommensurability of the infinite and the finite. For whatever else is the case, God and his creation in a straightforward sense occupy the same temporal order.

If God as being situated within the temporal order understood as an A-series does compromise his infinity, it is open to the A-series theorist, in defense of that view, to say that as with the case of time, so with the case of causality, God must "respect" the logical consequences of the existence of what is created by him, distinct from himself, even if what is distinct is continuously dependent on him. Such causality requires God to respect what he has created *ad extra*. It is a logical consequence of his decree to create. Yet for God to respect these differences is not to be bound by them as God himself is said to be bound by time. God remains the primary cause, whereas the flow of time affects God and his creatures similarly.

It may be that it is possible to construct a view of God's infinity that takes account of the contours of the A-series, but being infinite is not the first thing that comes to mind in relation to articulating the being of such a God. God may be a priori infinite (no concept of God could fail to have infinity ascribed to him), but the infinitude in question in the case of such a God has to be designed to fit his relation to time.

By contrast, the eternality of God is best thought of as entailing a reference to any temporal series that is best captured as a B-series. On the B-series view of the temporal series, the view according to which the significant temporal relations, as far as time itself, or times, is concerned, are three: "simultaneous with," "after," and "before." Indexicality is necessary for conscious and intelligent beings to cope with the temporal series, but not necessary for setting out and appreciating that series as such. In order for me to prepare for Christmas it is of importance that that day is tomorrow and not yesterday. For although the relation of Christmas Day to Christmas Eve is immutably fixed, I cannot prepare for Christmas Day if both it and the day before it are in the past, over and done with. They may be over and done with, with respect to present action, but they are every bit as real as the present. Temporal indexicality is indispensable to enable us to handle and steer by time, but not indispensable to the nature of time itself. This feature regarding the nature of time enables us to use the temporal series in connection with divine eternality.

The B-series view is satisfactory for thinking about God's relation to temporality, because it allows God to have the same metaphysical relation to each moment of time. But it does not follow that God is himself a part of any B-series. There is need to make the distinction between God's *timelessness* and the *tenselessness* of the B-series. To conflate the two, and work only with tenselessness, would be to compromise God's atemporality.

TU QUOQUE ARGUMENTS

Since Arthur Prior wrote "Thank Goodness That's Over" (1959) and "The Formalities of Omniscience" (1962), there have been those who have argued along with him that someone who says truly, "It is now 29th August," refers to the "leading edge" of his personal history, something he does not do when he utters a truth by saying, "The 29th August is earlier than the 30th August." And therefore such temporal indexical truths cannot be expressed by an atemporal individual, such as God. In this spirit, William Hasker (2002, 200) writes, "That this fact [the characteristic "leading edge" point] cannot be expressed from the standpoint of a timeless eternity reveals the inherent limitations of such a standpoint; it does not cast doubt on the reality and significance of the fact itself." We shall comment on this outlook later on.

Of course, an infinite knower would know that *for me* on August 31 at August 31, that day is the leading edge of my personal pilgrimage in time. The next day, September 1, is the leading edge on September 1, and so forth. And he would know the correct indexical beliefs that I have on August 31 are transposed into a corresponding set of correct indexical beliefs (assuming my beliefs are accurate) when I reach September 1. He knows how my memory enlarges in some respects at each successive time and (later still) that it probably shrinks, as my brain shows definite signs of wear, and further he knows that the expectations I had for the September 1 are fulfilled or disappointed, and so on. These may all be definite and concrete facts for me, and God knows the series of such facts that make up each human life. But they do not guide the Eternal Being in the same sense that they may guide individuals in time.

I'm not sure why the fact that all this cannot be expressed from the standpoint of a timeless eternity reveals or may have revealed the inherent *limitations* of such a standpoint, as Hasker claims, and so have adverse implications for God's infinity. What it does undoubtedly show is that existing in time is necessarily different from existing in eternity. Such differences therefore have modal consequences for all concerned in each mode of existence. And these consequences could be thought of as "limitations" for an eternal being. But why? A being that exists in time cannot understand facts from a standpoint that is "outside" time. Would this not be a limitation on God if he exists sempiternally (Leftow 1991, 321)?

Whatever we may think of the last question, it may be thought that atemporality has its share of other compromises. For consider an analogous example, the distinction first made in the Middle Ages, between God, the primary cause, and creaturely secondary causes. God's power is limited or conditioned by the character of such secondary causes, such that he must "respect" these. For example, if Joe at a certain time ties his shoelace, then it is not God as the primary cause who ties Joe's shoelace. Even if we suppose that the action is not indeterministically free, the distinction between what God can do and what Joe does remains, on a straightforward theistic view. There is occasionalism, of course, according to which God is the only efficaciously causal agent in his universe, but I shall not consider this possibility further here.

Richard Gale tells us that although he can stand such ideas as that of God not being able to tie Joe's shoelaces, he is unable to stomach the paradox of perfection that an eternal God cannot know what time it is now. He says, "While we are not crushed by the thought that God, in virtue of his absolute perfection, is barred from playing football, committing suicide, or performing a morally wrong act, we are by the thought that it precludes his knowing any A-propositions" (Gale 1991, 90). Perhaps Gale ought to have spoken more cautiously at this point. In any case, here the prospect of "bespoke" metaphysics shows itself. Some of us may prefer a God before whom we can prostrate ourselves and tremble at his sovereignty, and think that God's existing atemporally best fills this bill. Gale reveals a different set of preferences in his claim on behalf of God's sempiternity, when he says that timelessness contributes to God being religiously unavailable, but this preference by itself is not an argument for God's temporality (90). Gale continues:

> God's inability to know A-propositions in virtue of his timeless eternality, not only is a blight on his omniscience, but also is another factor that contributes to his being religiously unavailable. A religiously available God is one with whom we can commune. But we can commune only with beings who share our tensed perspectives, who know how it goes with us right now, and not just how it goes timelessly with us at t_7, even if now is t_7. (Gale 1991, 90)

Gale may well have a different view than some others about what religion is. For there are different views of religion as there are different views of time. Still, if metaphysical arguments about God can be settled by religious preferences, if, in effect, "being available" or "being open to negotiation" are to count as divine great-making properties, then this will severely constrain the prospects of philosophical argument, and the possibility of philosophical agreement at such points.

There is more philosophical mileage in considering God's relation to time than there is over whether or not he can play football, but there is still a point of principle over the paradoxes Gale dismisses or plays down. In virtue of who God is, the essence he has or the set of his essential properties, it is trivially true that there are sets of things that he can and cannot do. Gale displays his expertise in the discussion of God's relation or lack of relation to temporal indexicality. He thinks that there are all sorts of tangles that we get into if God is not in time. But then others have become entangled by the thought that God does not have a body, for (they claim) such a God cannot understand human embodiedness. Or that God does not have fits of emotion, and so does not understand human passion. Others have thought that a God that is not utterly transcendent cannot be worth much. Still others have thought that a God who is not in control of everything that comes to pass is unworthy of the name, and so on.

Hasker is among those who regard the future as ontologically open to God, and as a consequence the omniscience of God does not extend to the entire future, at least to that part of it that is effectible by human indeterministic choices, and so long as there are such choosers the entire future cannot be foreknown by him. Parts of it have no truth value now and so are unknowable now, even to God. Is this not a more obvious limitation? By contrast, others, such as Alvin Plantinga (1993, 43), think that it is monumentally difficult to see how any proposition could fail to have a truth-value.

Each view of God's relation to time engenders limitations, which become visible as the logical consequences of that view are recognized. Every philosophical or religious view that has a definite meaning has such limitations. Perhaps one pays one's money and one takes one's choice of "limitation," depending on what one believes is (in Leftow's [1991, 321] term) well supported or well backed elsewhere in one's understanding.

Let us return to the task of considering how these limitations work out in the case of divine atemporality. To begin with, an atemporal God can know how it goes with us right now. For God to know facts about his creation is necessarily for God to know what is going on at a time. And that knowing includes how it goes with us right now by knowing what goes on at that time. Eternally knowing that a particular time is now for Joe, God may eternally will to disclose something to Joe at that time, the knowledge of which will have an effect upon Joe at that time. In this connection Brian Leftow expresses the opinion that the denial of God's knowledge of essentially tensed facts may not reduce the religious value of the concept of God. He goes on to say, "But this denial may well not do so, for as we have seen, God does not need to know which events have present-actuality to answer prayer or to act providentially or salvifically" (Leftow, 1991, 327).

So regarding limitations there seems to be nothing to choose between the sempiternalist and the atemporalist. There is cost and benefit in each view. Any definite philosophical view seems to be in the same plight. What is a limitation in the one case is not in the other, and vice versa. Perhaps in philosophical theology and in other areas of philosophy bespoke metaphysics is after all inevitable.

CONSISTENT ATEMPORALITY

So far I have argued for the possibility of divine atemporality by means of a *tu quoque* argument against charges made by temporalists, and therefore have argued that atemporality is an aspect of divine infinity. As I mentioned at the outset, the traditional view of God's eternity is that it is timelessness, what I am calling "atemporality." Since Augustine, perhaps, but certainly since Anselm, this has been expressed negatively, as God the Creator being apart from temporal duration, and sometimes more positively, as God's life encompassing all times in one eternal present, time without succession or duration. The two kinds of expression often sit uneasily side by side.

As part of a defense of atemporalism and of its coherence with divine infinity, I shall try to show that some current objections to it rely upon these more positive expressions of divine eternity, but that such expressions are not necessary to express atemporalism. I shall argue that a would-be atemporalist ought to take what is a more principled line than that conveyed by some of the expressions we find in Augustine and later. Expressions such as this one by Anselm:

> For in eternity a thing has no past or future but only an eternal present, though in the realm of time things move from past to future without any contradiction arising. Similarly, that which cannot be changed in eternity sometimes, before it occurs, proves to be, without involving incongruity, changeable because of free will. Moreover, although in eternity there is only a present, nevertheless it is not a temporal present as ours, only an eternal one in which all periods of time are contained. (Anselm 1998, 442–43)

Or consider the Reformed Orthodox theologian Francis Turretin: "For the eternal duration of God embraces indeed all time—the past, present and future; but nothing in him can be past or future because his life remains always the same and immutable" (Turretin 1688–90, I. 203). This language may be said to express atemporality imperfectly, since an eternal *time*, an eternal *duration*, is implied.

So it is important to reiterate at the outset that, despite such flirtings with the idea of a "divine time" on the classical theistic view of God's eternity, that aspect of the expression of the traditional view ought to be eliminated. In my view, God is in no sense temporal. As Howard Robinson has put the matter:

> Philosophers have spoken as if even the notion of atemporality has to be explained within the framework of temporal concepts, so that the challenge becomes how one can, so to speak, stretch the natural topology of time so that it can accommodate the eternal. Atemporality gets explained in terms of a certain kind of now or of a peculiar duration, namely a limitless one without succession. It is as if an atemporal being is the limiting case of a kind of temporality. (Robinson 2007, 59)

When writers such as Augustine and Anselm referred to all times as being "present" to God they cannot have meant being temporally present, but sometimes their language conveys the opposite impression.

For all the elegance of Anselm, and what is said later in the tradition, there is in such language the potential to seriously mislead. Anselm's characteristic way of stating the relation of God's infinite life to the temporal series of the creation is to say that eternity is that "in which all periods of time are contained," and that eternity "encompasses" or "envelops" time. Even the repetition of the seemingly innocuous phrase "in eternity" easily lends itself to a temporalizing of eternity and of time, if not in the mind of Anselm himself, then in the minds of some of those who endeavor to understand him. It encourages them to think of God's eternity as lived out in a "divine time," which embraces all other times. The presence of this danger is one reason why it is seriously misleading to think of God himself occupying the B-series of time, perhaps by performing the amazing feat of occupying all times of the B-series "at once." This temporalizing tendency may be aided and abetted from another classical source of atemporalism, from Boethius's understanding of the relation between infinite eternity and time on analogy with that of a hilltop spectator and what he sees of the goings-on in the valley beneath him.

In the same vein it is said that all times are present to God. But this language is not intended, or in my view should not be intended, to imply that the temporal order is *contained in* something more temporally capacious than itself, in eternity. Just as God's omnipresence is not spatial and cannot literally encompass space, so his eternality is not temporal, and in my view it is not wise to consider atemporality in quasi-temporal terms.

But at least in the case of Augustine, temporality comes with the creation and is a mark of creatureliness. So the changeless God who is the Creator of the created order must be apart from temporality, which is a sequence and a mark of change. This gives us the warrant to assert a more consistent atemporalism than we possess if we include Anselm's talk of it as involving God encompassing all times in his eternal present.

Expunging these suggestions of an eternal duration, how are we better to understand atemporalism, then? I suggest, by thinking of Augustine and other eternalists as (maybe despite themselves) referring to the mind of God, which exists atemporally. The mind of God does not contain all times, by God "visualizing" them all in his consciousness, but rather by his immediate intellectual access to all times, and his immediate voluntary control of what occurs at a time. And he would have had such control of any alternative universe he might have chosen to create. As we saw when discussing the *tu quoque* argument, such control must be qualified by or refracted in the distinct natures of what is created. If the language of "containing" continues to be used in connection with the mind of eternal God containing all times, it must be thought of as more like the way in which valid conclusions are "contained in" their premises or the connotations of a word are "contained in" that word.

Thus the mind of an infinite God may be said immediately to know (not immediately in a temporal sense, but purely intellectually, without the need for any intermediate ratiocination), and thus immediately to "comprehend," the temporal order, and to support that order from a point atemporally distinct from it. All times are equally accessible to God's infinity in a way that does not require his own existence in time, much less his temporally simultaneous awareness of all times. Indeed, as I am suggesting, it requires that he *not* exist in time.

One might wonder if God's atemporality, understood in this way, provides him with the leverage of being able atemporally to affect events that are now past to us. This would be to impute to him a lordship of a different kind, that of having a uniform relationship to all parts of time. Perhaps he could eternally decree that part of the temporal series that is past to us is extinguished. Perhaps it's only his wisdom that prevents him doing so, if indeed he does not do it. Did I undergo a tragedy that is now some years in the past? That past tragedy is certainly as real as my present pleasure. That is a possible state of affairs on this understanding of atemporality. But perhaps that part of the series in which my pain exists is to be extinguished. Perhaps the B-series under theistic auspices can have some of the features of some variants of the A-theory in that certain periods are not simply over and done with but deleted from the record. However this may be, this speculation serves to remind us that the exact topography of a B-series, like the entire series, is metaphysically contingent matter.

So all times are equally accessible to God because they are all equally in the mind of God. All periods of time are "contained" or "encompassed" in the mind of God, not in some other "temporal place" but in the only atemporal "place" there could be, namely, God's mind. To exist "simultaneously in an eternal present," as Anselm in *De Concordia* (1998, 443) puts it, surely with an air of deliberate paradox, is for two or more events to exist together in the mind of God, but not in any way that confuses or conflates the divine understanding of them. This mind contains matters, Anselm tells us, that do not exist for us in time, and that some things exist in time which do not exist in eternity (here it is tempting to insert an "already"). Expressed in this way there is less temptation to insert an "already" or any other temporally modifying adjectives and adverbs (cf. Helm 2010, 36–37).

I shall attempt to work this out a bit further. Delmas Lewis says this: "Strictly speaking, then, it cannot be said that God *did* exist in the past or *will* exist in the future, because he does not exist pastly or futurely: he simply exists in the timeless mode of existence peculiar to an eternal thing" (quoted in Hasker 2002, 183). That is well said. We can say that for any feature in our past it was true then that the atemporal God exists, and in our future it will be true that he exists, but not, on either occasion, that it is true that he *has* existed or *will* exist. Lewis thinks that these states of affairs present a problem for the atemporalist, the limitation problem we discussed earlier, but I think they offer a solution to a problem for him.

God cannot have immediate knowledge of the temporal order of creation in the sense that such knowledge is in the least sense temporal, nor may he set out in his mind's eye or in his "consciousness" or "experience" of time as a B-series. The atemporality of God is not a matter therefore of the character of God's experience. He does not experience temporal events immediately—what would that be like?—but rather he has an atemporal intellectual grasp of them. Hasker (2002, 185) claims that as a consequence of asserting divine atemporality "temporal beings also exist in eternity and as such are present to God." But here he seems to have in mind eternity as a timeless duration that he finds in early medievals such as Anselm. It is one thing to say that according to the B-series times have a tenseless existence, another to say that time "exists in" God's atemporality.

Such arguments as Gale's and Hasker's against an atemporal God having immediate knowledge appear to turn out to be the indexical objection all over again, that God does not have knowledge as a creature does when the creature at a particular time, say, knows by seeing. God is thus deprived of indexicality, particularly of any ability to employ the temporal "now" as an agent in time employs it. As a consequence God is hamstrung, for he is on the wrong side of the barrier that the atemporalist has erected, and maybe in that position he is not a fit candidate to be infinite; he has so to speak disqualified himself (Hasker 2002, 198). But that does not mean that he exists in time understood only as an A-series. We have already seen the lameness of this objection as an argument for the superiority of sempiternity over atemporalism, it being open to countercharges against temporalism that result in a stalemate. So I shall concentrate on my chief claim, namely, that God immediately possesses and exercises the Creator's control of all times by willing or by some other activity implies and is implied by a clear sense of his infinity.

To do this I shall consider some other features of the argument of Hasker's in his article "The Absence of a Timeless God" that have already been referred to. And first I shall consider the claim that God's atemporality expressed in some words of Anselm's means that "while God performs actions that have temporal consequences, the acts themselves are not temporal" (Hasker 2002, 183). The assumption here seems to be that to act in time, time has to be considered an A-series in which the agent is situated and in which he operates. But why? As we have seen, it is true that "an atemporal God bears no temporal relations to any object or event whatever" (Delmas Lewis, quoted by Hasker 2002, 183). But what does this show? It shows that an atemporal God may nonetheless atemporally will that an event P occurs in time and that an event Q occurs, which follows P in time. Why is this atemporality a "barrier"? What does its fact prevent?

Hasker also says that it is a consequence of atemporalism that evanescent events become eternalized. Atemporalism, he states, "is seen to be dubious because it assumes that temporal objects exist only for a brief period. In time, to be sure their existence is so, limited, but that they also possess an existence that knows no temporal bounds" (Delmas Lewis, quoted by Hasker 2002, 196). Due to the infelicities of expression of Augustine and Anselm, for example, and all who slip into thinking that God's eternity is durational, their view carries the consequence, according to Hasker, that in addition to their temporality, states and events have an eternal duration in God's eternity. Perhaps some of Anselm's language does carry such an implication, but a more consistent atemporalism need not.

The price that has to be paid for this inconsistent temporalism is "[the extravagance] involved in the idea that all the familiar things and events of everyday life do not readily pass away, as we seem to see them do, but instead persist eternally. . . . The chalk-stub that I wore down on the chalk-board and eventually threw away—that this chalk-stub literally exists in eternity . . . is a mind-boggling thought, and one might well wonder whether eternal existence is not being dispensed a bit too freely" (Hasker 2002, 196).

Indeed. But the chalk stub does not literally exist in an eternal temporality—how could it?—and the thought that it exists in eternity can only come about because the B-series is being construed as a form of temporal duration. But if God exists truly atemporally, temporal objects and events continue to hold their place in the temporal series, the entirety of which is sustained atemporally by him. Such comments as Hasker's seem to treat God's atemporality as if it is a state of temporal unboundedness, which "persists." He says, "Now we see that the temporal realities are available to God in eternity" (Hasker 2002, 196). But what exists atemporally is God's immediate knowledge of the entire temporal series, a series that contains all things that exist for a time within that series, having beginnings and endings that are eternally known to the divine mind.

So Hasker's characterizations of divine atemporality seem to me to be infected with elements of that "eternal temporalism" or "eternal duration" that I have suggested are present in Anselm's idioms and elsewhere. But I have also argued that those idioms need constant watchfulness in order to maintain an account of atemporalism in an unsullied form.

One cannot parse atemporality by thinking that "future events of the world, including your and my future, always exist in the timeless eternity of God" (Hasker 2002, 197). One cannot parse God's eternal control using temporal expressions such as "always" or "already" or "simultaneous with" to say that they "always exist in the timeless eternity of God" (197). It is not possible to express strict atemporalism by referring to the actions that take place in the temporal order as "actions [that] are already present in the divine eternity" (197). There is no need to understand the "foreknowledge" of God in the sense that God knows temporally before the temporal order of the creation, or some part of it (198). And so on. Rather, the use of expressions like "God foreknows" refers to a temporal knower's recognition that what "God eternally knows" is true before the present moment. It was just as true yesterday, and in 1066, that God eternally knows (cf. Helm 2010, 98–99).

Against the impure or inconsistent atemporalism that I suggest can be found in Anselm and others, Hasker's points may well be valid. So this is not a comment on the soundness of his arguments but on the satisfactoriness of one of his premises.

It may be that I have misunderstood Hasker at this point. If so, I reiterate that any B-series (at least from a theistic point of view) is contingent, in the part or in the whole. Hence my earlier speculation regarding God's lordship over it. And if Hasker's thought is that the series has an ontological necessity, and hence its content exists eternally, this does not take account of God's ability to engage in annihilation. (In any case, what's so awful about there being a chalk stub in the B-series, so long as there is such a series?)

Earlier I referred to Boethius. In his article "The Eternal Present," Leftow argues that it is possible that God's life has for its temporal aspect one permanent event, and that this was Boethius's view (Leftow 2002, 40). Such would be a concept of atemporality without temporal duration, for being temporally present as distinct from being eternally present means that the temporally present passes, but the eternal present does not pass. The eternal now is a permanent occurrence, not transient. For according to Boethius and Augustine, for example, only what is present is real.

If one holds such a concept of the eternal now, having the character of one necessary event, it would seem not to help with the idea of such an eternal being experiencing, say, the activity of something in time distinct from itself, or possessing a stream of consciousness. I dare say that such a concept of eternity may avoid some of the difficulties of ascribing duration to eternity. It is not my idea to rule out all other concepts of divine eternity than what I have called pure atemporalism, since I have no wish to claim exclusivity for this view. It may be that Boethius's view, as articulated by Leftow, would also give us a satisfactory account of the eternality of an infinite being. Nonetheless Boethius's notion as understood by Leftow is a curious notion. It is not clear what the advantages of such eternalism might/would be. Perhaps the thought is that unless God dwells in time, unless he "occurs," however this is qualified, he cannot be alive. But what vivifying power is conveyed by an eternal being's occupancy of an eternal event such as Leftow attributes to Boethius's view is not clear.

The objection may be made that strict atemporalism is a piece of "negative theology." But, as we have seen, quite a bit that is positive can be said of it: that God is outside time, that he has the same relation to all times of the created order, and (as befits an infinite being) that he has immediate access to all such times. But of course it is true that we cannot begin to know what it is like to be God, what sense if any it makes to say that God is conscious. Nor could we have access to the atemporal *verstehen* by which God understands what it is like to be an agent or a patient in his creation. This is, after all, God *in se*, about whom our discussion must perforce be somewhat tight-lipped. And such a recognition, I have to admit, cannot contribute positively to my peroration here.

SUMMARY AND CONCLUSION

I have tried to make a case for what I think is a consistent atemporalism, a sense that does not imply temporal duration for God of even the sort that we find such as Anselm employing, or which some of his language suggests. Prior to that I argued that on any account of the Creator-to-creature relationship, whatever the details, there are things that the Creator cannot do that his creatures can do, and so such inabilities cannot in themselves be a serious barrier to God's infinity. Instead I have proposed as a test of infinity a principle about immediate accessibility, cognitive and volitional, to all moments, where immediacy has a purely intellectual sense. There seems to be no reason why an atemporal God should not have that ability. However, I do not claim exclusivity or sufficiency for this strict atemporalist view, and it may be that other views will also do the trick.

At the same time, I realize that such consistent atemporality imposes on its proponents the discipline of saying less rather than more about the details of the divine life. So I shall not try to say more.

NOTE

Thanks to James Dolezal for comments on an earlier draft.

REFERENCES

Anselm of Canterbury. 1998. *De Concordia*. In *Anselm of Canterbury: The Major Works,* edited with an introduction by Brian Davies and G. R. Evans, translated by Thomas Bermingham. Oxford: Oxford University Press.

Aquinas, Thomas. 1969. *Summa Theologiae*. Edited by Thomas Gilby. Garden City, NY: Image Books.

Augustine. 1992. *Confessions*. Translated by Henry Chadwick. Oxford: Oxford University Press.

Craig, William Lane. 2006. "Pantheists in Spite of Themselves." In *For Faith and Clarity,* edited by James K. Beilby, 135–56. Grand Rapids, MI: Baker Academic.

Gale, Richard M. 1991. *On the Nature and Existence of God.* Cambridge: Cambridge University Press.

Hasker, William. 2002. "The Absence of a Timeless God." In *God and Time: Essays on the Divine Nature*, edited by Gregory E. Ganssle and David M. Woodruff, 182–206. New York: Oxford University Press.

Helm, Paul. 2010. *Eternal God.* 2nd ed. Oxford: Oxford University Press.

Leftow, Brian. 1991. *Time and Eternity.* Ithaca, NY: Cornell University Press.

———. 2002. "The Eternal Present." In *God and Time: Essays on the Divine Nature*, edited by Gregory E. Ganssle and David M. Woodruff, 21–48. New York: Oxford University Press.

Plantinga, Alvin. 1993. "Divine Knowledge." In *Christian Perspectives on Religious Knowledge*, edited by C. Stephen Evans and Merold Westphal, 40–65. Grand Rapids, MI: Eerdmans.

Prior, Arthur N. 1959. "Thank Goodness That's Over." *Philosophy* 34:12–17.

———. 1962. "The Formalities of Omniscience." *Philosophy* 37:114–29.

Robinson, Howard. 2007. "The Self and Time." In *Persons: Human and Divine*, edited by Peter Van Inwagen and Dean Zimmerman, 55–86. Oxford: Oxford University Press.

Rogers, Katherin. 2000. *Perfect Being Theology.* Edinburgh: Edinburgh University Press.

Swinburne, Richard. 1979. *The Existence of God.* Oxford: Clarendon.

Turretin, Francis. 1688–90. *Institutes of Elenctic Theology.* Translated by George Giger, edited by James T. Dennison Jr. Philipsburg, NJ: P & R Publishing.

CHAPTER 15

Infinite Goodness

BRIAN LEFTOW

A chapter on infinite goodness should say something about infinity and something about goodness. What I have to say about infinity is historical: I show how Aquinas understood it. What I say about goodness is constructive: having shown a connection between one understanding of divine infinity and perfection, I give part of an account of divine goodness that incorporates this.

AQUINAS ON DIVINE INFINITY

There is nothing mathematical about Aquinas's thinking about divine infinity. He is explicit that the only concept of infinity that can apply to God is nonquantitative (cf. Aquinas, *Summa contra Gentiles* [*S.c.G.*] I, c. 43). Aquinas approaches the concept etymologically: to be in-finite is to be without *finis* (end or termination). What ends or terminates something places a limit on how far it extends. So for Aquinas, to say that God is infinite in some respect is just to say that He is in unlimited in that respect (I, c. 43). We can see how he thinks of divine infinity by examining some moves he makes with the concept.

Thomas holds that God's power is infinite in a number of respects, but the key one is that there is no limit on how many effects He can have (Aquinas, *De potentia Dei* 1, 2). Thomas infers from this lack of limit that He is omnipotent, that is, that He can bring about all states of affairs such that the claim that someone produces them does not entail a contradiction (see Leftow 2012, 187–95). Behind the inference is the thought that if God could bring about less, His power would be not enough to accomplish something, and that would constitute a limit on how many effects He can have. Unlimited power, then, is for Aquinas all possible power: being unlimited entails having all that is possible in this respect. All is, of course, the maximum possible, and so it also entails having the greatest possible power. Again, the link is plausible. Any power less than the greatest possible is by comparison limited, and the more limited, the more it falls short of this.

Thus here Thomas takes being unlimited in a respect to entail having all that could be had in that respect—the greatest possible amount—of what there is to have in that respect. This doesn't entirely escape quantity, as we see it: "all" is a quantifier, and the notion of a maximum—the most—is quantitative. But Thomas would insist that here we are still outside the realm of quantity: as an Aristotelian, he saw quantity as a property of matter, and so no divine property could be strictly speaking quantitative.

Another move Thomas makes is likely to seem strange to us. Thomas reasons thus:

> IW: If there were an infinite whiteness, one not limited by a matter that receives it, it would have all that could be had of the perfection of whiteness. (*S.c.G.* I, c. 43)

It may seem odd to think that matter limits the perfection of whiteness: Isn't being material instead a necessary condition of being white? We can make sense of (IW) in extensive and intensive terms. Matter limits the extent of whiteness: there cannot be more whiteness in the world than there is matter to reflect light. Matter also limits the intensity of whiteness: the whiteness is only so bright and no more because paint can only be mixed in certain ways, or light can only be so intense and so on. Necessary conditions can also be limiting conditions. But whiteness *not* limited by matter would be immaterial, and what could immaterial whiteness be? Perhaps (IW) is just a counterpossible about a Platonic Form of whiteness. If it is not, one answer that springs to mind is the content of God's perfect idea of perfect whiteness. God has the maximum possible whiteness in mind, the standard by reference to which white things count as white, Thomas's analogue to the Platonic Form of white. This is perfect whiteness. So here for Thomas, again, being unlimited or infinite in a respect implies having all that is possible in that respect, being perfect in that respect (as omnipotence is perfection of power). The kind of lack of limitation differs, and that difference matters if we are evaluating the cogency of (IW). But I am not claiming that (IW) is true. I am merely pointing out what it tells us about Thomas's thinking. For Thomas, it seems, being unlimited in a respect implies having all that is possible in that respect, and so the most that is possible in that respect, and so perfection in that respect.

Thomas also thinks he sees a connection running the other way. Thomas infers from the claim that God's nature contains the perfections of all kinds of thing—that is, somehow contains all possible perfection—that it is unlimited in perfection (*S.c.G.* I, c. 43). Again, Thomas infers from the claim that a greater goodness than His cannot be conceived—that is, given Thomas's approach to possibility, that a greater is not possible, and so He has the greatest possible—that God is unlimited in goodness (*S.c.G.* I, c. 43). Thus "greatest possible" implies "unlimited" for Thomas. Thomas seems to take these connections as conceptual and so a priori. We see, then, that for Thomas, to say that God is infinite in a respect is a priori equivalent to saying that He is the greatest possible, maximal, or perfect in that respect, and to saying that He has all that is possible in that respect when there is a sensible "all" to be had.

I now offer part of an account of perfect goodness, whose key definition trades on the connection between being unlimited—infinite, in Thomas's root sense—and being perfect.

DIVINE VIRTUES

Western scriptures ascribe virtues to God: He is benevolent, just, merciful, compassionate, faithful. As a result, Western theist philosophers have held that God's moral perfection includes being perfectly just, merciful, and so on. I first defend the claim that God can have virtues. I then define the perfect degree of a single virtue, and support the definition by showing how it resolves two puzzles about God's virtues. Finally, I argue that the perfect degree of a virtue would have a strong stability property.

Can God Have Virtues?

I think the claim that God has virtues emerges simply from the claim that He is perfectly good. Suppose that God is perfectly good. Then as part of this, when it is appropriate to be just, He is, unless perhaps He deems it more appropriate to be merciful. When it is appropriate to be merciful, He is. Benevolence is always in season, and He is always benevolent, willing to all that exists the good of existence and much more. Now virtues are just dispositions of a certain sort, and dispositions are the *in re* equivalent or correlate of functions from circumstances to effects: whatever else the virtue of mercy is, it is something that takes us from inputs of certain sorts of beliefs about the world to outputs of attempts to do characteristically merciful acts at least some of the time, the proportion being what is suitable to the level of virtue. Whatever God's perfect goodness is in itself, be it one thing or many and be it or not something that somehow transcends the moral sort of goodness, it takes God from inputs of completely accurate beliefs about the world to outputs of just, merciful, and suchlike acts—acts that would be characteristic of agents with the virtues of justice, mercy, and so on if they were divinely equipped and informed. God's perfect goodness realizes the right functions. It walks like a duck and quacks like a duck. So either He has one thing that somehow does duty for all virtues and makes a suitable range of virtue-predicates applicable,[1] or He has many distinct virtues. I will speak the latter way for convenience.

Objections to Divine Virtues

Yet it does not sound quite right to say that God is virtuous. Perfect goodness, one wants to say, should be above and beyond mere virtue. Perhaps this is because "virtuous" conjures to mind someone who has had to work to be good and labors to remain so. The idea that virtue is essentially something to exercise against contrary tendencies has had advocates. Von Wright characterizes virtues as forms of self-control, concerned essentially with the rational control of passion or desire—to the extent of saying that one cannot act virtuously unless one is at least prey to the sort of temptation the virtue deals with (von Wright 1963, 146–47).[2] A perfectly good being, of course, would not be prey to temptation (leaving incarnational complications aside). Philippa Foot has a thought like von Wright's in calling the virtues essentially correctives, so that "it is only because fear and the desire for pleasure often operate as temptations that courage and temperance exist as virtues at all" (Foot 1978, 9). But Foot adds that although this is the virtues' general nature, it does not follow that each individual, to possess a virtue, must actually face the temptation it masters: she grants, as von Wright should have, that it is possible to be generous without having selfish impulses to master (10–11). If God is morally perfect and has virtues, all of them are ways of tending to the good that have no tendencies to the bad to counteract. God does not have to work at virtue. In this way His perfect virtue takes Him beyond being virtuous as we see it in human cases.

Another sort of objection to God's having virtues comes from a certain sort of Thomist. For Thomists, God is "pure actuality." He has no potentialities. Virtues are dispositions, potentialities to act. So, some Thomists conclude, God cannot have virtues, though biblical virtue-terms might apply to Him analogically (Davies 2011, 60–61). But Thomists are Christians. If God is perfectly good and can become incarnate as a man, He can in some way or other have virtues. For as incarnate, He is human. As perfectly good, He could not be incarnate as a morally imperfect human, and no human is morally perfect without virtue. The most Thomists can insist here is that God the Son's virtues are realized in His human natural endowment (which, of course, can have dispositions).[3] But if they are not thereby *God the Son's* virtues, if these dispositions are not thereby God's, then they are someone else's and there are two persons in the incarnate Christ—which is the Nestorian heresy. So for orthodox Christians, at least, that God can in some way have virtues is not in doubt.

Further, Aquinas distinguishes virtues that can be ascribed to God only metaphorically from others like justice and generosity (Aquinas, *Summa theologiae* [*ST*] I, q. 21, a. 1, ad. 1). The proper contrast to "metaphorical" is "literal." So even Aquinas holds that God literally has virtues, albeit how these literal virtues are realized in an atemporal, simple being differs from how they are realized in us. Dispositions are compatible with atemporality. Actually, God does this and that just act; due to something about Him, faced with different circumstances, He might have done others instead. That something about Him is a disposition that might have manifested differently, though neither actually nor counterfactually does the manifestation involve growth in virtue or change of any other sort.

Virtues include a characteristic motive. One is not compassionate unless one habitually acts from the motive of compassion; someone who acts as the compassionate would, but always from desire for praise, is not compassionate. We often think the motive here to be an emotion, and compassion is often understood as including *feeling* a certain way. Thus any who deny that God can have emotions also question whether He can have some or perhaps all virtues. Anselm bit this bullet, denying that God is literally merciful. Yet Anselm also gave us a way to ascribe the virtue literally to God, though he did not think he was doing so:

> You are impassible [and so] you do not have any compassion [and so] your heart is not sorrowful from compassion with the sorrowful, which is what being merciful is. [But] in act you are merciful according to our way of looking at things . . . when you look upon us in our misery . . . we . . . feel the effect of your mercy, but you do not experience the feeling. Therefore you are both merciful because you save the sorrowful . . . and . . . not merciful because you do not experience any feeling of compassion for misery. (Anselm 1965, 125)

Anselm requires for literal possession of mercy the experiencing of certain qualia. This is why he denies that God is merciful. But he notes that there is nonetheless in God the characteristic input/output pattern of the merciful: given inputs of perceptions of occasions to be merciful, there are for that reason outputs of merciful behavior. If we think of the mind of God as functionalists think of minds generally, as causally linked states realizing characteristic functional patterns, we might well find there not just the characteristic input/output pattern of mercy, but the patterns of intramental state causation characteristic of the merciful. If some virtues are states whose essences are functional rather than qualitative, we can simply bypass the "feeling" objection. And some appear to be of this sort. Perhaps to be brave requires mastering fear and so requires feeling fear, but there is nothing particular it need feel like to be just, or faithful, or benevolent. Perhaps even compassion can be based on a desire to alleviate suffering rather than an emotion about it. I do not myself have a problem with the claim that God has emotions. But those who do, I have argued, do not have sufficient reason to deny that God can have virtues.

If God has virtues, He has perfect cases of all those He has—perfect justice, benevolence, and so forth—unless it is impossible to have them all perfectly. I now give a simple account of what it consists in to have a virtue perfectly, and show that it dissolves two puzzles about God's virtues. First, the puzzles.

Two Puzzles

One is a cleaned-up version of Rowe's a priori argument for atheism. Rowe's argument hinges on what he calls "Principle B":

> B: If an omniscient being creates a world when it could have created a better, then it is possible that there exist a being morally better than it. (Rowe 2004, 91)

(B) speaks of creating worlds. But Rowe's worlds are possible worlds, and by "creating" Rowe really means actualizing a possible world (Rowe 2004, 77n4). Further, though Rowe doesn't speak specifically of divine virtues, the respect in which someone would be morally better is plainly being benevolent. So Rowe's (B) should really be this:

> B2: If an omniscient being actualizes a world when it could have actualized a better, it is possible that someone be more benevolent.

Now for a version of Rowe's argument. It is standard perfect-being theism that

> 1: If God exists, God is omnipotent, omniscient, and perfectly benevolent.

Rowe assumes that

> 2: If God exists, God has actualized the actual world.

The argument needs (2)—if God has not, (B2) doesn't apply to Him. Further,

> 3: If God exists, for every world He can actualize, there is a better world He can actualize. (Rowe 2004, 88)

Given (2) and (3), if God exists, He has actualized a world, but could have actualized a better. But then *per* (B2), if God exists, it is possible that someone be more benevolent than God actually is. *Per* (1), if God exists, God is perfectly benevolent, and

> 4: Necessarily, no one can be more benevolent than someone perfectly benevolent.

So if God exists, no one can be more benevolent than God actually is. And so if God existed, a contradiction would follow, namely, that it is and is not possible that someone be more benevolent than God. So given (1) to (4) and (B2), God does not exist. Like arguments are fairly widespread in the literature (see, e.g., Quinn 1982, 197–213; Grover 1988, 244; Conee 1994, 815–25). Now (4) seems certain, (3) seems almost as plausible, and theists insist on (1). An obvious place for a theist to jib is (2): if we have libertarian freedom, as most theists think, then God does not actualize the actual world. Rather, we and He actualize it together. But one can reboot the argument in terms of actualizing all of history before the first free creature or free material creature existed. That neutralizes (2). So the theist needs a way to reject (B2). A good understanding of perfection in benevolence should provide this.

Our second puzzle is of long standing in Christian theology: How can God be both perfectly just and perfectly merciful? Here is Anselm's (1965, 123) formulation: "How do You spare the wicked if you are . . . supremely just? For how does the . . . supremely just do something . . . unjust? Or what kind of justice is it to give everlasting life to him who merits eternal death?"[4] To the extent that one is just, Anselm thinks, one applies due penalties. If one is merciful, one sometimes remits due penalties. So acting mercifully requires not acting justly. But presumably someone perfectly just would always act justly when justice is due, and someone perfectly merciful would always be merciful when mercy is possible. Thus it seems that no one could be perfect in both justice and mercy. A good understanding of perfection in virtue should deal with this. I now try to provide one.

Defining the Perfect Degree of Virtue

Erik Wielenberg (2000, 42) defines omnipotence this way:

> For every being *x*, *x* is omnipotent if and only if it is not the case that there is some state of affairs p such that *x* is unable to bring about p at least partially because of a lack of power in *x*.

This won't do, for reasons I give elsewhere (Leftow 2009, 167–98). But if we restrict p to possible states of affairs, we can see that Wielenberg's basic thought was that to be omnipotent is to have power enough to bring about anything possible. It does not for Wielenberg entail having knowledge or rationality enough, so it is compatible with having a restricted range of feasible effects, but it is (so to speak) having raw strength enough. Note that to lack some power is to be limited in power: one reaches one's limit where one's lack begins. So the basic thought in Wielenberg's definition is that omnipotence is unlimited strength, strength enough to bring about any possible state of affairs.

A parallel thought would be that to be maximal in (say) benevolence is to have virtue enough to do any possible benevolent act, whether or not other things are conducive to performing such acts, so that perfect benevolence is compatible with having a restricted range of benevolent effects. So I suggest as a definition of perfection in any virtue V that

> For all *x*, *x* is perfect or maximal in V just if for any act A, if *x* did A and A were surpassable as an expression of V, this would not be due even partly to a lack of or defect in V.

Again, one reaches one's limits where one's lacks or defects begin. So here the basic idea is that to be perfect in virtue is to have unlimited virtue. But this broad sort of definition is available for virtue of any degree:

> *x* has V in degree *n* just if for any act A, if *x* did A and A were surpassable as an expression of V, this would not be due even partly to a lack of or defect in at least *n*-degree V.[5]

Applications

This sort of definition lets someone perfect in V assemble only a surpassable record of V-acts. It allows that someone might do more and/or better V-manifesting acts than someone perfect in V. It insists only that the reason the V-perfect person's record could be surpassed not be defect in or lack of V. On this definition, that Smith gives more or better gifts than Jones doesn't entail that Smith is more benevolent than Jones. Thus too, on my definition, that in one possible world God gives more gifts (by creating a better universe) than He does in another doesn't entail that He is more benevolent in the one than in the other. My definition entails only that in any world God makes, if some goods are lacking, it won't be because some benevolence is lacking.

If this is the right way to define perfect benevolence, perfect benevolence is compatible with a surpassable record for benevolence. If it is, the move from (B2)'s antecedent to its consequent is just a non sequitur. From the fact that someone's record for benevolence can be surpassed, nothing follows about whether the degree of someone's virtue of benevolence can be surpassed. Perhaps the record is surpassable because the degree of virtue is. But perhaps the degree of virtue is maximal and the record is surpassable for some other reason. The truth of (B2)'s antecedent is compatible with the falsity of its consequent. That is so only if (B2) does not state even a valid material implication. But if it does not, (B2) is false.

Perfection in virtue entails nothing about quantity of acts or effects. Still, being perfect in a virtue could have entailments about acts' or effects' quality, as I now suggest.

Qualitative Perfection

If I have the power to whittle wooden dolls, I cannot ever use this power as much as it can be used. However many I make, I could always have made more: quantitatively, my actual whittling-record cannot be unsurpassable. But I can be a perfect whittler qualitatively. It can be the case that each doll I make is flawless, however few they are. Of course, one whittler who produces only perfect dolls can be better than another in other ways—if I need a year to produce one and you only a day, clearly in one respect you are better at it than I am, and ditto if I get it right only by rectifying many mistakes while you make not one imperfect stroke.

Still, in respect of quality of eventual product, there is an obvious maximum not affected by how many I whittle, or how long it takes: getting them all perfect. Perfection in benevolence can have entailments about quality of benevolent acts. If I am perfect in benevolence, none of mine will deviate even slightly from the Aristotelian mean toward either excess or stinginess. If it is not obvious that there is a vice of excess in benevolence, consider this: Would you rather have a rich aunt with awful taste or a poor one who gave you just the right thing for Christmas? In gift-giving, more need not be better. Perfect benevolence entails giving the most appropriate gifts, not the largest.

So too, I suggest, God's justice can be perfect qualitatively: His penalties can be perfectly suited to the criminal and the crime. Two judges can be perfectly just even if one decides fewer cases. Per my account above, that they are perfectly just implies only that if any of their decisions is surpassably just, this is not due to a lack or defect of justice, but has some other explanation: they decide whatever cases they get as justly as the evidence, argument, and the functioning of their faculties permit. Thus if each always has full evidence, is always at peak efficiency and so on, both decide all cases just as they should. If both always decide this way, each compiles a record for justice that is *qualitatively* maximal: it contains only perfectly just decisions. If being perfectly just were a matter of doing more rather than fewer just acts, a judge with a short career could not be in any sense perfectly just. This is implausible. Judges can be equally just even if they produce different quantities of justice. If God makes a world with ten moral agents, and all get just what they deserve, He has not (I submit) been more just than He would have in making a world with just nine and so treating *them*. He has been just to more people. There are more cases of justice. But this doesn't amount to having been more just. The ten-world and the nine-world are equally, perfectly just; a world is not more perfectly just for containing more cases of perfect justice, for all it takes for a world to be perfectly just is to contain some cases of perfect justice and none of imperfect justice or injustice. God has been equally just in a world with more or fewer cases of justice. If this is plausible, then plausibly a perfect record for justice is a matter of quality, not quantity of acts.

I incline to treat benevolence similarly. Consider a world of ten persons, to all of whom God gives the gifts best for them, and one with just nine, treated the same way. If God actualizes the ten-world rather than the nine, He has been perfectly benevolent to more, but His record for benevolence may be qualitatively perfect in both worlds. A perfect record for benevolence may just consist in being perfectly benevolent to each being benefited, however many or few there are. Further, perfect benevolence to each is a matter of the most appropriate gift, not the largest.

I have suggested that there is a qualitatively defined maximum record for recipient-relative benevolence, namely, qualitatively perfect benevolence to each recipient. That the most apt gift some can receive is larger than the most apt gift others can receive doesn't entail that God is more benevolently disposed to those who receive the larger gifts, though He realizes more good by giving larger gifts. Thus there *are* expectations about effects that go with the claim that God is perfectly benevolent. They are just not the ones Rowe's argument requires.

Justice and Mercy

Let me now take up the justice/mercy issue. Mercy is better viewed as a-just rather than unjust: in doing mercy one simply steps outside that particular order of moral considerations. (Granted, to the dutiful son or those who have worked for eleven hours in one day, mercy to the prodigal or those hired at the last hour don't *look* simply a-just.) On my account of perfection in virtue, being perfectly just does not require maximizing a quantitative justice-record. It entails only a maximal qualitative justice-record, and that where an act does not express the virtue of justice, the reason is not any lack of that virtue. We can add that someone perfectly just will deviate from justice only for a morally suitable reason. But my definition allows there to *be* morally suitable reasons. A desire to be merciful is compatible with perfection in justice; one need not lack in justice just because one feels the urge to mercy. This desire can be a morally suitable reason not to exact a due penalty in a given instance, and so one compatible with overall moral perfection. Thus on my definitions, although *expressing* maximum justice in a given case is not compatible with *expressing* maximum mercy, *possessing* the one is compatible with *possessing* the other.

Saying only this leaves other puzzles that exercised Anselm untouched—for example, why God might choose to be merciful to some and strictly just to others. But that puzzle arises only within a strong predestinarian framework. Theists are free to reject that, and say that God wills to be merciful to all, and will apply strict justice only to those who do not choose to accept His mercy.

Perfect Virtue: Stability

There is much more to say about perfection in even one virtue. But space constraints limit me to one issue: stability. Most writers about virtue agree with Aristotle that virtues are stable character traits, not easily dislodged and so tending to persist. If I am easily caused to become unjust, by (say) one reverse in my life, then I was not very just at all: that seems plausible. But if it is true, then if I am very just, I will not be easily caused to become unjust, and so my justice will be a fairly stable character trait.

Some, however, demur. Thomas Hurka defines virtues simply as intrinsically good attitudes to goods and evils, and wants to let these be transient (see Hurka 2001, 20, 42). This sounds odd. If I am habitually cruel but suddenly have a rush of merciful feeling and spare a potential victim, it doesn't seem sensible to say that I was a merciful man for a minute, save as a colloquial way to say that I briefly was motivated and acted as merciful men are and act. Virtues are part of character, and character doesn't change that way: instead, it seems right to say that we sometimes act out of character. Yet stability is only a tendency to persist. A tendency can be overridden. Suppose that at t I achieve virtue of a certain degree—stable, but able to be dislodged by a challenge of a particular strength. Perhaps I have become moderately courageous, but still, one truly awful scare could provoke an act of cowardice that would leave psychological scars and strongly influence me to run away ever after. Then one minute after t, I have just such an awful scare. Ever after, I am a coward. Should this really entail that I did not in fact become virtuous at t? If it does entail this, then whether I am ever virtuous, for any length of time, is always dependent on how my future will go. I am virtuous at t, say, only if my virtue lasts till $t + 1$; but I am virtuous at $t + 1$ only if what I have at $t + 1$ similarly lasts till $t + 2$; thus no one is virtuous at any point before death, and, of course, death is too late.

So perhaps Hurka has a point: virtues can be transient. Perhaps the truth lies somewhere between Aristotle and Hurka: virtues are character traits at least intrinsically apt or tending to be stable, but this does not preclude losing them quickly, because any tendency can be overridden, if only by an omnipotent God. Still, for present purposes I needn't take sides. I am after an account of perfect virtue. Even if one can be virtuous for just a minute, by a sudden shift of attitude, it would be better to be more stably virtuous. The more stable the virtue, the more deeply engrafted it is into one's character. Virtue more deeply engrafted into character makes one more admirable morally. The stronger or more intense the virtue, the more it should resist the pressure of sudden reverses, and the stronger the virtue, the more admirable *and* the more stable. By definition, the more stable a disposition, the harder it is to efface—that is, the less likely that an attempt to dislodge it will succeed. So someone more stably virtuous is more likely to retain virtue in the face of (say) a given temptation, and so *ceteris paribus*, someone more stably virtuous will retain virtue longer, in the face of more adverse circumstances. Thus again, it is better to be more stably virtuous. These are the only points my account requires.

Stability so far may sound like a probability-property, but it is really modal: an attempt to dislodge a disposition is less likely to succeed the more of the closest worlds in which it fails. A disposition has maximal conceivable stability in the face of an attempt to erase just if in all worlds in which it meets the eraser, the disposition persists. The maximal conceivable stability of virtue, then, would be a modal property: if an agent attains it at *t*, then in every causally possible continuation of history after *t*, the agent is never less virtuous than at *t*. Such virtue would be proof against diminishment and could not be lost.

It might seem that maximal-degree virtue is possible only if this maximal degree of stability is possible. For *ceteris paribus* one is more virtuous the more stable one's virtue. If the sort of stability just sketched is not possible, there is always a higher degree of stability. So there is always a higher possible degree of virtue, and so no maximum. I now consider two responses to this.

It might seem that the following would be a fix: say that one has maximal-degree virtue at *t* just if at *t*, one's virtue is *meta-stable* at some high point: it is proof against ever being less stable than that, and perhaps is able to occupy any degree of stability above that point. But there would be no nonarbitrary way to fix the lower bound of the range of permissible degrees of stability. More importantly, given any lower bound, a being with a higher lower bound would be possible, and *that*, *ceteris paribus*, would be a being with a higher degree of that virtue.

A better and simpler response is just this: even if the modal property just described is not possible, there might be other modal properties that are possible and provide natural maxima of stability and so of virtue (in this respect). Even if, for any sort of virtue one has, one causally can thereafter be less virtuous, there might be a sort of virtue such that one is thereafter less virtuous only in one causally possible world, or only in two: Why not? Any such property could provide a maximum *possible* degree of virtue stability, and so a maximal possible degree of a virtue, even though a greater degree is *conceivable*. If conceivability entails possibility, then the maximal conceivable degree of stability is possible. If the maximum conceivable degree of stability is impossible, then conceivability does not entail possibility. But if it does not, that we can conceive of greater stability than anything we claim to be the maximum possible degree does not entail that a maximum degree is not in fact possible, and is not in fact the one we claim it to be. So I submit that maximal degree virtue including a maximum possible degree of stability might be possible even if the maximum conceivable degree of virtue stability is not.

This is a good thing, because there is at least some reason to think that a maximal degree of virtue involving maximum conceivable stability is *not* possible. Let us ask whether someone maximally virtuous with maximal conceivable stability can act out of character, that is, voluntarily fail to do an appropriately virtuous act when one is called for, available, believed to be available, and so on.[6] It would seem not, on two grounds. One is that it is hard to see how acting out of character could fail to diminish one's virtue a bit: do it once, and surely it's a bit easier to do it the next time there is pressure to do so, even if perhaps a sense of shame engendered the first time around is there to fight that tendency. But maximal-degree virtue with maximal conceivable stability is proof against diminishment. Further, if we suppose that someone thus stable in virtue could occasionally act out of character while remaining perfectly virtuous, still, if someone is able to act out of character, a maximality problem threatens about the difficulty of acting out of character. The harder it is to do, surely, the more virtuous one is, *ceteris paribus*. But if the maximum—simply being unable to act out of character—is not available, then for every degree of difficulty in doing so, there is a greater, and so a greater degree of virtue.

Suppose on the other hand that someone maximal in a given virtue with maximal conceivable stability cannot act out of character with respect to it: the maximally brave, say, can do no cowardly act. They can move as cowards would, but they can do so only (say) in the belief that (paradoxically) that is the bravest thing to do, or that this is best overall for some other reason—and these motives keep their acts from being cowardly. Then apparently, although the maximally brave may be free with respect to which brave act to do, or how brave to be, they are not free not to act bravely. Perhaps they are responsible for their pick of brave action. But they are not responsible for acting bravely rather than not. They cannot get moral credit for being brave, though they can for *how* brave they are, or which precise way they choose to be brave. But seemingly something is the virtue of bravery only if by exercising it one does acts that bring one moral credit for being brave. To be morally valuable, virtues have to be habits of responsible action, actions for which we get moral credit.[7] If one has a property that necessitates brave actions, one might think, one is not responsible for being brave. If adding maximal conceivable stability to the description of a case of virtue yields something that cannot play the role of that virtue, it is not possible that some case of that virtue be maximally stable. It seems, then, that adding maximal conceivable stability to our conception of maximal-degree virtue generates a problem: someone with such a property could neither have nor lack the ability to act out of character. If we cannot solve the problem, a maximal degree of virtue involving maximum conceivable stability is *not* possible.

Some authors argue broadly on one side of this dilemma. Thus Bruce Reichenbach suggests:

> If it is logically impossible . . . that [God do] wrong . . . then despite the fact that [God] is free to perform a variety of acts, each one of which [would be] right . . . it would seem . . . impossible morally to commend him for performing any of those . . . [For] though [God] is free with respect to performing any given action . . . it is logically necessary that what he does is right. But then what ethical significance can there be to his "choosing the right." . . . Since he is not free to choose . . . other than the right, moral commendation becomes . . . impossible. (Reichenbach 1982, 134)

Stephen Davis has a related thought: "I see no good reason to praise God for His moral goodness if God is unable to sin; likewise I see no good reason to thank God for His benefits to us and answers to our prayers if it is not in His power to be malevolent to us" (Davis 1983, 95).

These arguments do not target divine virtues explicitly, but the second speaks of moral goodness generally, of which virtue is a prominent component, and has in view actions expressing His benevolence, while the first targets a property God has in virtue of His overall set of virtues.

I do not think these arguments succeed. Davis and Reichenbach make a significant slide, from praise simply for doing good—that is, doing some good act or other—to praise for the particular good things God does. Even if God necessarily does some good thing or other, I submit that He can be thanked, praised, and a fortiori credited for the precise goods He chooses to do: for doing this, as versus doing some good or other. God has a choice of good acts. So He can be thanked and praised for doing the ones He chooses. If it is inevitable that He do some good act, still it is not inevitable that He do the particular ones He does.

Suppose I'm in a group of hostages. A terrorist puts a gun to my head and says, "I am going to kill either all of you or all but one. You can pick the one I save. Whoever you pick goes free; if you do not pick, I kill you all." At this point it is right that I pick someone, and if I am perfectly rational, it is inevitable that I will do so, for there could be no reason to refrain. (Even if, e.g., I incline to be partial to white people, and so likely to unacceptably ignore the worthiness of nonwhites I might save, it is better to save someone for reasons including a bad one than to save no one.) Perhaps I save someone I know is about to find the cure for cancer, and let a notorious murderer die. Doesn't the world owe me praise for both choices? Perhaps I pick you. Should your parents not thank me for doing so? Would it be right for them to say, "So what—he had to pick someone?" Surely they owe me gratitude—for though I had to pick someone, I didn't have to pick you. This is true even if I had no special reason to pick you: I still lighted on you and selected you rather than continuing to cast my eyes about the group. I formed an intention that you be saved and acted on it, even if I had no real reason to do so.[8] God doesn't have to pick any specific nonobligatory good act to do, nor any specific obligatory act as long as He has some choice about how to fulfill His obligations—and there is always such choice. So He is still owed thanks for picking the ones He picks. And when we thank Him, we're giving Him credit—and we're right to do so, because He could have refrained from doing these particular goods. The only area of conflict here might come with goods He is obligated to give us. But even here, He could have avoided the obligation, if only by avoiding our existing, and so we can thank Him for letting Himself in for doing what He does, and derivatively, therefore, for doing it.

The most Reichenbach and Davis can suggest, then, is that God might not deserve moral commendation merely for doing something or other good. I now argue that He can, even if He cannot fail to do something or other good. God is morally responsible for doing some virtuous act iff it is appropriate to thank or praise Him for it. Inability to act other than virtuously need not keep these from being appropriate (*pace* Rowe 2004, 114). Your kind aunt gives you gifts. You have always thanked her. But whether you ought to thank her depends on why she does it. If you find out that someone brainwashes her into doing so, you needn't thank her. She is just the brainwasher's tool to give you gifts; if you thank anyone, it should be the brainwasher. If you do thank her thereafter, it might simply be because (say) you know she doesn't know about the brainwashing, and would feel hurt if you didn't. But suppose instead that you find this out: she is not brainwashed. She is not coerced. Nothing like that is true. She is not subject to irrational internal compulsions either. She always gives you gifts because she loves you. In fact, she loves you so much that you are literally irresistible to her. She just can't help herself; she loves you so much that she just has to give you the gifts. If inability to do otherwise is necessary for deserving thanks, upon learning this you should then cease thanking her.[9] You should say that Auntie would deserve thanks only if she loved you less.

Suppose that Auntie has always loved you just below the threshold that would make you irresistible to her, and then one day she comes to love you even more and crosses the threshold: if ability to refrain is necessary to deserve thanks, loving you more would make her cease to deserve your thanks. Here we're in the realm of moral intuitions, and my own, at least, are that caring for me more cannot be a reason for thanking her less, *ceteris paribus*.

I infer that thanks can be appropriate even if the one thanked could not have done otherwise. That is enough to suggest that moral responsibility does not require this. But in addition, it is just a small step from thanks to praise: "Thank you—what you did for me was great!"[10] It remains to ask whether there is anything distinctively moral in these thanks and praises. Well, I thank Auntie for being kind and good to me. Having received her kindness many times and so believing that this expresses her character, I might well say, "Thank you—you have always been kind to me. You're a truly good person!" Such thanks seem appropriate, and include giving her moral credit for what she has done. So if it is appropriate to thank her even if I'm irresistible to her, it's appropriate to give her distinctively moral credit even so.

Now it's a bit odd to think of thanking someone simply for choosing good: "Thanks for doing something or other good" isn't something we typically say, let alone say to someone for whom we know doing evil was never a live option, and in cases where we know exactly what good was done. All the same, it's imaginable. If I know that God has done something good for my friend, and do not know what, this is just what I might say. So I think we can credit God morally for doing something good rather than something evil, and can do so even if He cannot do otherwise in this respect. All of this defeats the dilemma argued above. Even if God cannot act out of His virtuous character at all, He can be morally responsible for doing good, and doing the precise goods He does. His virtues can be habits of responsible action, and so genuine virtues, even if they have the maximum conceivable stability.

I have considered one argument against the claim that maximal-degree virtue includes maximal-conceivable stability. The argument fails, and as the stability property seems perfectly possible and intuitively ought to go with maximal virtue, we ought to conclude that maximal degree virtue would indeed have it. Thus I suggest that all of God's virtues have this property.

NOTES

1. Or, is that one thing, on a strong doctrine of divine simplicity.
2. The remote cause here is Aristotle; see, e.g., *Nicomachean Ethics* 1102b, 1103a.
3. In traditional if unhelpful language, He has them qua man.
4. Anselm 1965, 123.

5. Of course, it may be due to a lack of even more V than *n*.

6. One could also raise a question about acting out of character with respect to the perfect degree of a virtue: Can the perfectly brave fail to do a maximally brave act if one is available?

7. Otherwise (paradoxically) cowards would get more moral credit for their occasional, out-of-character brave acts than the brave do for their characteristic bravery. This is a paradox, even though in one respect, it is harder for a coward to do a brave act than for someone brave to do the same act, and so we might credit the coward for more moral effort.

8. At least, this follows if it's truly I who make the choice. If your parents learn that I simply drew straws, and you lucked out, they would not owe me thanks, because I had not chosen. (My thanks to Tim Mawson here.) But as long as I chose, the following seems (at least to me) a thing it would make sense to say: "I don't know why you did it. I don't care why you did it. I'm just glad you did it. And you did do it, and mean to do it. So thanks!"

9. My Auntie is a variation of Senor's (2008, 186); the arguments I base on her differ from his, but wind up where his do. See also Bergmann and Cover 2006, 400.

10. Thomas Senor (2008, 185) suggests that theists praise God only for who He is, not what He does, for being both all-powerful and disposed by nature to treat us with love. This example suggests that theists also praise Him as a moral agent freely doing good.

REFERENCES

Anselm of Canterbury. 1965. *Proslogion.* In *St. Anselm's Proslogion*, translated with an introduction and philosophical commentary by M. J. Charlesworth. Notre Dame, IN: University of Notre Dame Press.

Bergmann, Michael, and Jan Cover. 2006. "Divine Responsibility without Divine Freedom." *Faith and Philosophy* 23 (4): 381–408.

Conee, Earl. 1994. "The Nature and Impossibility of Moral Perfection." *Philosophy and Phenomenological Research* 54 (4): 815–25.

Davies, Brian. 2011. *Thomas Aquinas on God and Evil.* Oxford: Oxford University Press.

Davis, Stephen. 1983. *Logic and the Nature of God.* London: Macmillan.

Foot, Philippa. 1978. *Virtues and Vices.* Berkeley: University of California Press.

Grover, Stephen. 1988. "Why Only the Best Is Good Enough." *Analysis* 48:224.

Hurka, Thomas. 2001. *Virtue, Vice and Value.* Oxford: Oxford University Press.

Leftow, Brian. 2009. "Omnipotence." In *The Oxford Handbook of Philosophical Theology*, edited by Thomas Flint and Michael Rea, 167–98. Oxford: Oxford University Press.

———. 2012. "Aquinas on Omnipotence." In *The Oxford Handbook of Aquinas*, edited by Eleonore Stump and Brian Davies, 187–95. Oxford: Oxford University Press.

Quinn, Philip. 1982. "God, Moral Perfection and Possible Worlds." In *God: The Contemporary Discussion*, edited by Frederick Sontag and M. Darrol Bryant, 197–213. New York: The Rose of Sharon Press.

Reichenbach, Bruce. 1982. *Evil and a Good God*. New York: Fordham University Press.

Rowe, William. 2004. *Can God Be Free?* London: Oxford University Press.

Senor, Thomas. 2008. "Defending Divine Freedom." In *Oxford Studies in Philosophy of Religion*, edited by Jonathan Kvanvig, 168–95. Oxford: Oxford University Press.

Von Wright, Georg Henrik. 1963. *The Varieties of Goodness*. London: Routledge and Kegan Paul.

Wielenberg, Erik. 2000. "Omnipotence Again." *Faith and Philosophy* 17 (1): 26–47.

CHAPTER 16

Divine Infinity and Personhood

KEN PERSZYK

My main interest in this chapter is with the notion of an infinite person. In particular, can God be literally and metaphysically both infinite and a person? The short answer, I think, is no. The road required to reach that conclusion, however, is long and difficult. There are many different senses or conceptions of infinity, and of divine infinity. Trying to sort them out might remind one of a limerick:

> There once was a chap in divinity
> Who got very interested in infinity,
> But his mind became so muddled
> And his tongue got so befuddled
> That he had to be content with the Trinity.
> (Sweeney 1992, 539)

And, of course, discussions of personhood also constitute a philosophical thicket.

Christians have widely maintained that God is both (in some sense) infinite and (in some sense) a person. If infinity is a divine attribute, it has been one of the least controversial.[1] The claim that God is infinite goes back at least as far as Clement of Alexandria (*Stromata* 5.12) at the end of the second century AD.[2] It is affirmed in the Augsburg Confession (article 1 asserts God's "infinite power, wisdom, and goodness"), the Belgic Confession (article 1 lists infinity as one of the attributes of God), and the Westminster Confession (chapter 2 states that God is "infinite in being and perfection"). And it is *de fide* for Roman Catholics (that God is "infinite in intellect and will and in every perfection").[3]

Analytic philosophers of religion, however, have (with few exceptions, e.g., Graham Oppy) said very little about the infinity of God. One reason might be that they do not really consider infinity to be a divine attribute, but a modifier of other attributes—especially power, knowledge, and goodness (see, e.g., Craig 2006; Harrison 1999, 420; and Hick ^{2}2004, 258). However, if that were the case, we should not expect to find them discussing (divine) simplicity either, but they often (still) do.[4] Of course, the doctrine of divine simplicity has been among the most controversial in recent times. So perhaps the general lack of controversy over divine infinity best explains its near absence from the contemporary analytical literature. But yet it has often been unclear, despite its apparently wide acceptance, just what the claim that God is infinite is supposed to mean, and so one would expect analytic philosophers of religion to have given more attention to it.

So far as I am aware, the claim that God is a person is not *explicitly* found in any of the historical creeds or confessions. In the Trinitarian formula, God is not "a" person, and the claim is not that there are three persons *of* God but *in* God (see, e.g., Webb [1918] 1971, 61–62; Aulén 1960, 135; and McConnell 1924, 263).[5] Personhood may be essential to the Godhead; it may apply analogically and literally to God in relation to the Trinity, but it does not follow that God is a "person," especially in the modern sense of the term: the three "persons" of the Trinity are not discrete individual substances or centers of consciousness or psychological atoms that stand next to each other; if they were, it would be hard to avoid tri-theism. Yet, that God is a person has widely been thought to be absolutely essential to Christianity, if not to theism itself.[6]

Analytic philosophers of religion tend to assume, rather than argue, that God is a person. This is not to say that arguments cannot be or have not been given.[7] One likely explanation for the relative lack of explicit arguments may be the conviction that it is or ought to be simply obvious to any "rational" person that God is a person. If this is not by itself obvious, surely it is entailed by other key religious claims and practices? For example, scripture abounds in talk of a God who acts—a God who creates us, who continually provides for us, who calls us into loving personal relationship with him, who intervenes occasionally in our history, who answers our petitionary prayers, and so on. If such claims are literally true, these appear to be the sort of thing that only a person can do. Another likely explanation for the relative absence of arguments for the personhood of God is perhaps the thought that nonpersonal (or impersonal) accounts of the divine are hopelessly inadequate, religiously if not philosophically, if they are even coherent to begin with. But even if so, given the putative centrality of God's personhood to Christianity, and the obvious suspicion among critics that the idea that God is literally a person is, for example, guilty of naïve anthropomorphic projection, it is odd that it has drawn so little attention from analytic philosophers of religion.

They have given even less attention to the claim that God is both infinite and a person.[8] Whether God is or can be both depends in large part on what is meant by the terms "infinite" and "person." These terms have long and fluid histories. What are the necessary and/or sufficient conditions for being a person, and are they compatible with being infinite? Whether an infinite God could be a person also seems to depend on what other attributes God may have (and whether he has them essentially or only contingently). In particular, what is the relationship between infinity and other putative divine attributes? Might being infinite entail an attribute that is incompatible with being a person?

Let us try to make our way sure-footedly through this minefield.

INFINITY AND GOD'S INFINITY

The English word "infinite" and its cognates derive from the Latin *infinitas* (*in* = "not" and *finis* = "limit," "boundary," "end"). What is infinite is what is not limited, not bounded, without end. On this reading, the word "infinite" seems primarily to have a negative or apophatic use. One could try to give it more positive content by specifying the respect(s) in which a thing is not limited. To say, for example, that something is unlimited in power *might* tell us something about what that thing is, and not just what it is not. Insofar as a limit is itself negative, some see the double negation ("not limited") as a positive affirmation—indeed, "the infinite" is the highest unsurpassable reality or the most perfect thing (see Zimmerman

1910).[9] Another historical sense of "infinity" is that which is unclear, indeterminate, indefinite, undefined (Oppy 2012, 436).[10] And still another sense is mathematical and quantitative—what Oppy labels "the generic modern conception of the infinite: *that which exceeds all finite measure*" (431; emphasis in original).[11]

There are also different conceptions of divine infinity or of "the theological infinite." One big question here is whether the mathematical sense of infinity, and in particular the concept of measure or cardinality, is applicable to God. There have been differences of opinion on this.

Some want to allow the mathematical sense of "infinity" to apply to God or at least to some of God's attributes (see, e.g., Swinburne [1977] 1993; Flavour 2004; Achtner 2005; and Steinhart 2009).[12] To say that God is infinite is to say that he is infinite in at least certain respects, such as power, knowledge, and goodness. To say that God's power is infinite is or may be to say that it exceeds all finite measure. And the same applies to God's knowledge and goodness. This might mean that God is able to bring about infinitely many states of affairs, that God knows infinitely many (true) propositions, and that God is able to bring about infinite value/good, where the extent of God's power, knowledge, and goodness may be unlimited in the sense of being unquantifiably infinite, or it might be possessed to a given infinite cardinal degree. On one reading, infinity does not apply so much to the divine attributes themselves but to their extent or scope, for example, not to God's power itself but to its (possible) effects, which may be infinite in number.

Although perhaps few would deny that an omniscient God would know aleph-10 true propositions, if that is how many true propositions there are,[13] the idea that the mathematical sense of infinity applies to God or God's attributes is not unproblematic. One reason is that the mathematical infinite is an extension of the finite; it is constructed out of a finite collection. However, there is an important difference between the notion of the finite infinitely extended and of the infinite that is divine.[14] In addition, suppose that God is potentially infinite and that the magnitudes of at least some of the divine attributes are potentially infinite. This seems to imply that just as for any infinite number there is a number bigger than it (any transfinite ordinal is limited in the sense that there is a greater or larger one), God is open to possible improvement in certain respects.[15] Anselmians at least will find such a conception of God to be woefully inadequate. Some others reject the idea that the mathematical infinite applies to God on the ground that God is not a number or collection or set (see Le Blanc 1993, 52; Harrison 1999, 420; Craig 2006).

If one thinks it is literally true to say that God is infinite, but the mathematical sense is not applicable to God, then "infinity" would seem primarily to be understood in the negative or apophatic sense mentioned earlier—as what is not limited or bounded. Alternatively, to say that God is infinite is shorthand for saying that God is maximally perfect; it appears to be a hyperbolic way of saying that God (literally) has maximal power, knowledge, goodness, and so forth.[16] However they understand these attributes, the idea seems to be that it is impossible for anything or anyone to be more powerful, more knowledgeable, or better than God. Le Blanc (1993, 57) calls this the "upper bound" conception of divine infinity.[17] This differs from the mathematical sense of infinity. So too does the sense of the infinite as what is inexhaustible. What is inexhaustible may be extremely large but quantitatively finite. If the quantitative, mathematical sense of infinity is primary or taken to be the strict sense of the term, it would probably be better to say that God is nonfinite, that is, does not have a finite measure, because the concept of measure does not properly apply to God.[18]

So, there seem to be a number of closely related, nonmathematical senses of divine infinity at play here. They seem to be attempts to flesh out some sort of "qualitative" sense of infinity based on the core notion of the infinite as what is unlimited or unbounded. From the middle of the thirteenth century it was common to say that God's being or essence is infinite.[19] This was grounded in God's being Pure Act, free from (not limited by) any matter or potentiality. This was closely related to, if not indistinguishable from, the idea that God is infinite, that is, unlimited, in the sense of being totally independent of others or of things outside himself or his nature. This is the classical doctrine of God's aseity. There is also the related idea of the infinite as what is complete or full or perfect.[20] On this reading, to say that God is infinitely powerful, knowledgeable, good, and so on seems to be a way of saying that God is completely or fully or perfectly powerful, knowledgeable, and good. Infinity appears to be the mode in which God possesses these properties, or it refers to the completeness of God's being or nature. This seems to be just another way of saying that in his being, power, knowledge, goodness, God is not limited by anything (or at least by anything external to him or his nature). Moreover, God's properties seem to be understood qualitatively—infinite (or maximal) goodness, for example, may only require a single right or good act. In fact, it seems that God would be infinite in the sense of being (perfectly) complete without doing or bringing about anything.

These notions of the infinite as what is unlimited in the sense of being unsurpassable, inexhaustible, free from any potentiality, self-sufficient, and complete do not seem to have anything to do with a mathematical sense of the term. More importantly, for my purposes, are these senses of being infinite compatible with being a person? More generally, is the notion of an unlimited person coherent, and (if so) can God be both unlimited and a person? Before we address this, we must look briefly at what it might mean to be a person, and in what sense(s) God is supposed to be a person.

PERSONHOOD AND GOD'S BEING A PERSON

The English word "person" and its cognates derive from the Latin *persona*, which originally referred to a mask worn by an actor in a theater. It came to refer to an actor's character or dramatic role or part, but it could also mean a party to a legal contract or suit.[21]

There have of course been different (philosophical) conceptions of what a person is, and especially of necessary and/or sufficient conditions for being a person.[22] Thankfully, for my purposes here I need only extract a few of the main and less controversial themes. Thought or intellect—the capacity, if not exercise—seems essential to personhood on most accounts. Boethius's (*Liber de Persona et Duabus Naturis*, chap. 3) classic definition of a person is "an individual substance of a rational nature." Though it was presumably implicit in the Boethian definition, consciousness, and in particular self-consciousness, came to the fore after Descartes. Locke defines a person as "a thinking intelligent being, that has reason and reflection, and can consider it self as it self" (Locke [1690] 1979, bk. 2, chap. 27, §9). Action (or volition) also seems essential.[23] Several accounts of a person require an agent who freely carries out its purposes or plans in the light of knowledge or beliefs, and who interacts with other such agents (which presupposes the ability to communicate, and form relationships, with others).[24]

The claim that God is a person has meant different things. Some speak of God as "a person." Others say that God is "personal." Some speak of the "personality" of God. Still others say that God is "like a person" or "more like a person than anything else (we experience)." These are different notions. Being a person is different from, and stronger than, being personal, having a personality, and being (more) like a person. Being (more) like a person (than anything else) is arguably the weakest notion, and too weak to be of genuine interest here: whatever the necessary conditions of being a person are, a thing could fail to satisfy *any* of them and still be more like a person than anything else (or a nonperson). Being "like a person" also suggests that personhood is a matter of degree, but that seems wrong; though there may be gray areas regarding what counts as a person (and the identity of persons over time), being a person seems to be an all-or-nothing affair. Some deny that God is *a* person (because he is *three* persons), but they say God is personal (see, e.g., Peterson et al. [3]2003, 64n13). We can avoid Trinitarian controversies if by "a person" we mean "at least one person." Even so, some still prefer to speak of God being personal rather than being a (at least one) person, where "being personal" seems to mean "having some properties of persons."[25] Some who speak of God being "personal" might want to say or suggest that God is more than a person (or not merely a person); if so, this implies that God is at least a person. For others, to say that God is personal just is to say that God is a person for whom it is appropriate to ascribe various mental- and action-predicates.[26]

What is at issue here is whether God's infinity is compatible with the claim that God is a person, where the latter is understood metaphysically, that is, as the claim that God belongs in the metaphysical category of a person. Or, if this is distinct, whether God is personal or a personal agent, where this is understood as a claim about the nature of God.[27] The basic or minimal idea (widely endorsed or assumed by analytic philosophers of religion) seems to be that God is a person or personal agent at least in the sense that he freely and intentionally acts and interacts with us in the light of knowledge or beliefs. It is important to stress the metaphysical interpretation of the claim that God is a person or personal agent because one might admit (leaving aside, e.g., certain Indian views) that it is *necessary as a matter of religious psychology* that we think of God as a person or as a personal agent—how else could we, who are persons, try to relate ourselves theistically to whatever is the highest reality? But from this, of course, it would not follow that God is in fact a person or a personal agent at the level of religious understanding or metaphysics.

What is also at issue here is the claim that God is *literally* a person or personal agent.[28] We need not, I think (hope?) plunge into the depths of theories of metaphor or figurative language, or into the perhaps even murkier depths of the doctrine of analogical theological predication. Even if the claim that God is a person or personal agent is metaphorical, this does not necessarily rule out there being a literal core. So if to say that God is a person or personal is taken to mean that he is "like a person" in certain respects, for example, he thinks and acts, the likeness statements are themselves supposed to be literally true or false (see Wainwright 2010, 6). Sometimes I think people use "the analogical" as a contrast with "the literal," but so far as I am aware, Aquinas and others who take the doctrine of analogical predication seriously do not intend it as nonliteral predication. So if in saying that God is "a person" or "personal agent," or that "thought" and "will" apply to him, these terms are understood analogically, this is still supposed to say something literally true (and not something purely figurative) (6).[29]

CAN GOD BE BOTH INFINITE AND A PERSON?

Although "the folk" probably equate "person" with "human being," I do not think it is necessary or analytic that persons are humans. If it were, this would be a much shorter chapter, for the vast majority would surely agree that human beings are essentially finite or limited.[30] Some have wanted to say that certain nonhumans, such as dolphins or chimpanzees, are persons; some are willing to entertain the possibility that sophisticated computers are persons. We also have the legal notion of "corporate persons." Any angels or demons, and any alien life-forms there may be, might also be persons. So it seems that there could be different kinds of person.

Can an infinite being, and in particular God, be a person? I think it is reasonable to answer no, though I do not believe there are knockdown arguments here (or elsewhere in philosophy). Either the notion of an infinite person is incoherent, because various kinds of limitation are essential to persons, or the sense in which an infinite being is "a person" or a person is "infinite" is purely figurative (or more figurative than literal).[31] Put another way, being a person is a "mixed perfection" because it implies some limitation, and so an infinite God can be a person only metaphorically or figuratively. There are some direct and more indirect lines of argument for this. Let me begin with a more indirect approach.

Arguably, infinity is or has been one of the core divine attributes, or—if not strictly a divine attribute—the core notion from which most if not all of the attributes derive (cf. Toner 1909). The indirect approach I have in mind here in effect focuses on "the classical conception of God," and in particular on the relation between infinity and other putative divine attributes—especially what have been called the "metaphysical" or "entitative" or "incommunicable" attributes. The strategy is to argue that infinity entails at least one divine attribute that is incompatible with being a person. (Alternatively, one might mount an argument for the implausibility of thinking that God can be both infinite and a person. We might get a cumulative case argument based on a number of considerations, some involving divine attributes.) Some of these arguments may be better than others. Replies to them will either challenge the entailment claim or deny that the entailed attribute is incompatible with being a person. Some of these replies may be less plausible than others.

I said earlier that God's aseity seems entailed by or just is equivalent to the claim that God is infinite. If what is infinite is not limited by anything (or at least by anything external to it), what is infinite would seem to be independent of any person, thing, condition, or influence outside it for its existence or nature. But persons seem essentially dependent in ever so many ways on things outside themselves for their (continued) existence, nature, and activities. These are limits or constraints. If so, an unlimited, totally unconditioned or self-contained God cannot be a person.

Closely related to God's aseity has been God's alleged impassibility, that is, inability "to suffer" in the old-fashioned sense meaning "to undergo." God is not the kind of being to which "things happen." And so an impassible being also cannot experience or be moved by emotions such as anger or sorrow. Arguably, the ability to "undergo" things or to be the kind of being to which "things happen" or to be affected (negatively or positively) by various emotions is essential to personhood. And so, an infinite, impassible God cannot be a person.[32]

In addition, it seems to follow from God's infinity that God is immaterial or incorporeal, since being corporeal would seem to be a limitation for obvious reasons, especially for a creator *ex nihilo*. If being corporeal is essential to being a person, as some have argued, an infinite immaterial God cannot be a person. Of course, many Christian philosophers seem happy to deny that persons are essentially corporeal.

Leaving aside whether the notion of immaterial, supernatural personal agency is deeply problematic, it appears that being infinite entails being timeless. A sempiternal (everlasting) being would not be limited insofar as it would be without beginning or end, but it looks like such a being would still be limited internally. According to Leftow, "Any life in time contains at least one limit, an inner boundary dividing its past from its future. This inner limit is as real a constraint as outer limits are. For it walls us off from parts of our lives" (Leftow 1997, 258). Is being (essentially) timeless compatible with being a person? There has been a fair amount of debate in the literature over whether the notion of a timeless person is coherent.[33] One point to highlight from this debate is that at least many of the seemingly important things (the psychological- and action-predicates) that are true of ordinary persons—the only uncontroversial examples of persons we are familiar with—would (could) not apply to a timeless being. Anything that requires or implies temporal duration or being in time, a "before" or "after," would be excluded. For example, a timeless being could not anticipate, deliberate, remember, make choices (if choosing implies earlier indecision), desire (what it does not already have), make inferences, and so on.[34] It is hard to see how a timeless being could act in time and respond to things that happen in time. And so it is difficult to see how an infinite, timeless being could be a person in any robust sense of the term.

It would also seem that being infinite implies the absence of spatial limitations. An infinite being would not be limited by, that is, contained in or confined to, a particular spatial location or region, but would in some sense be present everywhere, that is, to every spatial location/region and thing. An infinite being would be "omnipresent." But, arguably, omnipresence is incompatible with being a person. Persons seem essentially limited to a particular spatial (spatio-temporal) perspective.[35] They are not present to every spatial location and thing. Each person has a unique particular perspective on reality, a "first-person point of view"—

one among indefinitely many other possible perspectives. It might be possible to be aware of all different perspectives at once, but that would seem to reduce omnipresence to omniscience.[36]

Or consider change. People change—or so it would appear. If something cannot change but is still a person, many will think they have lost a grip on what a person is.[37] An (essentially) infinite being could not change if a change implies the possibility of limitation, or of an increase in or loss of perfection. If being (essentially) infinite implies being immutable, but persons are essentially mutable, we cannot have an infinite person (cf. Gale 1991, 92).

Being infinite also seems to entail an absence of any complexity or composition, given that these are or imply the possibility of limitations. A "simple" being without limits would thus lack temporal and spatial parts or extension, intrinsic accidental properties (and strictly speaking any distinct essential properties either), and lack any real distinction between essence and existence. To say that there is no real distinction in God between essence and existence is presumably to say that God is not a member of some general kind. On the other hand, persons are complex; they possibly are or are composed of spatio-temporal parts; they appear to have intrinsic accidental properties; and there is a real difference between essence and existence in them. So, once again, it does not seem that there can be an infinite person.[38]

There is a more direct line of argument against the notion of an infinite person (personal God), "direct" in the sense that it does not proceed via what (divine) attributes may be entailed by being infinite, which are then claimed to be incompatible with personhood. Rather the focus is on personhood itself, and the key claim is that persons are essentially limited in various ways, and hence we cannot have an infinite person. There are different ways one might develop this line of argument. Here I will sketch just two, and then consider some objections to both the "indirect" and "direct" arguments.

I pointed out earlier that the capacity to enter into relationships is often thought to be essential to persons. For some, to be a person or personal is to be in relationship. An emphasis on persons as beings-in-relation (as opposed to a focus on persons as discrete individual substances) goes back at least as far as the early Church Fathers, such as Gregory of Nyssa, and it is closely connected with the etymology of "person." It seems to me that one of the main reasons for thinking that God is a person is precisely the idea that we are called into loving personal relationship with God, together with the idea that one cannot have such a relationship with what is not itself a person or personal. However, it is very hard to see how an (essentially) unlimited being could enter into genuine, two-way, reciprocal personal relationships. Being in a personal relationship necessarily involves some constraints or limits, perhaps the most basic being that one party is not "the other." Genuine personal relationships necessarily seem to involve some give-and-take, some risk and vulnerability, on both sides. Are the best interpersonal relationships such that one of the parties could be essentially unlimited in the sense of being unable to be affected in any way by the other or the relationship?[39]

In addition, on most accounts personal agency requires at a minimum acting for reasons. Having a mind (or consciousness) is thus required. But the notion of an infinite mind, a mind not limited or constrained in any way, strikes me as incoherent because everything we seem to know about what it is to be or have a mind involves capacities that are exercised in the context of various limitations. At the least, we seem to lose a grip on what could literally be meant by "having a mind" if we try to remove or strip away all limitations. If God is a personal agent, God must have choices. Free action is often said to be essential to personhood, and God is supposed to be the supremely free agent. But choice situations necessarily involve a background of constraints; otherwise there would not be options about which one deliberates and chooses to act or refrain from acting. It is unclear whether an infinite mind (agent) could "intend" anything, for there could never be a gap between the intention to act and the action, nor presumably between the intention to act and the occurrence of the intended result. An infinite mind would appear to face no conflicts in rational decision-making. There could be no possible weakness of will, no pull of emotions, and no wants or desires to begin with insofar as these are directed at what one lacks. Strictly speaking, it does not appear that "rationality" would apply to an infinite mind. It might be better to speak of an infinite being's "intellect." But even so there can be no cognitive capacities that develop and are exercised against a backdrop of various constraints and limits. An infinite being's knowledge or beliefs about the world would not seem to depend on what happens in the world; rather, what happens in the world would seem entirely dependent on what the infinite being wills. In short, the notion of an infinite person is deeply problematic, if not incoherent.

"But hang on, mate!" a friend of infinite personhood might interject. "Aren't you trying to pull a fast one here? Your arguments against infinite personhood seem to require God to do the logically impossible, but that's nonsense. No one in the history of philosophy, with the possible exception of Descartes, has thought that God can do the impossible. To give just one example, you seem to be arguing that since choice situations require options, and options are limits, choice situations are incompatible with being literally unlimited (infinite), but since choice situations are required for (free) personal agency, there is or can be no infinite personal agent. But that just turns on asking God to do the logically impossible, that is, have choices without options."[40]

In response, I say that this objection comes perilously close to completely begging the question. My point is that limitations, logical and perhaps nonlogical, are imbedded into the very concept of a person or personal agency. You cannot get the concept of personal agency going without placing constraints on it. If one were already convinced that God is a supreme personal agent, God's power, for example, would necessarily be limited by logical possibilities and the kind of being he is. If he is immaterial, it would, for example, be no legitimate objection to complain that he cannot ride a bicycle. There would be lots of things God cannot do as a powerful agent if he were or must be a person. But that, or whether God could be both infinite and a person, is precisely what is in question.

"But hang on!" friends of infinite personhood will shout from all corners. "Don't focus on what may be essential to *human* persons! Human persons, as you yourself admitted, need not exhaust the kinds of person there are. A divine person, if infinite, would be unique and the way it has properties (their "mode of signification," to use Aquinas's phrase) would be vastly different from the way we exemplify them. Of course many psychological- and action-predicates true of ordinary persons could not apply to an infinite being. But that does not mean that no person-predicates are applicable. Of course human (or created) persons are finite, that is, limited, in ever so many ways. That isn't front-page news! But we must not take these limitations to be essential to personhood itself."[41]

Despite all the admitted differences between ordinary finite persons and an infinite being, there is (following Alston 1989b) allegedly enough of an abstract common core (or partial overlap) of "functional concepts" (personal-predicates) applicable to both infinite and finite beings. To resolve this decisively we would need to look much more closely than I can do here at this core of functional concepts (person-predicates). Are the concepts or predicates that are applied to an infinite God, and the functionally interrelated psychological states they describe, too thin or sparse to constitute a personal system or agent?[42] Arguably, the differences between the psychological states of a so-called infinite mind and a human finite mind are so great that "mind," "intellect," 'thought," "will," "purpose," and so on cannot literally apply to both. For followers of Aquinas these terms are used "analogically,"[43] for Alston they are used (partially) univocally. Either way, their ascription to infinite God is supposed to be literally true. But arguably they are purely figurative or symbolic, or more figurative than literal.

Suppose, however, that there are grounds to allow a coherent notion of an infinite person. Pike (1970, 129) asks, "How much of a person would a timeless person be?" Not much, he answers. Here too we can ask: How much of a person would an infinite person be? Could it be God? Could it be worthy of worship? Though Alston defends the coherence of the notion of an infinite (and timeless and immaterial) person, he himself (Alston 1989c, 101–2) concedes that his functional account is not enough for religious or devotional purposes.[44] He clearly recognizes (101) the important distinction I drew earlier between what may be required as a matter of religious psychology and what may be sound metaphysics (or theology). Whereas he sees his highly abstract and austere functional account as sufficient for a literal sense of an infinite personal God, I think the claim that God is infinite, if literally true, signals that God is not literally a person. Personhood seems to be an inappropriate category for that which ultimately explains all that is.

Now, one might think that in arguing against the notion of God as an infinite person I have made heavy weather over a number of the more esoteric divine attributes, but have conveniently given short shrift to the "personal divine attributes," such as omnipotence and omniscience. There does not seem to be anything incoherent about the notion of a person who can do or bring about everything it is possible for a person to do or bring about.[45] And can we not also agree that having this property might deserve to be described as being omnipotent? Similarly, there does not seem to be anything incoherent about the notion of a person who knows everything it is possible for a person to know. And can we not also agree that having this property might deserve to be described as being omniscient? But surely, then, it will be argued, there can hardly be anything incoherent about the notion of an infinite person (i.e., a person who is unlimited in the above ways)!

But are we talking about omnipotence and omniscience in the classical theological sense? I do not think so. Consider omnipotence. It is problematic—some might even suggest scandalous—to see God's power as merely or even primarily the ability to do things, with God's power at the extreme end of a continuum where he has vastly more power than other persons by virtue of being able to do everything it is possible to do. This seems to conflate the important distinction between the finite infinitely extended and the theological infinite (cf. Van den Brink 1993, 178). And it is doubtful that an appeal to analogy really helps here, that is, that God's way of doing things, of knowing things, and such is very different from our way. Omnipotent God is not a superpowerful agent among other agents. Arguably, actualizing everything that is *ex nihilo* is on an entirely different level or plane of power altogether, not reducible to the causal power of an agent—however great they might be. Davies (2006, 61–62) warns against treating God as "nothing but an inhabitant of the universe . . . more powerful or more knowledgeable than other persons. Yet . . . nothing more than that—an extraordinary person, one of a kind, hardly the Creator of the Universe, hardly that which accounts for there being any world at all."

It is true of course that omnipotence has been understood as the capacity to do all things (to bring about everything it is logically possible to bring about, etc.). This sense of omnipotence has been at the fore in the contemporary analytical literature. I agree that there is historical evidence supporting this sense of omnipotence. Van den Brink (1993, 48–49) distinguishes three senses or meanings of "omnipotence" (*pantokrator*) based on a close examination of the history of words denoting God's power in different classical languages. One of these is the capacity to do all (possible) things. This sense, he says, was built into the Latin word *omnipotens*, but is foreign to the Greek *pantokrator*. The other two senses or meanings are "God's power as universal dominion and authority over all and everything" and "God's power as shown in the creation and preservation of the world." The former, says Van den Brink, goes back to the Septuagint and is clearly found in the early patristic literature. He points out that the latter, especially the notion of sustaining power, was present in the first century BC but became prominent among the Church Fathers. I would not then go so far as to say that the "doing all things" sense is a distortion of Christian belief, at least as long as God does not become another item in the world, albeit more exalted.[46] But it isn't (wasn't) the only, or even the primary, sense. And if or when classical theists have a "doing things" sense of God's omnipotence in mind, there's also an important qualitative sense of "omni" at work—omnipotence may only require a single act and God would be perfect in power without actually doing anything.

Let me put the issue another way. Perfect-being philosophical theologians often start with a set or series of goods of finite beings (i.e., us), which they also think apply to God, such as power, knowledge, and goodness. They then strip off any limitations or imperfections that may be associated with us (or our mode of exemplifying them), and they say that what is left of this cleansed (maximally consistent) set of attributes gives us the best picture of God's nature that we can reasonably hope to get. A different picture, however, emerges if one starts with the notion of an absolutely infinite God. Arguably, such a God is "too big" to (literally) be "a person."

But have I not then gotten things the wrong way around? Of course nothing can have the classical divine attributes and be a *finite person*. But God, it might be urged, is the *perfect person*. He is a person by nature, and it is his nature that sets what is essential to personhood.[47] We are but imperfect, pale imitations of the real thing. Reply: The *imago dei* idea might well give us a legitimate sense in which God is *the* Person, without it following that God is *a* person (cf., e.g., Plato's "third man" argument and Aquinas's claim that Being Itself is not a being). Theists will agree that our personhood is dependent on and derives from God. But this needn't imply that God is a person. Perhaps more importantly, the objector's move isn't going to deliver a robust enough sense of God's being a personal agent-cause doing or producing things in the way ordinary agent-causes produce things. We end up with a very stretched or thin sense of "personhood," and I think it is better to reject the idea that God is literally and metaphysically a person.

NOTES

I wish to thank the participants in the Infinity of God workshop at Ruhr-Universität Bochum, August 8–11, 2013, and especially the organizers, Benedikt Göcke and Christian Tapp. I also thank John Bishop for discussions on this material.

1. There have been few Christian finitists. Origen may have been one of the first, apparently due largely to the pejorative connotations of *to apeiron* in Greek thought. Canon 8 of Justinian's *Book against Origen* condemns anyone who thinks that the power of God is finite (Denzinger 1946, no. 210). See Sweeney 1992 (chap. 16) for some early thirteenth-century theologians who rejected the ascription of infinity to God's being or essence. Process thinkers say that God is infinite in some sense(s) but finite in some other(s). See, e.g., Griffin 1973 (187–89). Hartshorne has been one of the harshest critics of the classical view of divine infinity. See Hartshorne 1984 (8) for a particularly strong statement. Some have recommended God's finitude as a way to try to reduce the force of the Argument from Evil. See Dilley 2000. Hume directed the Logical Argument from Evil at an infinitely powerful and praiseworthy God. See Holden 2010 (154–58) for a discussion of this.

2. Philo of Alexandria (first century), who influenced many of the early Christian writers, may have thought that God is infinite, but it is apparently not found explicitly in any of his works. It is found in the earliest Latin writers, e.g., Novatianus, *De trinitate* (e.g., chaps. 2 and 4) and Minucius Felix, *Octavius* (e.g., chap. 18) in the early third century. God's infinity takes center stage in Gregory of Nyssa in the fourth century, and John Duns Scotus in the late thirteenth century.

3. First Vatican Council (Denzinger 1946, no. 1782). Ott (1955, 31) says it is *de fide* that "God is actually infinite in every perfection." Some qualifications on "every" seem needed here—e.g., all those pure perfections that are mutually consistent.

4. Simplicity is arguably not a divine attribute but a doctrine concerning the way God's attributes are related to each other and to God or God's nature (essence) or being.

5. Webb claims that the first talk of God being "a person" is found in theologians of the sixteenth century, who *deny* Trinitarianism. Davies (2006, 59) thinks the first occurrence of "God is a person" in English is found in the 1644 trial of (the Unitarian) John Biddle on the charge of heresy. Webb has no doubt that the first occurrence of the "personality of God," outside Trinitarian controversies, is to be found in the late eighteenth century.

6. For the latter claim, see, e.g., Owen 1967 (97) and Wainwright 2012 (1). There are too many examples of the former claim to cite here.

7. Taliaferro (1998, 100–102) lists nine arguments for a personal God. Yates (1990, 165–67) gives six reasons why the category of personhood as ascribed to God is indispensable. My aim here is not to consider the merits of these arguments, or whether God is in fact a person. In a series of papers, John Bishop and I reject "personal omniGod theism" in favor of a nonpersonal alternative that employs a

"euteleological" conception of God as the ultimate supremely good purpose (*telos*) for the sake of which the universe exists.

8. The most notable exception has been Alston 1989b.

9. As Oppy (2011, 234–35) points out, this seems closer to the original sense of *to apeiron* in Anaximander—the "boundless, imperishable, ultimate source of everything"—though it gets understood personally in Christian thought. Oppy (2011, 244) sees this as a "loose or metaphorical sense" of infinity. For an excellent discussion of the history of the connection between infinity and perfection, and why it might be important to keep them distinct, see Clayton 2000.

10. This has been associated with a more epistemic sense of God's infinity, viz., his incomprehensibility. Others, e.g., Descartes, divorce entirely "the indefinite" from the notion of "the absolute or theological infinite." See Le Blanc 1993 and Clayton 2000.

11. This conception of infinity was first analyzed by Aristotle and received its most systematic treatment by Cantor. Whether Aristotle's actual infinite is Cantor's actual infinite is another matter.

12. See Oppy 2011 and Oppy 2012 for an excellent discussion of this issue.

13. If there are true propositions about our future free actions, some deny that an omniscient God knows them on the ground that it is logically impossible to know them. They say that this is not incompatible with omniscience because an omniscient being (still) knows everything it is logically possible to know.

14. Clayton (2000, 311) points out that Descartes and Leibniz saw the infinite as what is absolutely different from the finite or limited. If I am not mistaken, the idea that infinity is what distinguishes the divine nature or essence from anything created goes back at least as far as Gregory of Nyssa's *Against Eumonius*. If we equate Cantor's "absolute infinite," Ω, with God, as Cantor may have thought, Cantor's absolute infinite is not a being or person.

15. See Oppy 2006 (xii–xiii) and Oppy 2011 (249–51) on this and potential problems for the idea that God is actually infinite. Also see Nagasawa 2008 (593–94). Hartshorne (1948, 20) prefers to define God as "that than which no *other* individual being could conceivably be greater, but which itself, in another 'state,' could become greater." Those who think of God as (essentially) timeless or immutable will have to reject this.

16. Craig (2006) says that "'infinity' serves as an umbrella-term for capturing all those properties which serve to make God the greatest conceivable being. . . . God's infinity is constituted precisely by these great-making properties [necessary, self-existent, omnipotent, omniscient, holy, eternal, omnipresent, etc.]." Gellman (1994, 59) says, "To say that God is 'infinite' is to say that God is omnipotent, omniscient, all-good, eternal, and the like."

17. Also see Oppy 2011 (247) regarding the connection between this sense of infinity and Anselm's "that than which no greater can be conceived."

18. See Oppy 2012 (437–38) on distinguishing the nonfinite from the infinite.

19. See Sweeney 1992 (chaps. 18 and 19) for a discussion of this, especially in Fishacre and Aquinas.

20. See Bochner 1973 (614) on Aquinas, and Clayton 2000 (261) on Leibniz.

21. For an interesting history of "person," or *persona*, as a philosophical term, and of its relation to the Greek *hypostasis* and *prosōpon* and the Latin *substantia*, see Webb ([1918] 1971, Lecture II). The *Lewis and Short Latin Dictionary* includes the following definition of *persona*: "Law t.t., *a being having legal rights and obligations* (including the state, etc.; not including slaves)."

22. See Teichman 1985 for a helpful discussion of some of the prominent philosophical (and nonphilosophical) conceptions of a person.

23. Locke goes on to say that "*Person* . . . is a Forensic Term appropriating Actions and their Merit; and so belongs only to intelligent Agents capable of a Law, and Happiness and Misery" ([1690] 1979, bk. 2, chap. 27, §26). The forensic sense of "person" seems to be our modern sense of the term: a person is an agent capable of bearing moral responsibility for outcomes that occur through his or her agency. This sense of "person" did not, however, originate with Locke but goes back to the Romans: slaves, who had no legal rights and obligations, weren't "persons." See note 21 above. If one rejects the idea that God is a moral agent with obligations, etc., and that God's perfect goodness is to be understood as moral goodness, I think one should reject the idea that God is literally a person.

24. Among analytic philosophers of religion, Kretzmann (1997, 197), Peterson et al. (2003, 64), Alston (1989a, 39), Swinburne (1996, 4), and Wainwright (1988, 10), to name just a few, include these in a minimal description of what it means to say that God is a person or personal being.

25. Thatcher (1985) argues that it is essential to Christianity that God is personal but not that God is a person. His main target (section II) is the "bodiless person theism" of Swinburne, and in section IV, he argues that the belief that God is a person is "foreign to Christian theology."

26. See Swinburne (1977) 1993, Swinburne 1996, and Alston 1989a (41), among many others.

27. Cf. Legenhausen 1986 (308) on these two points. The idea that God has a personality or character suggests that God develops over time, which will be problematic for some but congenial to others. Mander 1997 (410) argues that the idea of a person without a personality makes no sense or would be of no interest or value. I suspect that talk of God's having a personality, though certainly in line with biblical language and devotional practice, is too unbecoming of the ultimate source of all that is.

28. Burns (2009) takes this as purely metaphorical. Le Blanc (1993) is strongly inclined to think that the attribution of infinity to God is more figurative than literal.

29. I overlook some complications here. E.g., what Alston (1989a, 43) means by "the literal use of a predicate term" does not seem compatible with the doctrine of divine simplicity. Of course, Alston and many others in recent times reject the latter. If one accepts divine simplicity, I think one should deny that God is a person.

30. I leave aside any qualification that might be needed if we were considering the Incarnation.

31. I am of course hardly the first to doubt or deny that there can be an infinite person! Historically, it seems to be found primarily in the German and British idealists—Bradley, Bosanquet, Fichte, and (probably) Schelling. In more recent times, see, e.g., Hartshorne 1948 (26) and Legenhausen 1986 (315). Mander (1997, 404) does not think personality and infinity are compatible. Clayton's "minimal personalist theism" holds that "analogs of what we mean by personhood should be attributed to God" (Clayton 2000, 425), yet "one cannot speak literally of God as agent since a number of the features of agency as we know it are inapplicable, or at least problematic, in the case of an infinite being" (427).

32. For a defense of a doctrine of divine impassibility, see Creel 1986. Many Christian thinkers these days reject divine impassibility, and so they will deny that being infinite implies it (if they think God is a person). But is that plausible?

33. Those who argue for incompatibility include Coburn (1963), Swinburne ([1977] 1993, 214), and Gale (1991, 52–53). Pike (1970, 128) argues that a timeless knower could not act in ways required to be knowing or aware of anything, and if there were a coherent sense of a timeless person, it "would surely not rank very high on the personality scale." Those who reject the incompatibility claim include Alston (1989b), Craig (1998), Leftow (2009, 298–99), and Helm ([2]2010, chap. 4). Others maintain that while the notion of a timeless person may not be incoherent, we should not think of God as a timeless person. See, e.g., Hasker 1989 (chap. 9).

34. See Alston 1989b for a particularly good discussion of these matters.

35. If a thing has or occupies more than one spatial (spatio-temporal) perspective at once, it would seem to be a scattered object. See Helm [2]2010. There is an interesting and important related issue of whether divine infinity implies pantheism or panentheism. Unfortunately, I do not have the space to get into this here.

36. For more on omnipresence, and whether it is really distinct from God's power and knowledge, see, e.g., Leftow 1998. Omnipresence is often understood as, or as reducible to, God's power over everything that occurs anywhere in space, and his knowledge of what occurs everywhere.

37. But see Mann 1983 for a defense against the objection that an immutable being could not be personal.

38. The majority of philosophers of religion in recent times seem happy to jettison the doctrine of divine simplicity. However, that God is absolutely simple ("an absolutely simple substance or nature") is *de fide* for Roman Catholics, as defined by the Fourth Lateran Council and First Vatican Council (Denzinger 1946, nos. 428 and 1782, respectively). Few are likely to endorse the idea of a "simple" person, though Mann (1983) is an exception. Those who deny divine simplicity will presumably deny that being infinite implies being simple, but is that plausible?

39. If we ask whether an unlimited being could limit itself, this raises the paradox of omnipotence. It also raises, as Oppy (2006, xii) notes, the issue of whether an actually infinite God could become incarnate.

40. Brian Leftow vigorously pressed this line of objection against an earlier version of my chapter.

41. This line of objection is based on the approach taken by Alston (1989b).

42. See, e.g., Block 1978 (esp. sec. 3.2). Alston (1989c, 99n18) is sensitive to this worry but thinks he has adequately addressed it.

43. My impression is that Aquinas's doctrine of analogy has fallen out of favor in recent times, though that may depend on where you hang out and who your friends are. Hughes (1989, 57–59) argues that Aquinas's doctrine of analogy is based on his doctrine of divine simplicity. The latter, I said earlier, has fallen on very hard times.

44. Leftow (2009, 299) seems to take the opposite view.

45. Let's leave aside the stone paradox and whether, e.g., omnipotence is compatible with impeccability.

46. I find it revealing that the repeated biblical assertion that "nothing is impossible for thee [God]," where this does imply that there is nothing that God cannot do, always seems to occur in the context of prayer and prophecy, or in a liturgical setting. I cannot get into this here, but this suggests to me (once again) that it may be practically necessary, from the point of view of religious psychology, that we think of God as a person—even though I think it is false at the level of religious understanding or reality.

47. I am certain I came across this claim in Leftow 2012, but cannot recall the page reference. Leftow did not recall saying it there, though he agreed it is something he probably would say.

REFERENCES

Achtner, Wolfgang. 2005. "Infinity in Science and Religion." *Neue Zeitschrift für Systematische Theologie und Religionsphilosophie* 47 (4): 392–411.

Alston, William. 1989a. "Can We Speak Literally of God?" In *Divine Nature and Human Language: Essays in Philosophical Theology*, 39–63. Ithaca, NY: Cornell University Press.

———. 1989b. "Functionalism and Theological Language." In *Divine Nature and Human Language: Essays in Philosophical Theology*, 64–80.

———. 1989c. "Divine and Human Action." In *Divine Nature and Human Language: Essays in Philosophical Theology*, 81–102.

Aulén, Gustaf. 1960. *The Faith of the Christian Church*. London: SCM Press.

Baker, Lynne Rudder. 2000. *Persons and Bodies: A Constitution View*. Cambridge: Cambridge University Press.

Block, Ned. 1978. "Troubles with Functionalism." In *Perception and Cognition: Issues in the Foundations of Psychology*, edited by C. Wade Savage, 261–325. Minneapolis: University of Minnesota Press.

Bochner, Salomon. 1973. "Infinity." In *Dictionary of the History of Ideas: Studies in Selected Pivotal Ideas*, edited by Philip P. Wiener, 2:604–17. New York: Scribner's.

Burns, Elizabeth. 2009. "Must Theists Believe in a Personal God?" *Think* 8 (23): 77–86.

Clayton, Philip. 2000. *The Problem of God in Modern Thought*. Grand Rapids, MI: Eerdmans.

Coburn, Robert C. 1963. "Professor Malcolm on God." *Australasian Journal of Philosophy* 41 (2): 143–62.

Craig, William Lane. 1998. "Divine Timelessness and Personhood." *International Journal for Philosophy of Religion* 43 (2): 109–24.

———. 2006. "Pantheists in Spite of Themselves? Pannenberg, Clayton, and Shults on Divine Infinity." In *For Faith and Clarity*, edited by J. K. Beilby, 135–56. Grand Rapids, MI: Baker Academic. http://www.reasonablefaith.org/pantheists-in-spite-of-themselves.

Creel, Richard. 1986. *Divine Impassibility*. Cambridge: Cambridge University Press.

Davies, Brian. 2006. *The Reality of God and the Problem of Evil*. London: Continuum.

Denzinger, Henricus. 1946. *Enchiridion Symbolorum*. Barcelona: Herder.

Dilley, Frank. 2000. "A Finite God Reconsidered." *International Journal for Philosophy of Religion* 47 (1): 29–41.

Flavour, Fivel Minty. 2004. "Divine Infinity." PhD diss., Emory University.

Gale, Richard. 1991. *On the Nature and Existence of God*. Cambridge: Cambridge University Press.

Gellman, Jerome I. 1994. "Experiencing God's Infinity." *American Philosophical Quarterly* 31 (1): 53–61.

Griffin, David Ray. 1973. *A Process Christology*. Philadelphia: Westminster.

Harrison, Jonathan. 1999. *God, Freedom and Immortality*. Aldershot: Ashgate.

Hartshorne, Charles. 1948. *The Divine Relativity*. New Haven, CT: Yale University Press.

———. 1984. *Omnipotence and Other Theological Mistakes*. Albany: SUNY Press.

Hasker, William. 1989. *God, Time, and Knowledge*. Ithaca, NY: Cornell University Press.

Helm, Paul. [2]2010. *Eternal God: A Study of God without Time*. Oxford: Oxford University Press.

Hick, John. [2]2004. *An Interpretation of Religion: Human Responses to the Transcendent*. New York: Palgrave.

Holden, Thomas. 2010. *Spectres of False Divinity: Hume's Moral Atheism*. Oxford: Oxford University Press.

Hughes, Christopher. 1989. *On a Complex Theory of a Simple God*. Ithaca, NY: Cornell University Press.

Kretzmann, Norman. 1997. *The Metaphysics of Theism: Aquinas's Natural Theology in Summa Contra Gentiles I*. Oxford: Clarendon.

Le Blanc, Jill. 1993. "Infinity in Theology and Mathematics." *Religious Studies* 29 (1): 51–62.

Leftow, Brian. 1997. "Eternity." In *A Companion to Philosophy of Religion*, edited by Philip Quinn and Charles Taliaferro, 257–63. Oxford: Blackwell.

———. 1998. "Omnipresence." In *Routledge Encyclopedia of Philosophy*, edited by E. Craig, 7:103–7. New York: Routledge.
———. 2009. *Time and Eternity*. Ithaca, NY: Cornell University Press.
———. 2012. *God and Necessity*. Oxford: Oxford University Press.
Legenhausen, Gary. 1986. "Is God a Person?" *Religious Studies* 22 (3/4): 307–23.
Locke, John. (1690) 1979. *An Essay Concerning Human Understanding*. Edited by Peter N. Nidditch. Oxford: Clarendon.
Mander, William. 1997. "God and Personality." *Heythrop Journal* 38 (4): 401–12.
Mann, William E. 1983. "Simplicity and Immutability in God." *International Philosophical Quarterly* 23 (3): 267–76.
McConnell, Francis John. 1924. *Is God Limited?* New York: Abingdon.
Nagasawa, Yujin. 2008. "A New Defence of Anselmian Theism." *Philosophical Quarterly* 58 (233): 577–96.
Oppy, Graham. 2006. *Philosophical Perspectives on Infinity*. Cambridge: Cambridge University Press.
———. 2011. "God and Infinity: Directions for Future Research." In *Infinity: New Research Frontiers*, edited by Michael Heller and W. Hugh Woodin, 233–54. Cambridge: Cambridge University Press.
———. 2012. "Science, Religion, and Infinity." In *The Blackwell Companion to Science and Christianity*, edited by J. B. Stump and Alan G. Padgett, 430–40. London: Blackwell.
Ott, Ludwig. 1955. *Fundamentals of Catholic Dogma*. Cork: Mercier.
Owen, H. P. 1967. "Theism." In *The Encyclopedia of Philosophy*, edited by Paul Edwards, 8:97–98. New York: Macmillan.
Peterson, Michael, William Hasker, Bruce Reichenbach, and David Basinger. [3]2003. *Reason and Religious Belief: An Introduction to Philosophy of Religion*. Oxford: Oxford University Press.
Pike, Nelson. 1970. *God and Timelessness*. New York: Schocken.
Steinhart, Eric C. 2009. "A Mathematical Model of Divine Infinity." *Theology and Science* 7 (3): 261–74.
Sweeney, Leo. 1992. *Divine Infinity in Greek and Medieval Thought*. New York: Peter Lang.
Swinburne, Richard. (1977) 1993. *The Coherence of Theism*. Oxford: Oxford University Press.
———. 1996. *Is There a God?* Oxford: Oxford University Press.
Taliaferro, Charles. 1998. "Personal." In *Philosophy of Religion: A Guide to the Subject*, edited by Brian Davies, 95–105. London: Continuum.
Teichman, Jenny. 1985. "The Definition of Person." *Philosophy* 60 (232): 175–85.
Thatcher, Adrian. 1985. "The Personal God and a God Who Is a Person." *Religious Studies* 21 (1): 61–73.
Toner, Patrick. 1909. "The Nature and Attributes of God." In *The Catholic Encyclopedia*, edited by Charles G. Herbermann, 6:612a–14a. New York: Robert Appleton Company. http://www.newadvent.org/cathen/06612a.htm.

Van den Brink, Gijsbert. 1993. *Almighty God: A Study of the Doctrine of Divine Omnipotence*. Kampen: Pharos.

Wainwright, William. 1988. *Philosophy of Religion*. Belmont: Wadsworth.

———, ed. 2010. *Philosophy of Religion*. Vol. 1. New York: Routledge.

———. 2012. "Concepts of God." In *Stanford Encyclopedia of Philosophy*, edited by E. N. Zalta. http://plato.stanford.edu/entries/concepts-god.

Webb, Clement C. J. (1918) 1971. *God and Personality*. London: Allen & Unwin. Reprinted 1971. New York: Kraus.

Yates, John C. 1990. *The Timelessness of God*. Lanham, MD: University Press of America.

Zimmerman, Otto. 1910. "Infinity." In *The Catholic Encyclopedia*, edited by Charles G. Herbermann, 8:4a–7a. New York: Robert Appleton Company. http://www.newadvent.org/cathen/08004a.htm.

CHAPTER 17

Divine Infinity and the Trinity

THOMAS SCHÄRTL

Combining classic theism with the Trinity remains a sticky business. At first glance, there seems to be a painful contradiction between divine simplicity and the fact that the Godhead consists of three different persons. Whatever we understand by the term "person" does not fit nicely the idea that God is an entirely simple, somehow "homogeneous" entity, which must not be understood as the instantiation of a variety of properties or as subject to certain categories of being. Social Trinitarianism makes the problem of divine simplicity more pressing and yet more painful, but the problem itself originates from the Trinitarian doctrine as such.

MIRACULOUS INFINITY?

Although there have been different and maybe even convincing approaches to reconcile divine simplicity with some sort of complexity within the divine life (cf. Hughes 1989), Trinitarian faith still puts a burden on divine simplicity,[1] if the doctrine of simplicity prohibits *any kind of composition* within the Godhead (cf. Kelly 1994). In Aquinas, many types of logical or ontological relations count as composition (*Summa theologiae* [*ST*] I, q. 3, a. 1–4), so whatever we consider to be a case of constitution or coincidence and so forth falls under the prohibition of composition. It is not easy to see why the Trinity should be regarded as an *exception.*[2] However, the Latin tradition of Trinitarian theorizing underlines that the three persons of the Trinity are not mere parts of the Godhead, and the parthood relation is seen as an entirely *inappropriate* tool for spelling out the relation between God's Godhead and his Trinitarian existence.

INFINITY AND INDIVIDUATION: GREGORY OF NYSSA

Nevertheless, there is a notion of infinity that is actually meant to serve the Trinitarian doctrine. We can detect it in the writings of the Cappadocian Fathers. For instance, Gregory of Nyssa's *Ad Ablabium* spells out the relation between infinity and the Trinity. Wondering why the Trinitarian Creed does not lead to the idea of having three different Gods, Gregory points to the notion of infinity: as instantiations of the divine essence, the distinction of the Trinitarian persons cannot imply any kind of separation, which we have to face whenever we are confronted with finite entities. For the divine essence is infinite in its very own nature so that separation has no place in it. In other words, the mechanisms of individuation, which are at work within a metaphysics of finite beings, are entirely inapplicable to the Godhead. The mechanisms that individuate divine persons cannot be paralleled to whatever principle is responsible for the individuation of finite entities. Though the individuation of finite beings leads to temporal, spatial, and other forms of separation, the individuation of divine persons does not destroy the unity and uniqueness of the divine essence (cf. Gregory of Nyssa 1958, 43–47). But what does "infinity" mean in this context? Apparently, "infinity" is used to mark an ontologically significant difference between what is finite and whatever is the opposite of being finite. However, this whole *marking the difference* aspect looks like an *ad hoc invention* of specific rules to immunize the Trinity from being regarded as a cloaked version of polytheism. Does it not look as though infinity is being used as a trump card to make us believe that a multitude in God does not destroy His unity and our belief in monotheism? If metaphysics were a poker game, the infinity card would end it. Still, how could infinity do the trick? Simply by *marking the difference* between the finite realm and the infinite sphere? This not only sounds circular but also somewhat empty.

In fact, Gregory's concept is much more elaborated. It entails the thesis that God's life does not have *any limitations*, and, as a consequence, there are no spatial or temporal sequences in the Godhead. So, whatever the divine persons are, they cannot be treated like individuals that are separated by their positions in space and time or by causal relations. Furthermore, they cannot be seen as limiting each other, since there are no limitations in an infinite deity. For any finite entity individuation entails limitation, but this is not the case within the Godhead. Infinity even *ensures* the unity of the Godhead, because it overrides whatever might be regarded as limitation at first glance: once the different relations between the eternal Father and the eternal Son are put into the framework of the infinity of the divine essence, their difference becomes reconcilable with the unity of the Godhead. The Cappadocian Fathers had the somewhat vague idea that in infinity anything that implies a limitation or a boundary is taken away. So, infinity serves as a means to spell out unity and identity within the Godhead. But again, how does this work? And, if it works, why does infinity not simply override or erase the Trinity? Gregory seems to allude to a Neoplatonic concept of the One, which is the role model for the infinite by definition, because it has unlimited life and comprises everything there is or could be in an exceeding, transcending manner. Within this framework, *being exceedingly one* and *being infinite* are the same. Since the infinite entails the highest degree of self-identity, there cannot be anything like composition or parthood within the One. If the Godhead of the Trinitarian God is the One, self-identity has to be ascribed to it also. And the Trinitarian nature of God must be seen as a way to secure the self-identity of God. Unfortunately, the Cappadocian Fathers do not tell us much about how infinity provides unity and self-identity. They simply say that it does, while insisting on the negative elements in the notion of infinity: based on divine infinity, the divine essence remains beyond our human comprehension. Infinity thus contains incomprehensibility and also immeasurability (cf. Gregory of Nyssa 1958, 47, 52–53, 54–55). Sometimes, the Fathers treat infinity as plain incommensurability. The important fact remains that individuation within the Godhead is not the same as individuation in the realm of finite beings. But the Fathers shied away from describing the reasons for this difference. They might have shown us the dots, but they did not connect them for us.

INFINITY AND UNIQUENESS: NICOLAUS CUSANUS

God's incomprehensibility is a strong starting point for Nicholas of Cusa. It is the very reason, for him, why we cannot have what is usually called "knowledge" with regard to God, since knowledge presupposes that we are able to identify the kind-membership of something we know. Nicolas presupposes a *principle of comparison* as the basis of his notion of knowledge, which leads to a theory of *conceptual knowledge*, because it is aimed at a conceptual determination of what is known.[3] However, in God's case, the principle of comparison is not applicable, which becomes obvious as soon as we subscribe to the following argument proposed by Cusanus:[4]

1. Any *comparison* presupposes a well-determined framework of comparison.
2. If there is no such thing as a well-determined framework of comparison for entities *x* and *y*, then to compare those entities is devoid of meaning.
3. There is no well-determined framework fit for the infinite.
4. Therefore, any attempt to compare the infinite with anything else is devoid of meaning.

According to Cusanus, any scientific, knowledge-based propositions presuppose unequivocally meaningful concepts, and since concepts are the outcome of comparison, we cannot achieve conceptual knowledge of the infinite God. Infinity, in this case, explains incomprehensibility. However, Cusanus does not leave us to the demons of *apophatic* theology but offers possibilities, which somehow might allow us to approach the infinite in an epistemologically significant way. Alluding to Kant's distinction between *concepts* and *regulative ideas*, we might argue in favor of Cusanus's proposal that we can come up with regulative ideas of divine infinity, which would allow us to speak of God in a rule-governed and epistemologically significant way and simultaneously to retain the doctrine of divine incomprehensibility. Based on this, Cusanus is able to offer a somewhat dialectical idea of God. As the infinite, God is maximal equality, that is, He is literally equidistant from any other entity. How can this be so? Again, Cusanus offers an interesting argument (cf. *De docta ignorantia* I, 4, n. 11):

(1) If x and y are totally equal, then they are specifically and sensitively distant from any z they are different from.
(2) Being specifically and sensitively distant means that any x or y that is different from any z is z_1-wise different from z_1, z_2-wise different from z_2, and so on.
(3) The same goes for equality: If x and y are different while they are equal to any other z, then they are z_1-wise equal to z_1, z_2-wise equal to z_2, and so on.
(4) If the infinite is beyond comparison, then the rules of being equal and being different do not apply to the infinite.

From (4) Cusanus derives the conclusion that the infinite is not only equally distant from everything, but also *equally equal to anything*, because being specifically and sensitively distant means to be unequally distant and so forth. The rules of comparison proposed in premises (2) and (3) are somewhat unusual, but it is possible to make sense of them once we take into account that any comparison is sensitive to the coordinates of a system of categories. So, for example, comparing a dog to a cat differs, in Cusanus's view, significantly from comparing a dog to a cow and, even more so, from comparing a dog to a stone. A Platonically conceived system of coordinates, which places higher beings above and beyond lower entities, could serve as the ontological framework of comparison. Within such a framework, Cusanus's notion of specific and sensitive comparison makes sense. Now in God's case, the situation changes dramatically. It is precisely because of God's infinity that we cannot say that God is, comparatively speaking, closer to certain entities than to others in a meaningful way. Although Cusanus doesn't offer a mathematical analogy in this case, we can easily come up with one that does justice to his intuition: Everybody agrees that the values of the numbers 1, 2, and 3 are different. We can also add, by applying Cusanus's notion of *sensitive comparison*, that whatever makes 1 different from 2 is not the same as that which makes 1 different from 3. Now, although this is true, the difference in value disappears once we divide the numbers by infinity: $^1/_\infty$, $^2/_\infty$, or $^3/_\infty$ equals zero in each case. Analogously, whatever we want to say about God conceptually transcends any conceptual content we gain by comparing finite entities and using finite categories.[5]

The equidistant closeness and distance of God opens the door to Cusanus's dialectical doctrine of God. For God is equally the supreme maximum and the supreme minimum. Since everything participates in God and God doesn't participate in anything else, He is the true maximum (defining "the ontological maximum" as the one entity in which everything else participates). And, as the ultimate constituent, God is part of everything, while nothing else could ever be a constituting part of him (defining "the ontological minimum" as that which has no logical or ontological parts but is part of everything else) (cf. *De docta ignorantia* I, 24, n. 76). *To participate in* and *to be part of* cannot be one and the same relation in this context. If so, this would lead into a straightforward contradiction. Cusanus embraces paradox, but not contradiction. Therefore, we have to clarify how *participating in* differs from *being a part of.* Participation entails *having being or existence from something that is not contingent*, whereas *being a part of* entails (in a way specific to what Cusanus wants to flesh out) *giving identity and singularity to something, as its constituent*. Eventually, Cusanus places God at both ends of the spectrum of metaphysical principles: God is the absolute maximum that provides existence to everything else by mode of participation. And He is the ultimate singularity that provides identity as the omni-engaged constituent, thereby fulfilling His task as an ontological minimum. Both tasks are *bound together* by divine infinity: because of his infinity, God is the Absolute that encompasses any contentful determination. He is the ultimate principle that represents the unique identity any finite identity must depend on. For Cusanus, God-talk has to be dialectical to do justice to both tasks God fulfills, at both ends of the spectrum. The horizon of a perfect being theology, therefore, has to be broadened to flesh out the idea of God as totally equidistant from everything and every determination. So, God is not just the most perfect *F* in any comparison of things that are *F*. He is also the least *F*-having entity, given that *F*-ness does not really apply to God (cf. *De docta ignorantia*, I, 4, n. 12).

But is this *F*-and-not-*F* rule not just weird? Does it entail a contradiction? Part of the problem might be that Cusanus seems to mix up the nonapplicability of *F* with not having *F*. If this is a failure on his part then this specific detail of his doctrine of God might just get thrown out the window. For cases of nonapplicability can be encountered, whenever we have to be aware of category differences and possible category mistakes. However, Cusanus wants to explain, by using the above-mentioned paradoxical expression, that to be infinitely *F* means both: being perfectly

F and—to a certain extent—not being *F*. So, are we stuck with a contradiction after all?

Let us try a somewhat different approach in order to make sense of what Nicolaus de Cusa is hinting at. Therefore, the notion of "*x* being infinitely *F*" comes, according to Cusanus, with two subnotions:

i. Whenever *x* is infinitely *F*, *x* is *unsurpassably F*.
ii. Whenever *x* is infinitely *F*, *x* is *exceedingly F*.

From (i) Cusanus could arrive at an interpretation which holds that whatever is unsurpassably *F* is also perfectly *F* since there cannot by any *y* which surpasses *x* in having *F*. In this case, infinity is connected to the notion of perfection. But there is also the second notion of (ii) that Cusanus seems to hold on to. From the fact that *x* is exceedingly *F*, Cusanus seems to conclude that *x* is beyond comparison with regard to *F*. It looks as if Cusanus would go on to insist that anything that is exceedingly *F* cannot be placed on a common ground with anything else that is said to have/be *F* in a more regular way. We can sum up this idea in the following manner:

> ($PRED_1$) The predicate *F* is applicable to an entity *a* if and only if *a* can be meaningfully and significantly compared to an entity *b* to which the predicate can be applied correctly.

Cusanus's view of divine infinity seems to be that ($PRED_1$) leads to the very paradoxical situation where a certain predicate *F* should be applied to God and must not be said of God; for if God is unsurpassably *F*, then He is realizing the maximum of having or being *F*, which seems to presuppose that He, indeed, is *F*. On the other side, if God is said to be exceedingly *F*, this insinuates that God is beyond comparison, from which—according to ($PRED_1$)—the nonapplicability of *F* would follow. What appears to be a contradiction is, at the very end, a result of Cusanus's holding on to two reasonable interpretations of infinity as a predicate whose business is to modify the meaning of more contentful predicates. Since Cusanus is not willing to favor one interpretation of infinity over the other, the paradox remains, which—for Cusanus's taste—is a good thing because it seems to justify the need of dialectical expressions within the theology of God.

Of course, the dialectics are to a certain extent the result of Cusanus's views on predication. This becomes even more obvious if we take a look at Cusanus's second "dogma" of predication:

> ($PRED_2$) The predicate *F* is applicable to an entity *a* if and only if we can conceive of cases in which Non-*F* is applicable to *a*.

From a nowadays perspective we can also deny that ($PRED_2$) is a justifiable principle. In Cusanus's eyes, ($PRED_2$) depends on the rule of comparison that, as we saw, is a prerequisite for developing so-called significant concepts. For Cusanus's rule of predication allows us to be equipped with such concepts. In the realm of finite entities, we will not face a problem relying on this rule: if we conceptually know that Black Beauty is a black horse, we can easily imagine cases in which Black Beauty is not a horse of that kind but, say, a white horse.

Still, as it stands, ($PRED_2$) looks plainly false. There are predicates like "being self-identical" or "being possible," which are applicable to finite entities although we cannot imagine the opposite. Furthermore, it makes Cusanus look like a prematurely lost member of the Vienna Circle. But we must apply a principle of charity here and say that Cusanus saw something important. Maybe he had in mind a difference between content-bearing and logical, between kind-membership-defining (or at least kind-membership-sensitive) and transcategorical predicates. For the latter, ($PRED_2$) as a rule is clearly false, whereas for the former it might make sense and could be taken to be true. Hence, we just need to modify ($PRED_2$) a little bit:

> ($PRED_3$) For all predicates *F** that are content-bearing insofar as they are either kind-membership-defining or kind-membership-sensitive, we have to note that predicate *F** is applicable to an entity *a*, if and only if we can conceive of cases in which Non-*F* is applicable to *a*.

If we try to connect ($PRED_3$) with Cusanus's idea of conceptual knowledge, we can probably understand why saying something about God might be a problem. For instance, if we conceptually know that Black Beauty is a celebrated racehorse, we can figure out cases in which Black Beauty is a pathetic loser. But in God's case everything is different, since ($PRED_3$) cannot apply. We cannot conceive of cases in which God, given that we understand the term "God" properly, is not wise, good, powerful, and so on.

So far, this sounds as if Nicolaus just tells us that God has his properties necessarily. However, he wants to take us a step further by insisting that we will not be able to conceive of God or to imagine God as not being good, as long as we understand the term "God" properly. But can we not imagine the opposite of divine goodness and bring God into the picture? Cusanus might answer that whatever we have conceived as divine goodness initially will turn out not to be divine goodness, even if we can just imagine the opposite to it. Once we add divine infinity to the picture, we have to admit that there is something immeasurable and incommensurable in anything we say about God.

Up to a certain extent, this view is supported by Nicolaus's paradoxical metaphysics: metaphysically speaking, God is the supreme substance for Cusanus, such that there cannot be anything that is "outside" or alien to him. Since this very substance is infinite and encompasses everything, there is (in its ideality) nothing left to be ontologically opposed to God (cf. *De docta ignorantia* I, 24, n. 78). Thus, divine infinity safeguards both God's unique substantiality and his omni-relationality.

As we can see, "divine infinity" does not just serve as a marker, indicating the crucial difference between the finite realm and God's nature. In Cusanus's thinking, divine infinity is also a means of nourishing the idea that God is a unique substance, which "reflects" or "mirrors" any finite entity through being equidistantly related to everything in a unique sense. Therefore, in his treatment of infinity the emphasis is on *divine uniqueness*.

But there is still more to it: infinity serves as a justification for speaking *paradoxically* about God, which allows us to use different ideas that seem to be contradictory at first glance. God is the most perfect unity for Cusanus, because he has no parts in any respect, but God is also a perfect complexity, since he is Trinitarian. To justify both lines of thinking, a concept of infinity is required, because infinity is able to turn unity into complexity and complexity into unity. Cusanus offers a geometrical analogy. Consider a line, a triangle, and a circle. These shapes differ significantly. But what can you say about an infinite line, an infinite triangle, or an infinite circle? The closer you approach infinity, the more these shapes will resemble each other, being approximately equal to each other, so that the infinite line *is* the infinite triangle, which, again, *is* the infinite circle. Adopting a phrase used by Eleonore Stump in order to describe Aquinas's notion of divine simplicity, we could speak of Cusanus's "quantum mechanics" of divine infinity (cf. Stump 2006).

Quantum mechanics, indeed, provides an analogy. In classical mechanics we are used to the idea that something is either a wave or a particle, but we have to give up this dogmatic distinction once we approach microreality. In the same way, we are used to the idea that something is either simple and homogeneous or complex and subject to constitution in the world of everyday objects, whereas we have to give up this distinction whenever we approach the *infinite*. Because of His infinity, God can be both absolutely simple and astonishingly complex.

This is where the Trinity enters the picture. God, as the infinite substance, is a *self-encompassing* substance, which therefore is related to itself. Based on this self-relational substantiality, we can distinguish unity, equality, and connection in God (cf. *De docta ignorantia* I, 8, n. 21; I, 9, n. 23). For Cusanus, "unity," "equality," and "connection" are the metaphysical names of the Trinitarian persons. They provide us with hints of their specific roles in the immanent and in the economic Trinity (cf. *De docta ignorantia* I, 9, n. 24). *Unity*, *equality*, and *connection* are metaphysical principles and the ultimate features of reality. Since finite entities cannot be a perfect union, and cannot be perfectly equal to each other or will not enter into a perfectly reciprocal connection (cf. *De docta ignorantia* II, 5, n. 120), we can see the difference between the finite realm and the infinite through the lenses of these principles—because of His infinity, God is perfect unity, equality, and connectedness. And because of His infinity, God as the perfect substance is the perfect self-relation. Therefore he must be Trinitarian.

UNWRAPPING DIVINE INFINITY

For Cusanus, and also for the Greek Church Fathers, infinity seems a trump card to reconcile opposite, if not contradictory, notions of God. Simplicity and complexity seem reconcilable, once we add the flavor of infinite to the mixture. But how is it possible for infinity to do the trick?

In Cusanus, we found the idea that infinity takes away conceptual differences—the infinite substance is also infinite omni-relationality. Infinite simplicity is infinite complexity, and vice versa. So, whatever is at the left or right end of the spectrum of metaphysical principles forms a unity when pushed into infinity. We could continue by claiming that the infinite thing is the infinite subject, the infinite universal is the infinite particular, and so on.

But is this sequence of peculiar identities not just some sort of wishful thinking, presumably, stemming from the fact that a concept of God can be developed from very different angles? Is not infinity, in this case, just a marker to tell us that within the realm of divinity everything is different compared to the realm of finite entities? When we use Cusanus's "quantum mechanics" to briefly describe his *coincidentia oppositorum*, is this anything more than a mere euphemism, hiding the fact that we do not have a clue what divine infinity is?

But before we surrender to the analogy and confess that we are clueless, two points seem to be helpful in spelling out divine infinity. Even if there are many more things to say, the notion of infinity clearly serves as a tool to understand *divine simplicity* and *divine uniqueness*. Therefore, we need to say more about the relationship between infinity, simplicity, and uniqueness. At the end of this journey, divine uniqueness might help us reconcile Trinity with divine simplicity.

INFINITY AND SIMPLICITY AS COEXTENSIONALITY

The notion of divine simplicity has some counterintuitive aspects. That the divine is not composed of spatial or temporal parts is the more understandable side of divine simplicity. However, a more complicated implication of divine simplicity is the thesis that *God is identical to his attributes* and that his *attributes are mutually identical* (cf. Mann 1982, esp. 452). Taken at face value, this idea seems to entail, as Alvin Plantinga notes, that God is nothing but a property (cf. Plantinga 1980, 47). But if God cannot be a property, it looks as though the doctrine of simplicity from which the conflation of the divine attributes is derived must be wrong. Is that so?

Among the strong defenders of divine simplicity, John Duns Scotus offers a remarkable approach to the whole problem. He would strictly deny that the properties *F*, *G*, *H*, applied to God in a literal and straightforward way, are identical. Their identity is the result of their being applied to God *in an infinite mode*. Thus, although being *F* is not identical to being *G*, or being *H*, being *infinitely F* is identical to being *infinitely G* and *infinitely H*.[6] So, Scotus takes a common idea in medieval God-talk, the notion of divine simplicity supporting the identity of God's attributes, and offers some clarification of it, while appealing to divine infinity.[7] But what does this mean in detail? Scotus simply states that properties become interchangeable, and really (but *not formally*) identical, once they are filtered through the mode of infinity, without ever saying how infinity does the trick.

Apparently having more than one property is somehow a sign of finitude: even if we have to use different concepts, which seem to refer to different properties, a *multitude of properties* would violate divine simplicity, as Scotus and various other scholars thought. However, *multitude* as such is not the core of the problem, since we could, in theory, defend a multiplicity of aspects of the divine being by pointing out God's overflowing being and goodness. Rather, for medieval metaphysics, the having of multiple properties is associated with limitation because multiple determinations imply specific limitations. We can presumably make sense of this doctrine if we regard determination as a procedure that places an entity *x* in a system of categories and, by doing so, in a network of logical and ontological relations. For example, to be a human being includes a logical relation to being an animal and another significant logical relation to being an irrational animal. In God's case, however, there is no such thing as being placed into a system of categories. Although, for Scotus, there is the framework of being and existence, a gap separates God's realm from the reality of finite beings. This gap is infinity as a *quality* of God's life. There is no limitation to God's life.

But how do we derive simplicity from infinity? We have to assume that infinity does this job as some sort of metapredicate. Infinity resembles perfection. Nevertheless infinity is not the same as perfection. But where is the difference? Again, we need to say that infinity fulfills its task of being a metapredicate by being a marker at the same time—it points at the difference between God's reality and the realm of finite beings. By saying that God is infinitely *F*, we would have to add that God's *F*-ness is strikingly different from any *F*-ness found in finite beings. But how does this lead to divine simplicity?

To explain this, we need two tools to describe consequences of divine simplicity and to illuminate property-ascription to a divine being. Both tools will help us to understand why God's properties are identical to each other and why God doesn't simply *have* His attributes but *is identical to* them. The first tool is the concept of *coextensionality*. Two properties are identical if and only if whatever fulfills their concepts is coextensional in every possible world.[8] Coextensionality is not automatically in place whenever we compare goodness and wisdom, by comparing the instances of being good and being wise in possible worlds. We can conceive of beings that are good but not wise, and so on. But once we add infinity to the picture, at least this is what Scotus would have us believe, the properties in question change significantly, so that whatever is infinitely good must be infinitely wise, and vice versa. But this is more than just saying that, whenever *x* instantiates infinite goodness, it necessarily instantiates infinite wisdom (as, for example, James Bond necessarily instantiates bravery and recklessness). The doctrine of coextensionality entails that the extension of "infinite goodness" is necessarily identical to the extension of "infinite wisdom," and vice versa.

This reconstruction might come close to what Scotus thought about the identity of divine attributes and could be the best way to understand him from a more contemporary point of view. In pointing out that we imitate the supreme simplicity of the divine by a number of perfections, such as being morally good, seeking wisdom and knowledge, Scotus dismisses an alternative explanation of the identity of properties, which entails that two properties *F* and *G* are the same if their instantiations in an entity *a* have the same effects (cf. *Ordinatio* II, dist. 17, pars 1, q. 2, n. 66–71). Since God has various effects on us, as we can see in the ways in which human beings try to imitate God or to engage in a salvific relationship with God, any effect-related explanation of property-identity won't do in God's case (cf. Cross 2005, 103–11). So, the best way to understand the identity of the divine attributes remains the doctrine of coextensionality.

But the doctrine of coextensionality has to be pushed to its limits. That the divine properties (or their conceptual equivalents) are coextensional in every possible world entails that they (or their conceptual equivalents) are cointensional. Since in God's case coextensionality is not random, but strictly necessary, there is no other possible solution. We have to face the fact that divine goodness (or any predicate ascribable to God) *turns metaphysically into* divine wisdom or any other attribute, once we qualify it as "infinite." Maybe the concept of identity is misleading in this context, especially if we use Leibniz's principle of identity as a basis for explaining any kind of identity. Consider the case of property-identity: Leibniz's principle of identity is inapplicable to properties themselves. And so we have to shy away from the concept of identity (giving credit to any critique on which coextensionality in every possible world does not entail property-identity) and turn to the idea of *inclusion*.

The least we can say is this: *infinity* makes properties mutually inclusive and their conceptual equivalents necessarily coextensional. On this interpretation, infinity is not a bomb you drop in order to indistinguishably conflate the properties ascribed to God, but the explanation of the strongest possible logical relation you can establish between properties. And if this relation cannot be named "identity," it might yet be coextensionality and inclusion. So, this is what infinity does: it *safeguards the strongest possible relation among divine attributes*.

INFINITY AND PARADIGMATIC PREDICATION

The other tool, which is needed to explain God's relation to his own attributes, is taken from Uwe Meixner's reconstruction of Platonic predicate theory (cf. Meixner 1992). Meixner alludes to an age-old problem in Plato's theory of predication, which becomes apparent at the stage of *self-predication*. How can we say that the idea of piety is itself pious without falling prey to a number of nonsensical derivations (such as, whoever is pious is immersed in prayer, therefore the idea of piety is immersed in prayer)? Plato's solution, pointed out by Meixner, consists in drawing a distinction. The "is" of self-predication has to be distinguished from the "is" of any other form of predication. In the first case, Platonic ideas are the subject of predication. As such, they serve as *paradigmatic subjects of predication*. They are the paradigmatic instances of the property they express. Thus, the idea of piety is the paradigmatic instance of being pious. However, any derivation has to be performed very diligently, since paradigmatic subjects of predication must not be confused with everyday subjects of predication. Everyday subjects of predication presuppose standards of predication, which, according to

Platonic reasoning, are given thanks to the *existence* of paradigmatic subjects of predication. Within this framework, paradigmatic subjects of predication perfectly instantiate the property they express, but they do not *have* the property they express in the usual sense of "having a property." In Platonic terms, "having a property" means "participating in the paradigmatic subject of predication," which most eminently and purely expresses the property in question. But, since a subject of predication cannot participate in anything else—while displaying the property the subject of predication is meant to express—there is no usual meaning of "having," or usual mode of having, relating the paradigmatic subject of predication and the relevant property.

Pretty much the same reasoning is at work when the medieval metaphysician considers the possible modes of God's "having" properties. If "having properties" is bound to a certain kind of participation, God cannot be subject to this mode of having properties. *Any predication has to be treated like the case of paradigmatic subjects of predication. So, whatever property and characteristic God may display, it has to be parallel to how we talk about paradigmatic subjects of predication.* In other words, the divine attributes are, so to speak, paradigmatic subjects of predication. Moreover, God himself is the ultimate and most eminently paradigmatic subject of predication. Since paradigmatic subjects of predication are "above" the instance-property distinction (and any distinction that depends on it, such as the distinction between universals and particulars), we have a solid reason to explain the "quantum effect" of divine infinity. If it is only divine infinity that makes God the paradigmatic subject of predication, God's nature is similar to the nature of a paradigmatic subject of predication, His nature resembles, paradoxically speaking, the nature of universals and also the nature of particulars. This paradoxical nature of God is proof of being beyond any system of categories that presupposes the distinction between universal and particular. *This result is not surprising at all for the medieval mind-set, since God, as the ultimate truth, serves as the model for being a universal and, as the first cause of everything, appears at the same time to be an ultimate and unsurpassable particular.*

In this context, "divine infinity" cannot imply that God displays innumerable attributes or that He "has" his attributes in a way that provides no form or no content. A less subtle idea is that an infinite mode of having the property *F* is like a borderless, so to speak, "gunky," or amorphous mode of being. But this is not true of God and his attributes. Their infinite mode of being has to be translated into the *mode of being F as a paradigmatic subject of predication*. Again, infinity serves as a marker to point to the crucial differences between the two modes of predication: the mode of participation and the mode of being a paradigmatic subject of predication. Only the latter can be true of God.

Now, what about the Trinity? For Scotus, the Trinity is only partly subject to this mode of predication. Whenever a divine attribute of the divine substance is ascribed to a divine person, the rule of infinity is in place. And Scotus uses it to strengthen a typical feature of Latin Trinitarianism, namely, that the divine persons and the divine substance are coextensional. It is not easy to make sense of this idea, which is common in other medieval authors (cf. *Summa theologiae* [*ST*] I, q. 39, a. 1). Maybe we come close if we reformulate the idea thus:

> (OT) All sentences referring to the divine nature, and also any sentence referring to one of the divine persons, have one and the same truthmaker: the divine substance.

The truthmaker approach—as expressed in (OT)—was adopted by Jeffrey Brower (cf. Brower 2009) as an additional way of explaining divine simplicity. This approach could help us make sense of the idea that the divine persons and the divine nature are essentially identical, which is emphasized by Scotus in his doctrine of the *infinite mode of predication*. Nevertheless, Scotus denies the applicability of this infinity rule to any predication that expresses inner-Trinitarian relations, that is, relations between the divine persons (cf. *Ordinatio* I, dist. 8, pars 1, q. 4, n. 215). The reason for this caution is more or less obvious: Scotus wants to avoid the conflation of the divine persons at all costs. On Scotus's view, infinite goodness and infinite wisdom are interchangeable, whereas divinity and paternity are not. Since being divine is infinitely true of God, it is coextensional with being father.

But paternity is not coextensional with divinity.[9] Since the properties of the divine persons are expressed in relative terms, they cannot be subject to the mode of infinity either. Given that identity is the symmetrical and transitive relation we know will not cope with Scotus's specifications, we need to distinguish necessary coextensionality from accidental coextensionality, and symmetrical from nonsymmetrical coextensionality. Infinite goodness and infinite wisdom are necessarily and symmetrically coextensional, while paternity and divinity are necessarily but nonsymmetrically coextensional. Symmetry implies *equal accessibility* from *every* possible world. But when we talk about how being an animal and being rational cohere in the essence of being human, we have to be aware of the accidental and nonsymmetrical coextensionality of the properties.

In order to avoid a conflation of the divine persons at the cost of invalidating the infinite mode of predication rule, Scotus's precaution obstructs what Cusanus offers a century later: making infinity itself the reason for combining simplicity and complexity in God. For Scotus, however, infinity has to safeguard divine simplicity and to mark the difference between God and the finite world. In Cusanus, on the other hand, infinity is a more positive, content-bearing predicate, announcing the unlimited, ever-fruitful, and everlasting divine life whose simplicity is mediated through the complexity of mental activities. But saying this presupposes an idealist metaphysics.

INFINITY AND UNIQUENESS

The problems of predication have already opened our eyes to the fact that infinity not only safeguards divine simplicity but, even more so, divine uniqueness. We left Scotus, giving the impression that his understanding of divine infinity gravitates around a rather negative interpretation of infinity. But this is just one side of a complicated story. Although Scotus cannot satisfy the needs we developed in dialogue with Nicolaus Cusanus, his notion of infinity plays a prominent role in his development of a *perfect being* theology.

Scotus tells us that the *existence* of an infinite being cannot be logically derived from the *concept* of an infinite being, since this very concept is beyond our epistemic capacities to grasp (cf. *Ordinatio* I, dist. 2, pars 2, q. 1–2, n. 15–38). This raises problems of religious epistemology, such as whether an infinite being can be an object of knowledge or the object of some kind of perception. For Scotus, as for many other medieval authors, knowing God has to follow a route that starts from concepts of the features of finite beings. Another side effect of divine infinity is monotheism (cf. *Ordinatio* I, dist. 2, pars 2, q. 1–2, n. 73). For if infinity entails unsurpassable perfection, there cannot be another unsurpassably perfect being. For, if there were two absolutely perfect beings, the sheer existence of one would limit the perfection of the other.

However, the core idea in Scotus is that infinity seriously changes the parameters for applying causal concepts to God. Not only *is* God the first, uncaused cause in the line of causation, but the entire distinction between cause and effect *cannot be applied* to God. This has consequences for God's inner life, which for Scotus, as for any other medieval theologian, consists in the activity of the intellect and the will. Saying that the activities of the intellect and the will are infinite means that they do not have any limitation. Therefore, they must have an unlimited and nonlimiting object, which cannot be anything but the divine substance through which everything is known or referred to by intentionality. But that is not enough: saying that God's intellect and his will are infinite means that *there is no distinction between cause and effect* in the activity of God's will (cf. *Ordinatio* I, dist. 2, pars 2, q. 1–2, n. 94–104). God's intentions are basic actions that form a unity as the *one divine activity and action,* and the same applies to the activity of the divine intellect. If there were memory or imagination in God, they would be one and the same intellectual capacity of pure self-awareness, in which the awareness of all things distinct from God is embedded.

This explication of infinity tells us that God is the most perfect unity, transcending any form of constitution appropriate to finite entities. This means that God is unique. His uniqueness, however, does not just point to divine singularity but also to *the mode* of God's existence: whatever God is, has, knows, or wills, he is and has it *all at once*. Infinity entails, so to speak, actuality (cf. *Ordinatio* I, dist. 2, pars 2, q. 1–2, n. 66–69). This itself points back to the limitlessness of God's life. Divine aseity, which follows from that, reveals the richness of God's life, while anything we would call "an activity" on God's part is entirely basic and bound into the unity of the one divine activity, which is identical to God's life as such.

But again, what about the Trinity? Scotus holds that the doctrine does not entail any contradiction with the simplicity of the divine substance, if we take the Trinitarian persons to be relations (cf. *Ordinatio* I, dist. 2, pars

2, q. 1–4, n. 196). His answer to the reconcilability of the Trinity with divine simplicity is very short, following the Augustinian lead that the divine attributes primarily belong to the divine substance. But if we treat the Trinitarian persons as relations of the one divine substance to itself, relations that are tied to the inner life of God as processions of intellect and will, we can develop a model to keep divine simplicity and the Trinity together. In idealistic terms—which, of course, remain hidden but not undetectable in medieval authors (cf. *Ordinatio* I, dist. 2, pars 2, q. 1–4, n. 279–80)—this is the model of the spirit. God as the perfectly self-transparent spirit mediates his own substance in the processions of mind and will and, therefore, is Trinitarian. But infinity as the all-at-once perspective God has on his very own substance safeguards the perfect unity within the full self-transparency of the divine substance.

THE TRINITY AND ITS SIMPLICITY

So we ask: How does infinity reconcile divine simplicity with the Trinity? Two hints come from classic theology. First, as a negative predicate, "infinite" reminds us of the crucial differences between God's reality and the realm of finite beings. Individuation within the Godhead is, because of infinity, entirely different from any mode of individuation of finite beings. And, as infinity implies simplicity, infinity entails that in God's case the truthmaker of property-ascriptions must not be a multiplicity of states of affairs but rather the one single divine substance.

The second aspect reflects the all-at-once instantiation of God's nature. If there is no unrealized potential in God's nature and if everything that is essential to God is also internal to him, the best model for describing God would be the idealist concept of Spirit: God is the absolute, unlimited life as the supreme Spirit, which has the highest self-transparency. So, if God is a substance, He has to be the substance of self-transparent Spirit. This model gives divine infinity a positive meaning. If God's life is truly unlimited, given that there cannot be any limit to infinity, then He must have the power to relate to every conceivable entity there is or could be in a way that does not make him dependent on these relations. Again, the model of Spirit helps here, because an unsurpassably self-transparent and cognitively fully self-aware spirit is also omnirelational.

There have been noteworthy attempts to make the Trinity the most significant consequence of the idea that God is the most perfect Spirit. Johannes von Kuhn, for example, regards God's Trinitarian being as the result of his "self-mediation" (*Selbstvermittlung*[10]) as Spirit.[11] Having a similar background in German idealism, Philip Marheineke takes this idea a step further: the Trinitarian persons must not be conceived as three individuals, but as three *instances* or *figurations* of the one divine Consciousness.[12] If being a person means *being an instance of consciousness* rather than being an individual, then we can retain the term "person" in Trinitarian theology. The notion of infinity is still influential, but cloaked under the Absolute. In Trinitarian theology, God has to be conceived as unlimited Spirit, as fully self-transparent. The Trinitarian persons are those relations that the divine substance has to itself.[13] In order to realize itself as the Absolute Spirit, the divine substance has to enter into those relations. Therefore, the Trinitarian persons must not impose any limit on God's divinity. At this point, the centuries-old idea that infinity reminds us of the different modes of individuation within the Godhead enters the picture again. Whereas individuation always includes some sort of metaphysical limit for finite beings, this is not necessarily the case within the Godhead. *Infinity rigorously excludes any form of limit.*

Johannes von Kuhn and Philip Marheineke favor Latin Trinitarianism, pushing forward St. Augustine's model of the Trinity within a Hegelian framework. At this point, however, we have to raise the question whether one Trinitarian model fits divine infinity better than another. If we agree that Kuhn or Marheineke have found a convincing way to reconcile the Trinity with divine simplicity, based on the idea of God being the Absolute and infinite Spirit, the social model of the Trinity would not be the way forward. Although one might be able to implement the idea of infinite love in this model, the truthmaker and the individuation problems still remain and threaten divine simplicity. It is hard to see how the one divine substance can be the truthmaker of all propositions about God (the Trinity included), if the Trinitarian persons are three different individuals (of any sort). If we cannot dismiss divine infinity and divine simplicity, the social model of the Trinity is truly under pressure. There might be other, more theological, reasons to endorse the social model of the Trinity. But in that case, divine infinity and simplicity are still at stake.

NOTES

1. So, we can find enough voices telling us to abandon the doctrine of divine simplicity; cf. Mullins 2013.

2. In fact, some recent models of Trinitarian theology conflict with the rule of excluded composition; for example, Keith Yandell's model of the Trinitarian persons as necessary and constituent parts of the Trinity is hardly reconcilable with the above-mentioned rule. See Yandell 1994, esp. 205, 211–16.

3. Cf. Cusanus *De docta ignorantia* I, 1, n. 2. Of course, if this was the only mode of knowledge, we would be seriously lost. But Cusanus is not considering alternative notions of knowledge at that point. This could mean that his argument won't hold water if confronted with another form of knowledge.

4. Cf. ibid., I, 1, n. 3: *Omnis igitur inquisitio in comparativa proportione facili vel difficili exsistit. Propter quod infinitum ut infinitum, propter omnem proportionem aufugiat, ignotum est.*

5. This analogy could take us into a different direction once we take a look at Scotus. For Cusanus the fact that the above-mentioned formulas point to one and the same value is a starting point for an interpretation of divine uniqueness in terms of identity. Scotus, however, might teach us to focus not only on the fact that the values are identical but that, on the other side, the content of the formulas remain different. In his view this leads to an interesting conclusion: God has different attributes. But thanks to infinity they are true in a unique way of a unique substance. Cf. Scotus, *Ordinatio* I, dist. 8, pars 1, q. 4.

6. Scotus, *Ordinatio* I, dist. 8, pars 1, q. 4, n. 215–22. For further details, see Cross 2005, 111–13.

7. Scotus, *Ordinatio* I, Dist. 8, pars 1, q. 4, n. 219–20: Haec ergo est falsa "animalitas est rationalitas," et e converso, et hoc quacumque praedicatione, quia non tantum extrema non sunt formaliter eadem, sed nec vere eadem; quiditas enim haec praecise est potentialis ad quiditatem illam, et non est eadem illi nisi propter identitatem ad tertium a quo abstrahuntur: ergo abstractio illa tollit causam veritatis affirmativae unientis ea.

Oppositum est in Deo, quia abstrahendo sapientiam a quocumque quod est extra rationem sapientiae, et bonitatem similiter abstrahendo a quocumque quod est extra rationem eius formaliter, remanet utraque quiditas, praecise sumpta, formaliter infinita, et ex quo infinitas est ratio identitatis eorum—in tali abstractione praecisissima—remanet ratio identitatis extremorum. Non enim haec erant eadem praecise, propter identitatem eorum ad tertium a quo abstrahuntur, sed propter infinitatem formalem utriusque.

8. For further discussion, see Mann 1982, 462–65.

9. Scotus, *Ordinatio* I, dist. 8, pars 1, q. 4, n. 219–20: quia bonitas et sapientia et cetera huiusmodi, sunt eadem quasi identitate mutua, quia utrumque est formaliter infinitum, propter quam infinitatem utrumque est idem alteri,—sed paternitas et

deitas non sunt eadem sic mutuo, quia unum eorum non est formaliter infinitum, sed tantum deitas est formaliter infinita, et propter istam infinitatem paternitas est sibi idem; et ex hoc . . . edoem est bonus et sapiens, "eodem"—inquam—identitate adaequata secundum perfectionem, quia utrumque infinitum; non sic paternitas habet ad deitatem identitatem adaequatam, quia non est infinita.

10. Cf. Kuhn (1857) 1968, 578: "Indem wir aber die Idee des persönlichen Gottes aus ihrer Unmittelbarkeit und Unbestimmtheit zu einer vermittelten und bestimmten Erkenntniß zu erheben trachten; so kann dieses, wie gezeigt, nur dadurch geschehen, daß wir über die abstracte Einheit und Einfachheit des göttlichen Wesens hinausgehend eine innere Unterscheidung, einen Selbstvermittlungsproceß Gott annehmen, kraft dessen er der lebendige Gott, der ewig sich aus seiner Substantialität zur Subjectivität erhebende absolute Geist, persönlicher Gott ist."

11. Cf. ibid., 586: Der Unterschied zwischen Vater und Sohn ist "ein Unterschied . . . in der immanenten Beziehung Gottes auf sich selbst als sich wissender Geist, mithin ein subjectiver oder rein persönlicher Unterschied. Gott weiß sich als in sich seiend (Gott Vater), sofern er sich in einem Andere, das er aus sich setzt, also in seinem Andern weiß, und er weiß sich als durch sich seiend (Gott Sohn), sofern er sich in einem Andern als seinem Grunde weiß."

12. Cf. Marheineke 1836, 222: "Es ist nicht schwer, einzusehen, daß Gott in sich selber seyend und aus sich selber seyend ein anderer ist, oder als Vater nicht der Sohn, als Sohn nicht der Vater ist; beide sind andere gegen einander. Daß dieser Unterschied aber im Wesen Gottes kein Unterschied sei, ist durch das Geistseyn Gottes bestimmt. Der Unterschied ist persönlich. Vater, Sohn und Geist sind als Personen bestimmt. Person ist Bewußtseyn und nicht mit Individuum zu verwechseln. In welchem Sinne aber von Personen im göttlichen Wesen die Rede ist, zeigt sich im Geistseyn Gottes; in der Liebe, welche Geist ist, läßt schon unter Menschen die starre Persönlichkeit sich los. So geht nun vollends das Eine göttliche Wesen durch das dreifach bestimmte Bewusstsein Gottes hindurch, ist das eine und selbige in demselben. Drei Götter hat man nur, wenn man die Persönlichkeit abstract und unbeweglich fixiert."

13. Cf. Marheineke 1819, 378–79: "Wollte man den personellen Unterschied zwischen Vater, Sohn und Geist . . . aus dem Grunde leugnen, weil das, was den einen vom andern unterscheiden soll, nothwendig als fehlend dem andern, mithin bei dem göttlichen Wesen als eine ihm fehlende Vollkommenheit angesehen werden müßte, was die Idee von Gott aufheben würde: so darf man dagegen nur bemerken, daß das, was die eine Person als solche, von der andern als solcher, in Gott unterscheidet, keineswegs als eine wesentliche Eigenschaft und Vollkommenheit Gottes, nicht als *attributum Dei*, wie Heiligkeit, Gerechtigkeit, Allwissenheit, sondern nur als eine im Wesen Gottes gegründete *Beziehung* des göttlichen Wesens auf sich selbst und die Welt . . . gedacht wird, die als solche . . . nicht das Wesen, sondern nur die göttlich eigenthümliche Art der Existenz . . . bezeichnet."

REFERENCES

Brower, Jeffrey E. 2009. "Simplicity and Aseity." In *The Oxford Handbook of Philosophical Theology*, edited by Thomas P. Flint and Michael C. Rea, 105–28. Oxford: Oxford University Press.

Cross, Richard. 2005. *Duns Scotus on God.* Aldershot: Ashgate.

Gregory of Nyssa. 1958. *Ad Ablabium.* In *Gregorii Nysseni Opera,* vol. 3.1, *Opera dogmatica minora,* Pars 1, edited by Friedrich Müller. Leiden: Brill.

Hughes, Christopher. 1989. *On a Complex Theory of a Simple God: An Investigation in Aquinas' Philosophical Theology.* Ithaca, NY: Cornell University Press.

Kelly, Charles. 1994. "Classical Theism and the Doctrine of the Trinity." *Religious Studies* 30 (1): 67–88.

Kuhn, Johannes von. (1857) 1968. *Die christliche Lehre von der göttlichen Dreieinigkeit.* Tübingen: Laupp.

Mann, William E. 1982. "Divine Simplicity." *Religious Studies* 18 (4): 451–71.

Marheineke, Philipp. 1819. *Die Grundlehren der christlichen Dogmatik.* Berlin: Nikolai.

———. 1836. *Lehrbuch des christlichen Glaubens und Lebens für denkende Christen.* Berlin: Nicolai.

Meixner, Uwe. 1992. "Eine logische Rekonstruktion der platonischen Prädikationstheorie." *Grazer Philosophische Studien* 43:163–75.

Mullins, Ryan T. 2013. "Simply Impossible: A Case against Divine Simplicity." *Journal of Reformed Theology* 7 (2): 181–203.

Nicholas of Cusa. 1985. *On Learned Ignorance.* Edited by Jasper Hopkins. Minneapolis: Banning.

Plantinga, Alvin. 1980. *Does God Have a Nature?* Milwaukee: Marquette University Press.

Stump, Eleonore. 2016. "Die Einfachheit Gottes und Thomas von Aquins Quantenmetaphysik." In *Eigenschaften Gottes*, edited by Thomas Maschler and Thomas Schärtl, 231–250. Münster: Aschendorff.

Thomas Aquinas. 1978. *Summa Theologiae.* Editio Leonina. Vols. I–V. Edited by University of Salamanca, Madrid. (= BAC Bde. 77, 80, 81,83, 87)

Yandell, Keith E. 1994. "The Most Brutal and Inexcusable Error in Counting: Trinity and Consistency." *Religious Studies* 30 (2): 201–17.

CHAPTER 18

(A)symmetries between God and World

Process Philosophy, Postmodern Theology, and the Two Families of Infinity Argument

PHILIP CLAYTON

THE PROBLEM OF TRANSCENDENCE

Process philosophers are famous for emphasizing the symmetries between God and world. The *locus classicus* for this view is the famous statement of six symmetries in Alfred North Whitehead's 1929 work *Process and Reality,* which represents Whitehead's strongest statement of a fully symmetrical relationship between God and world. It bears quoting in full:

(1) It is as true to say that (a) God is permanent and the World fluent, as that (b) the World is permanent and God is fluent.
(2) It is as true to say that (a) God is one and the World many, as that (b) the World is one and God many.
(3) It is as true to say that, (a) in comparison with the World, God is actual eminently, as that, (b) in comparison with God, the World is actual eminently.
(4) It is as true to say that (a) the World is immanent in God, as that (b) God is immanent in the World.

(5) It is as true to say that (a) God transcends the World, as that (b) the World transcends God.
(6) It is as true to say that (a) God creates the World, as that (b) the World creates God. (Whitehead 1978, 348)

One might infer from these six statements that the relations between God and world are *fully* symmetrical, and the writings of at least some process philosophers serve to deepen this suspicion. Thus David Ray Griffin advocates a complete symmetry: "God is internal to the world (as traditional theism said), [and] the world is internal to God: 'It is as true to say that the World is immanent in God, as that God is immanent in the World' (Whitehead 1978, 348)" (Griffin 2001, 141).

In an attempt to sort things out, I begin by analyzing Whitehead's six propositions and by exploring the fully reciprocal relationship between God and world that results. We'll then turn to consider the significant analogies between orthodox process philosophy and a class of positions on the God–world relationship known as *postmodern theologies*. As we'll see, the postmodern theologians share Whitehead's emphasis on the metaphysics of immanence, even though they rarely formulate their view in these terms. In fact, the Whiteheadian symmetries between God and world quite effectively summarize the central assumptions of this (usually unacknowledged) metaphysics of immanence. Should my criticisms of the symmetries be justified, they also open up serious concerns about the core assumptions of many postmodern theologies. Many of these theologies are guilty of "the domestication of transcendence" (to use the eponymous phrase from Bill Placher [1996]).

If these concerns are justified, how then could one de-domesticate postmodern reflection on transcendence? What concept or concepts can help overcome these limitations? I'll suggest that the concept of infinity has a crucial role to play. The notion of transcendence that reemerges when the overambitious symmetries are left behind is rooted in the classical metaphysical tradition and, in particular, in its heritage in the modern period, especially in German idealism. We will find that the notion of the infinity of God is essential for reestablishing an adequately robust notion of transcendence. Divine infinity offers a surprisingly effective bridge between classical metaphysics and the focus of contemporary thinkers on the Unnamable and Unspeakable, the apophatic Mystery that arises at the heart of all human reflection—but only in one of infinity's two traditional guises.

PROCESS SYMMETRIES

Whitehead's six symmetries, taken in the context of *Process and Reality* as a whole, represent a metaphysical proposal about how best to conceive God and world. Certainly Whitehead is not asking his readers to view the world without God, in the way that many postmodern theologians do. Whitehead's work is deeply metaphysical and—at least in this part of *Process and Reality*—he makes essential use of the concept of God (see Ford 1984 for details). But here, unlike in classical metaphysics, immanence and transcendence are delicately balanced.

Whitehead's symmetries represent a sophisticated means for disposing of the notion of the infinity of God. In this passage he lists a variety of classical attributes of God: God is permanent and One; God is eminently actual; the world is immanent in God; God transcends the world; God creates the world. In each case, he argues, it's just as true to affirm the converse—on the condition that one accepts Whitehead's own metaphysical proposal.

"It is as true to say . . ." we read six times. But *is* it as true to say these things? Let's take them one at a time, using the numbers and letters that I've inserted into the text above. (1) I take it that (1a) "God is permanent and the World fluent" is not a problematic assertion. But what about (1b), "the World is permanent and God is fluent"? For open theists, including the present author, there are specific senses in which God is fluent; that is, there is a sense in which God changes. Assume that future contingent propositions have no fixed truth value in the present and that *p* is such a proposition. Perhaps *p* is the proposition that you will order fish at the restaurant on Saturday night. Assume that on Saturday night you do in fact order fish. That means that there is a time (for example, now) when God cannot correctly affirm "*p* is true" and thus does not affirm *p*; and there is a later time (for example, Saturday at midnight) when God can and does correctly affirm "*p* is true." In this sense, I would say, God is fluent.

But is (1b) *as true* as (1a)? The world changes in a myriad of ways: new objects and agents come into existence; new types of ecosystems begin to function; new social realities are instantiated; and people think thoughts that were never thought before. Humans may shepherd the earth with wisdom and altruism, or we may drive it to destruction with selfishness and greed. Universal altruism toward others may triumph in the end, or it may all but disappear. The moral fiber of our being is at stake in the course of history, and the outcome is unclear. *None of these things is true of God.* For open theists, and for process theologians too, God's moral being, or what Whiteheadians call the "primordial nature" of God, is set for all time. I conclude that it is not "as true" to affirm (1b) as true as (1a). Actually, this conclusion should be reformulated, since it's problematic to speak of amounts of truth. We should say that "God is fluent" is true in only a very limited and circumscribed sense, whereas "the World is fluent" is very broadly true.

The same conclusion, I suggest, follows for each of the other alleged symmetries. To save time, I'll skip over the analysis of (3) and (4), since they turn on technical features of Whitehead's metaphysics that we don't have sufficient space to develop here. What about (2)? In (2a), "God is one" expresses the unity of the divine being, the unity of the divine character, the unity of the divine experience, divine simplicity, and divine aseity. "The World is many" reflects the vast plurality of existing things, classes, dynamics, types of agents, moral characters, and human goals. By contrast, (2b) "the World is one and God many" is true only in a carefully circumscribed sense. Those of us who are Trinitarians affirm the threefold nature of God. But because we wish to avoid being tri-theists, we formulate our Trinitarian assertions against the backdrop of the underlying unity of the divine nature. Likewise, there are limited senses in which physicists say that the world is one, for example, because there is a single space–time continuum or a fixed sum of matter-energy. Again here, there is not a full symmetry between (2a) and (2b).

This brings us to the final two propositions, and thereby to the most important debate between Whiteheadians and open theists (and classical theists even more so!). I reverse the order of (5) and (6), since it turns out that the position that Whiteheadians take on creation largely determines their position on the transcendence of God, and hence (as I will argue) on the infinity of God. So, *is* it "as true to say that (6a) God creates the World, as that (6b) the World creates God"? The analysis of these two propositions proceeds in line with the analysis already given of (1) and

(2). Open theists acknowledge a limited sense of (6b). The content of God's thought and consciousness is determined in part by the free actions of human beings (and perhaps by events in the physical world that are ontologically contingent—if indeed any such exist). If the world determines some significant part of the consciousness of God, and the consciousness of God is an important determinant of who God is, then, in at least a metaphorical sense, one might say that the world creates God.

For advocates of *creatio ex nihilo* such as myself, "God creates the world" is true in a very strong sense. No preexisting matter provided the sufficient ground for the big bang (or whatever was the origin of the physical universe). Instead, God willed that a universe should exist, and that universe came into existence. For advocates of *creatio ex nihilo*, (6a) and (6b) fail in a rather dramatic way to be symmetrical. Of course, process theologians do not generally affirm *creatio ex nihilo*, as the debates in Oord (2014) show. Orthodox process theologians affirm that, in all cosmic epochs, God was always accompanied by other centers of experience, which Whitehead calls "actual occasions." The creative responses that these actual occasions make to each other and to God (Whitehead's *concrescence*) influence the experience of God, and God's creative responses in turn influence the actual occasions. Hence for Whiteheadians—except for one caveat—there really is a full symmetry between world and God in this sense.

Here's the caveat: Whitehead (1978) famously said that God is not an exception to other metaphysical principles but their "chief exemplification" (22). Yet most Whiteheadian thinkers do not actually treat God as fully identical to other actual occasions. Ordinary actual occasions accomplish their process of becoming ("concrescence") in an extremely short period of time, perish, and thereby become available to be "prehended" by other occasions. One school (led by John Cobb) affirms that God is a single concrescence, not a chain of occasions that become and then perish. But the experience of an occasion only becomes available to others after it has perished, so in this case how could God influence any other centers of experience? This leads a second school of thought (led by Charles Hartshorne and David Ray Griffin) to affirm that God is in fact a society of actual occasions. It's just that there is a closer unity of vision and valuation across the divine society as a whole than in the case of any other society of occasions.

One could extend this argument in some detail, but the basic idea is clear. It's possible to conceive the divine experience as identical to the experience of finite occasions, and thereby to preserve the full symmetry expressed by (6). But it is costly to make this move. Thus one finds even some process philosophers pulling back from the full identification of the divine experience and normal human experience.

I suggest that affirming (6) is an important part of explaining why process philosophers would wish to affirm (5), "It is as true to say that (a) God transcends the World, as that (b) the World transcends God." For them, any actual occasion that prehends another transcends it in a very specific sense: it passes a creative judgment on the experience of the occasion that has just "perished." Your creative judgment of Mozart's music in some sense transcends Mozart, since he is no longer here to experience your judgment and to encompass it in some way in his own further musical experience. If one believes that *creatio ex nihilo* is false and that divine actual occasions and nondivine actual occasions have always been mutually experiencing and responding to each other in full symmetry, then one will of course affirm that God transcends the world as much as the world transcends God.

POSTMODERN THEOLOGIES

All of us are familiar with the strong criticisms of transcendence that lie at the basis of most postmodern and poststructuralist thought. In fact, late modern or "postmodern" thought can virtually be identified with the rejection of transcendence. One can trace late modern thought back to the 1840s, some ten years after Hegel's death, and to the so-called philosophers of suspicion whose work dominated the latter half of the nineteenth century. Late-modern attacks on transcendence took multiple forms: denying the existence of God; challenging all metaphysical reflection; arguing for the epistemic superiority of science; claiming that nonempirical language is strictly speaking meaningless; and, finally, construing all language about God as pragmatic, existential, hermeneutical, *Sprachspiel*-based, political, or cultural.

Those familiar with postmodern thought know how quickly incommensurabilities begin to arise when one tries to engage in rational dialogue with these attacks from a standpoint that is other than postmodern. For those of us who continue to be interested in metaphysics, a more fruitful dialogue opens up with those postmodern thinkers who explicitly advocate a metaphysics of immanence. In this section I have chosen three representative thinkers with whom to pursue this dialogue. The points could have been made by considering others—Mark C. Taylor or Slavoj Žižek, for example—although their positions push back more strongly against the constructive dialogue that I am seeking here.

Yirmiyahu Yovel's two-volume work *Spinoza and Other Heretics* offers an excellent first example. Here Yovel proclaims the adventures of immanence, the thrills of the denial of transcendence.[1] Yovel traces the history of this adventure as far back as the modern father of immanence, Spinoza, through the main philosophers of suspicion, and as far forward as the neopragmatist Richard Rorty and the deconstructive Derrida, whom he christens the true disciples of Spinoza. In his eyes, Spinoza is to be praised because he "rejected both the dualistic transcendence of the Christians and the denaturized, transcendence-ridden this-worldliness of his fellow Jews. A 'Marrano of reason,' he shed all historical religions (though not all religious concerns), and offered salvation neither in Christ nor in the Law of Moses, but in his own kind of religion of reason—naturalistic, monistic, and strictly immanent" (Yovel 1989, 2:169).

For Yovel, it's either metaphysics (along with transcendent beings and truths) or Rortyan immanentism (choices that are no longer grounded in Truth or God). Once we have dispatched with transcendence, we will be set free to enjoy our innate freedom to create values and truths without constraint.

Of course, Yovel would not be happy that I have christened him a metaphysician, even in this immanentist sense. He is less interested in working out the details of a naturalistic metaphysics and far more interested in propelling the reader through Marxian materialism and on to a joyful acceptance of whatever Nietzsche has to offer.[2] Still, he does single out Spinoza, the great metaphysician, to stand for his project of immanence, the denial of transcendence. By giving Spinoza pride of place, Yovel lends implicit support to a metaphysics of immanence, or at least to a more sophisticated critique of transcendence than one usually finds in postmodern texts.

Less sophisticated but more straightforward is the concept of "horizontal transcendence," first formulated by Michael Kalton and later popularized by Ursula Goodenough in *The Sacred Depths of Nature* (1998) and in a batch of other works. Kalton's definition is worth quoting in full:

> Horizontal transcendence finds its anchor in life rather than mind, thus displacing human consciousness from its privileged place. The movement from earth to cosmos, from biosystem to life, is a form of transcendence that is characteristic of degrees of abstraction rather than a movement towards some kind of Absolute metaphysical dimension. There is no cosmos posited apart from the historically ongoing one within which we find ourselves, nor is there life apart from ongoing living, at whatever level it is considered. . . . Instead of the typical vertical transcendence of the Greek inspired tradition, the movement of this kind of spiritual cultivation is horizontal, perfecting our relationship with the world of life about us. . . . In this kind of horizontally framed spirituality the question of belonging acquires a new kind of centrality. (Kalton 2000)

Horizontal transcendence is a form of transcendence without transcendence—transcending one's previous experience into new kinds of experience. Kalton intends the phrase "transcendence without transcendence" to sound paradoxical, presumably in the same sense that the phrase "metaphysics of immanence" will sound paradoxical to some. Using such a koan has a specific purpose—he hopes to foster a certain kind of "spiritual cultivation," namely, "perfecting our relationship with the world of life about us." Whereas Whitehead chooses becoming over being, Kalton affirms "belonging" over being.

Following Yovel, Kalton offers a second step into the metaphysics of immanence. The third step comes with the abolition of being. The notion of substance had been a major target since Descartes used it to dichotomize all existing things (*res cogitans* and *res extensa*) and the German idealists sought to replace substance with subject, a move whose consequences cascaded into the twentieth century. With the abolition of being, even more clearly, metaphysics transitions into ethics. This is the explicit outcome in the work of postmodern theologians such as Emanuel Levinas and Jean-Luc Marion. Whitehead's metaphysics of becoming cannot be equated with the theological ethics of relationships *sans l'être* (Marion 1995). But surely the theologians-without-being are extensions of the Whiteheadian logic one step further.

The fourth step—though certainly not the final one!—arrives equally under the aegis of Whitehead. It's the philosophy of the momentary event, of all reality as events and event-ing. One can find subtle and metaphysically sensitive works representing the fourth step, such as Catherine Keller (2003) and Roland Faber (2008). But for a more unrestrained exemplar of the fourth step in its radical form, it's better to turn to John Caputo's influential book *The Weakness of God* (2006).

Caputo claims that the name of God "is an event, or rather that it harbors an event, and that theology is the hermeneutics of that event" (2006, 2). The ontology of events represents a very close bond between Caputo and Whitehead. Consequently, the task of theology is "to release what is happening in that name, to set it free, to give it its own head, and thereby to head off the forces that would prevent this event" (2). Never one to eschew a poetic metaphor, Caputo uses the name "God" to express "electrifying event-ing disturbance, the solvent of the weak force of this spectral spirit who haunts the world as its bad conscience" (39–40).

Among the forces that would prevent the event-ing disturbance is classical metaphysics. A big chunk of Caputo's book is devoted to breaking with metaphysical notions of God through a process of "Unsaying":

> Suppose we imagine God, not as a prime mover unmoved, but as removed from the order of cosmic movements and cosmological explanations, removed from the onto-causal order altogether, from being, presence, power, and causality? Suppose we imagine God otherwise, not really having a seat in being at all, but below being and beings, simmering beneath the ontico-ontological difference, as the heart of a heartless world? (Caputo 2006, 36)

Unlike some postmodern theologians, Caputo is willing to make explicit use of the language of divine transcendence. Transcendence can't be opposed to this weak force of God, however; indeed, *it just is the weak force of God* (Caputo 2006, 36). What God's transcendence means is that "we are laid low by a call arising from on high, but whose heights are eventful, provocative, and arise from someone who, lacking the wherewithal to lay down his head, pitches his tent among the lowly bodies and nobodies below in the bodily bowels of hypo-being" (36). Traditional language of divine transcendence, by contrast, entails a

Superman-type God: "I treat God, not as an eminent omnipotent onto-power capable of leveling tall buildings and reducing his enemies (no need for gender-neutral language here) to ashes, but as the weak force of a call" (36).

In the end, Caputo suggests, the only way to avoid the Superman picture of God is to avoid the language of being or existence altogether. As he notes, "I prefer to say not that God 'is' but that God 'calls,' that God promises, not from beyond being but from below, without being or sovereignty" or, in the same context, "God is, without being, of unconditional import and the stuff of unconditional desire" (Caputo 2006, 36).

Unfortunately, Caputo decides that, in the end, talk of God's transcendence won't really do either—perhaps because God's event-ing has removed most of the semantic arrows that could have pointed the way to a meaningful use of the term transcendence. Caputo prefers to speak instead of God's "in-scendence" (Caputo 2006, 45). He is unsure whether there is a God (or a "God beyond God") behind our sense of call; in fact, the question doesn't really matter to him:

> Whether over and beyond what we might call the hermeneutics of the event, the lived experience of the call and of being on call, there is some entitative cause calling, some entity or hyper-entity out there with a proper name, verifiable by a metaphysical argument or certifiable by a divine revelation, is no part of my hypothesis, one way or the other (for or against). I leave that stock to fluctuate on the open market of existence. (Caputo 2006, 40)

Now, when it comes to our dislike of the hyperanthropomorphized Superman God, Caputo and I do not diverge. Both of us reject that God equally as vehemently, and for similar reasons. I also affirm the weakness of God, albeit in the form of divine kenosis. For Caputo, however, when traditional substance metaphysics has to go, everything else divine has to be thrown out with it. Here we part ways. The transcendence of God, decoupled from the very *un*transcendent anthropomorphisms of human, all-too-human projections, is exactly the antidote to the moral misuse of God-language that Caputo seeks. And, as I will argue, the concept of divine infinity provides an ideal means to protect God's transcendence from human projections.

COMMENCING THE METAPHYSICAL JOURNEY

Let's pause to take stock of where we now stand. We first explored the parallels between the Whiteheadian symmetries and the four key steps in the ongoing progression of postmodern theologies. Like Whiteheadians, the postmodern theologians put at least as much stress on immanence as transcendence. Both camps insist that all the kinds of things that God does, humans do also. There can be no talk of transcendence that is not equally about events in the world. That transcendence and immanence are (at least) symmetrical in this way turns out to be axiomatic for these thinkers. In fact, I suggest that what one perceives in the four steps of postmodern theology goes even further; it involves the *breaking* of the symmetry, the increasing dominance of the immanent over the transcendent, until transcendence begins to disappear altogether. God-language comes to be more about worldly events and, in the extreme cases, exclusively so.

What does it take to tip the scale back in the other direction? Whitehead himself never moved back from his six symmetries. Other process-influenced thinkers, however, including this author, do pull back from the demand for full God–world symmetry. The infinite divine Creator, we argue, can be involved in deep reciprocal relations with finite agents without being in nature the same as them. The importance of this realization cannot be stressed strongly enough. In the end, it turns out, *the notion of infinity is uniquely suited to move beyond the symmetries, to challenge the domestication of transcendence, and to explore the asymmetrical features in God's relation to the world.*

But note: exactly how one introduces the symmetry-breaking features of infinity makes all the difference. Placher may have taught us to worry about the domestication of transcendence, yet it's crucial to de-domesticate without producing a new enslavement, a belittlement of immanence. It is Whitehead's enduring contribution to have developed a metaphysics able to conceive the realm of immanence as mattering, even to God. (Open theism achieves a similar result by drawing more deeply on biblical resources.) If one is to end up with a philosophical theology that adequately conceives immanence, one has to retain an adequate place for the kingdom of finites: ourselves, living things, and the creative chaos that constitutes the realm of nature. Exactly how one pulls back from the precipice of the collapse of transcendence—how one actually commences the metaphysical journey—makes all the difference.

For a positive example, consider the recent work of Joseph Bracken. Fr. Bracken, a well-known Catholic process theologian, focuses on the twin themes of immanence and transcendence, very much as we have done here. That his solution moves beyond Whitehead's symmetries is clear from the fact that he remains a Trinitarian theologian, and from book titles such as *The One in the Many* (Bracken 2001). Consider this short excerpt from one of Bracken's early drafts for his book on systems theory and theology (Bracken 2014), which he has graciously allowed me to cite here. Bracken writes,

> The God–world relationship is an integrated organic reality or "super-system" in which the higher-order system proper to the life of the three divine persons of the Christian doctrine of the Trinity is by the free choice of those same divine persons symbiotically linked to the lower-order set of systems proper to the cosmic process. In this way, the three divine persons exercise downward causation upon the cosmic process even as the cosmic process exercises upward causation upon the inner life of the three divine persons.[3]

Like postmodern theologians, and like Whitehead, Bracken chooses to begin with the natural world, and with systems theory in particular, which is an important twentieth-century school of philosophy and has been one of the fastest growing areas in microbiology over the last several decades. He is clearly looking for analogies between the best account of the natural world that science can give, on the one hand, and a theology of the intra-Trinitarian life, on the other. But the point of the analogy is precisely *not* to establish the God–world relation as fully symmetrical. Instead,

> in line with this understanding of systems theory in the natural sciences, one can rationally claim that the higher-order system of the divine life possesses its own ontological integrity, its distinctive pattern of unchanging existence and activity, and the lower-order set of systems proper to the cosmic process possesses its ontological integrity, its own ever-changing pattern of existence and activity. In this way,

> the ontological distinction between the natural and the supernatural can be sustained even as a creative existential interaction between the two is constantly maintained. (Bracken 2014, manuscript, 7)

The final sentence of the quoted passage gives a perfect expression of what I am arguing is the appropriate goal: God and world exist in "creative existential interaction," an interaction that has real effects on both, even as the ontological distinction between Creator and created is preserved. Elsewhere in his manuscript Bracken puts this conclusion in Christological terms: "Just as the human Jesus possessed a significant measure of personal freedom in his unswerving obedience to the Will of God, so the cosmic process will find its own way through an ongoing process of trial and error to the final integration of the natural and supernatural, matter and spirit, intended for it by the three divine persons from the beginning" (Bracken 2014, manuscript, 8).

It is no secret that this position does not represent the dominant program in Roman Catholic thought today. Theologians such as Bracken and (for example) Denis Edwards in Australia are certainly not in the majority—even though they write in the spirit of Thomas Aquinas and the great thinkers of the Catholic Middle Ages. Far more typical among liberal Catholic theologians is the use of "Rahner's Rule" as the framework for linking immanent and transcendent. Followers of Catherine LaCugna, for example, use scripture, tradition, and experience as the basis for claims about the economic Trinity and hence the mode of divine presence in the world (LaCugna 1991). Rahner's Rule then allows them to infer that the immanent Trinity is similar in nature to these conclusions about the economic Trinity.

Bracken reverses the direction of inference, using these reflections on the immanent Trinity as the ontological foundation for conceiving the economic Trinity. As he describes it, "I interpret both the immanent Trinity and economic Trinity in terms of my own version of systems theory, in which a lower-order system can be incorporated into a higher-order system without losing its own ontological identity as a lower-order system (e.g., atoms as still atoms within the constitution of a molecule)."[34]

More generally, the position that Bracken has been developing over several decades simultaneously affirms the transcendence and immanence of God—yet without subjecting itself to Whitehead's six symmetries.

Bracken's thought represents an important move beyond the domestication of transcendence. Like the postmodern theologians and Whitehead, and arguably like Aquinas also, he begins with conceptual categories derived from the study of the natural world. But the structures he uses are not the structures of symmetry; they are open-ended, upwardly open—open to and toward a transcendence that cannot be fully grasped by concepts in their finite or limited form.

TWO FAMILIES OF INFINITY ARGUMENTS

Bracken employs an argumentative form classically known as the *via eminentiae*, the "way of eminence." In the *via eminentiae*, one begins with categories known through the natural world, extending them upward to the highest level. One then asks in what sense these categories could or should be affirmed of God. To do this, one works to leave aside the imperfect features of the qualities one is analyzing and to conceive each quality in its most perfect possible form. As Sauvage concludes in his *Catholic Encyclopedia* article, "it is easily understood that the perfections affirmed, in these two ways, of God, as First and Perfect Cause, cannot be attributed to Him in the same sense that they have in finite beings, but only in an absolutely excellent or supereminent way (*via eminentiae, excellentiae*)" (Sauvage 1907).

This argument form lies at the heart of much of what Fr. Bracken has written throughout his long career, just as it lies at the heart of my program in philosophical theology starting with *Mind and Emergence* (Clayton 2004) and further extended in *Adventures in the Spirit: God, World, Divine Action* (Clayton 2008). As should be clear by now, although this argument structure begins with mundane states of affairs, as Whitehead's symmetries do, in the end it points in a vastly different direction and achieves correspondingly different results. Instead of presupposing a full parallelism between God and world, it precisely undercuts any such parallelism.

One discovers, somewhat surprisingly, that this "way of eminence" (*via eminentiae*) can be used in two very different ways, thereby yielding two very different sets of results. I wish to show, in conclusion, that they represent two distinct families in the history of Western thought. For shorthand, let's call them the "Way of Unlimited Perfection" and the

"Way of Absolute Infinity." What one says about divine transcendence is significantly affected by which of these two approaches one adopts. In the second approach, the Way of Absolute Infinity, infinity limits human language about God more severely than in the first. As we'll discover in this final section, infinity's limiting function in the Way of Absolute Infinity connects back in interesting ways to some of the themes of the postmodern theologians.

Both of the two ways are already familiar to most readers, so I'll be brief. In the Way of Unlimited Perfection, one considers a set of goods, namely, the *kinds* of qualities that one thinks would apply to God, for example, knowledge, power, wisdom, and benevolence. Using the "cleansing mechanism" of the *via eminentiae*, one then seeks to purify them of their imperfections, which are due to the imperfections of their human source. (A man is wise, but he is not perfectly wise, or *how* he is wise introduces imperfections.) When the perfect-making qualities are maximized by this method and the imperfections are eliminated, what emerges is the highest level of perfection that one can conceive for any given quality. Each of these qualities helps one to understand an attribute of God. The sum of the attributes thereby attained, it is argued, offers the most accurate picture of God that philosophers can achieve.

The program just described exists in many forms, perhaps the clearest of which is the Anselmian program.[5] Formally, it proceeds by maximizing perfections, that is, by maximizing great-making properties and by conceiving them in their most excellent or "eminent" form. William Carroll sketches a second argument from Unlimited Perfection in his chapter in this book (chapter 4), as he has done frequently in discussions, and Brian Leftow's chapter (chapter 15) sheds further light on this mode of argument. As Carroll shows, St. Thomas addresses the question of the infinity of God in the *Summa theologiae* (I, q. 7). Before that, however, he discusses perfection in question 4, construing it as completeness or as being fully actual, in the well-known sense of Aristotle's metaphysics. Only when that framework is in place can Aquinas begin to discuss the sense in which God is good (qq. 5–6). The similarities and differences between these two forms of the argument are extremely interesting and important; they are well and ably presented in other chapters in this volume.

The Way of Absolute Infinity has also played an important role in the history of Western thought, and I am glad for the opportunity to include it in this volume alongside the other Way. This second *via*, I'd like to suggest, also employs a type of the *via eminentiae*. But here the emphasis lies not so much on magnifying perfections without limit, but rather on the privative *in-* in the term *infinity*: *not* finite. In the case of both the divine being and the divine attributes, God is without any limits whatsoever; nothing finite can be allowed to limit God. Hence we must question or eliminate all finite features and connotations with which our language shackles God. Any concept that one wishes to apply to God must be tested in terms of this criterion. In Neoplatonism after Pseudo-Dionysius, it produced the *via negativa*; in Meister Eckhart it produced the God above God; in Nicholas of Cusa, the *coincidentia oppositorum* and the doctrine of "divine ignorance"; and in mystics such as Julian of Norwich, Teresa of Avila, and Catherine of Sienna, a questioning of all language of God outside of the unitive experience itself, the moment of being overwhelmed by the divine presence.

The Way of Unlimited Perfection maximizes perfection, while the Way of Absolute Infinity maximizes infinity, which is to say that it minimizes finitude when it comes to language about God. The first way, we might say, is the route of infinite perfection, the second the route of perfect infinity. Needless to say, the results of the two programs have differed historically, and it's not hard to see why. It's one thing to minimize imperfections in our language of God and then to affirm the result of God; it's a very different thing to seek to eliminate all finite elements whatsoever from one's language about God.

CANTOR'S CONTRIBUTION

I spoke on the topic of infinity at the 2013 World Science Festival in New York City. Georg Cantor played a major role in my preparations for the talk, as he did in the presentations of the mathematicians and physicists with whom I shared the stage. To Cantor we owe the brilliant theory of transfinite sets, which allows one to differentiate between the various orders of infinity in a rigorous way and also to formulate and evaluate claims about infinite sets.

As you may know, Cantor was deeply concerned with the status and implications of his own theory. Is the series of orders of the transfinites—$aleph_0$, $aleph_1$, $aleph_2$, $aleph_3$, and so on—itself transfinite? In the process of establishing for the first time a way to calculate with infinite sets, had the very thing that he was after somehow eluded his grasp?

Later in his career, Cantor introduced a notion that transfinite set theory could never capture: *das absolute Unendliche*, the absolute infinite. He struggled with these issues in his 1885 article "Über die verschiedenen Standpunkte in bezug auf das aktuelle Unendliche" (Concerning Various Perspectives on the Actual Infinite). Here he concluded that there are three kinds of actual infinities: (1) the absolute infinite, which pertains to God alone; (2) the concrete infinite or the *Transfinitum*, which he believes is found in nature; and (3) the abstract infinite, which is expressed in mathematics through the formalism of transfinite numbers.[6] Of these three actual infinities, both of the latter two can be comprehended by the mind: "*Omnia seu finita seu infinita definita sunt et excepto Deo ab intellectu determinari possunt*" (All things, whether finite or infinite, are definite and, with the exception of God, can be determined by the intellect).[7]

Toward the end of his life, Cantor focused increasingly on an order of infinity that is beyond all the orders, a something that is infinite in the most absolute way possible. In 1899 he wrote to Dedekind: "Now I envisage the system of all numbers [*System aller Zahlen*] and denote it W. . . . Thus the system W of all ordinal numbers is an inconsistent, absolutely infinite multiplicity [*Das System W aller Zahlen ist eine inkonsistente, eine absolut unendliche Vielheit*]."[8]

By *W* Cantor meant the entire number sequence, the collection of all number-classes. (Note that the class of all numbers is not a Cantorian set.) This collection is not limited in any sense; hence it is absolutely infinite. But there is no more analysis that can be given from a set-theoretical perspective; the collection cannot be systematized in any other way. Here mathematicians must fall into silence.

In a January 1886 letter to Cardinal Johannes Franzelin, Cantor worked hard to show that the infinity of God is distinct from the reality of the transfinites. In their earlier correspondence, the cardinal had raised his suspicion that affirming the reality of the transfinites amounted to the heresy of pantheism, a charge that deeply pained Cantor. Cantor was also firmly convinced that the charge was false. He thus wrote:

> I distinguish between an "*Infinitum aeternum sive Absolutum*," that refers to God and his attributes and an "*Infinitum creatum sive Transfinitum*" that above all testifies to where in *natura creata* an

> actual infinite must be acknowledged, as for example, according to my firm conviction, in the actual infinite number of created individuals in the entire universe and also on our earth, and in all probability, in every smallest part of extended space itself, a matter on which I completely agree with Leibniz.[9]

In short: transfinites are created, whereas God, the Absolute Infinite, is eternal. No *wissenschaftlicher* framework, such as set theory, can contain the Absolute Infinite; hence it cannot be grasped with the rational mind—or, in the phrase from Cantor cited above, it cannot "be determined by the intellect." In this sense, the divine Infinity remains forever shrouded in mystery.

Of course, not all those who follow the Way of Absolute Infinity are driven to silence, or at least not immediately. In our writings we do, however, struggle on a regular basis with the limits of language about the infinite God. Does one set the bar too high by trying to avoid all finite categories in talking about God? Or is it that, though the project itself is acceptable, one inevitably falls into inconsistency by trying to speak about what, according to one's own project, cannot be spoken of? Is it not clear that one cannot seek to transcend the finite altogether and nonetheless continue speaking?

CONCLUSION

We begin by considering positions that pushed back against all asymmetries in the relationship between God and world. "Orthodox" process theologies—we considered Alfred North Whitehead and David Ray Griffin—defend six strong symmetries between God and world, symmetries radical enough as to rule out the infinite/finite contrast as applied to the God–world relationship. We then found that a number of postmodern theologies deny the asymmetry in a similar way. As we saw, they reject the metaphysical Other, preferring increasingly immanent approaches to theology as one moves through the four steps.

Finally, leaving the symmetrical theologians behind, we concentrated on approaches that affirm the infinity of God. Here we found significant differences in how the concept of infinity is applied to the

God–world relation—differences so great that one is required to treat them as two separate families. The Way of Unlimited Perfection and the Way of Absolute Infinity—or, most briefly, the way of infinite perfection and the way of perfect infinity—turn out to be two competing schools for conceiving divine transcendence.

Anselmians seek the level of greatest excellence or perfection, and Thomists seek to understand the divine as fullest completion, *ipsum esse subsistens*. Contrasting in important ways with their project, however, is another way of conceiving the infinity of God, the Way of Absolute Infinity. Both *viae* seek to describe the divine reality as the ground of all that exists. But as the paths vary, so too do the claims they make about the ultimate referent. As Cantor, the great modern exponent of the second Way, concluded, "All things, whether finite or infinite, are definite and, with the exception of God, can be determined by the intellect."[10]

NOTES

I wish to acknowledge the partnership in research with my research assistant, Matthew Fisher. Matthew's assistance with texts and sources contributed greatly to the quality of this chapter, and I happily give this public acknowledgment of his role as my co-researcher.

1. In Yovel's *Spinoza and Other Heretics*, see, especially, vol. 2, *The Adventures of Immanence*. For commentaries, see the symposium on his work in *Inquiry*, vol. 35, esp. Henry Allison (1992), Richard Schacht (1992), and Yirmiyahu Yovel (1992).

2. The distance from Spinoza increases in the epilogue as Yovel dismisses the notion of the universe as *causa sui* (1989, 2:173) and asserts that "there is no *natura naturans* in Spinoza's sense, no infinite substantive unity underlying finite things in the world" (2:175).

3. From an early draft of the introduction to Joseph A. Bracken's *The World in the Trinity* (Bracken 2014, manuscript, 6–7). Joe Bracken generously shared the introduction and the chapter on the Trinity from his book manuscript.

4. Joseph Bracken, e-mail communication, August 2, 2013.

5. The Anselmian program is classically expressed in *Anselmian Explorations* (Morris 1987), though it is present in one form or another in a vast ($\approx \infty$) number of publications in metaphysics and Christian philosophy.

6. Cantor, "Über die verschiedenen Standpunkte in bezug auf das aktuelle Unendliche" (Concerning Various Perspectives on the Actual Infinite), written 1885, later published in *Zeitschrift für Philosophie,* vol. 88. See Newstead (2009, 548).

7. Cantor 1980, 176; also in Ewald 1996, 2:891, quoted in Newstead 2009, 546.

8. Letter from Cantor to Dedekind, August 3, 1899; the text from Ewald 1996 (1) is now available online at http://www.logicmuseum.com/cantor/cantorquotes.htm (accessed November 15, 2013).

9. Cantor 1991, 254, quoted in Newstead 2009, 549–50; the translation is hers.

10. See note 7, above.

REFERENCES

Allison, Henry. 1992. "Spinoza and the Philosophy of Immanence: Reflections in Yovel's *The Adventures of Immanence*." *Inquiry* 35 (1): 55–67.

Bracken, Joseph A. 2001. *The One in the Many: A Contemporary Reconstruction of the God–World Relationship.* Grand Rapids, MI: Eerdmans.

———. 2014. *The World in the Trinity: Open-Ended Systems in Science and Religion.* Minneapolis: Fortress.

Cantor, Georg. 1980. *Gesammelte Abhandlungen mathematischen und philosophischen Inhalts.* Berlin: Springer.

———. 1991. *Briefe.* Edited by Herbert Meschkowski and Winfried Nilson. Berlin: Springer.

Caputo, John D. 2006. *The Weakness of God.* Bloomington: Indiana University Press.

Clayton, Philip. 2004. *Mind and Emergence: From Quantum to Consciousness.* Oxford: Oxford University Press.

———. 2008. *Adventures in the Spirit: God, World, Divine Action.* Minneapolis: Fortress.

Ewald, William, ed. 1996. *From Kant to Hilbert: A Source Book in the Foundations of Mathematics.* 2 vols. Oxford: Clarendon.

Faber, Roland. 2008. *God as Poet of the World: Exploring Process Theologies.* Louisville, KY: Westminster John Knox.

Ford, Lewis S. 1984. *The Emergence of Whitehead's Metaphysics, 1925–1929.* Albany: State University of New York Press.

Goodenough, Ursula. 1998. *The Sacred Depths of Nature.* New York: Oxford University Press.

Griffin, David Ray. 2001. *Reenchantment without Supernaturalism: A Process Philosophy of Religion.* Ithaca, NY: Cornell University Press.

Kalton, Michael. 2000. "Green Spirituality: Horizontal Transcendence." In *Paths of Integrity, Wisdom, and Transcendence: Spiritual Development in the Mature Self,* edited by Melvin Miller. London: Routledge. Accessed January 19, 2018. https://faculty.washington.edu/mkalton/green%20spir1.htm.

Keller, Catherine. 2003. *Face of the Deep: A Theology of Becoming.* London: Routledge.

LaCugna, Catherine Mowry. 1991. *God for Us: The Trinity and Christian Life*. San Francisco: Harper.

Marion, Jean-Luc. 1995. *God without Being: Hors-texte*. Chicago: University of Chicago Press.

Morris, Thomas V. 1987. *Anselmian Explorations: Essays in Philosophical Theology*. Notre Dame, IN: University of Notre Dame Press.

Newstead, Anne. 2009. "Cantor on Infinity in Nature, Number, and the Divine Mind." *American Catholic Philosophical Quarterly* 83 (4): 533–53.

Oord, Thomas, ed. 2014. *The Theology of Initial Creation: Creatio ex Nihilo or Something Else?* New York: Routledge.

Placher, William C. 1996. *The Domestication of Transcendence: How Modern Thinking about God Went Wrong*. Louisville, KY: Westminster John Knox Press.

Sauvage, George. 1907. "Analogy." In *The Catholic Encyclopedia*, vol. 1. New York: Robert Appleton Company. Accessed August 5, 2013. http://www.newadvent.org/cathen/01449a.htm.

Schacht, Richard. 1992. "Adventures of Immanence Revisited." *Inquiry* 35 (1): 69–80.

Whitehead, Alfred North. 1978. *Process and Reality: An Essay in Cosmology*. Edited by David Ray Griffin and Donald W. Sherburne. New York: Free Press.

Yovel, Yirmiyahu. 1989. *Spinoza and Other Heretics*. 2 vols. Princeton, NJ: Princeton University Press.

———. 1992. "Spinoza and Other Heretics: Reply to Critics." *Inquiry* 35 (1): 81–112.

CHAPTER 19

The Quantitative and the Qualitative Infinity of God

BENEDIKT PAUL GÖCKE

I distinguish quantitative from qualitative notions of infinity, and argue that quantitative notions of infinity are of limited use in the philosophy of religion: they apply at most in the analysis of some divine attributes and do not increase our understanding of the infinity of God itself. I then turn to qualitative notions of infinity and finitude and argue that they refer to different *modes of being* on a tripartite scale of ontological determinacy, reaching from infinite negative quality, through finitude, to infinite positive quality. I then turn to the question of whether there is an entity the mode of being of which is that of positive-qualitative infinity. I argue that models of ultimate explanation, which imply that there is a single ultimate source of everything, entail that there is such an entity and that this entity is what we call "God." Consequently, to assert that God is infinite is to assert that there is a positive qualitatively infinite entity, which is the single ultimate source of everything. I end by arguing that this entity is indistinguishable from the world, since any property that any finite entity exemplifies is a property that a positive qualitatively infinite entity also exemplifies. However, paradoxically, God is not only indistinguishable from the world but, by this fact alone, as Meister Eckhart observed, the more distinct insofar as He is indistinct.

THE QUANTITATIVE INFINITY OF DIVINE ATTRIBUTES

Hart (2011, 255) is right to say that "there is not—nor has there ever been—any single correct or univocal concept of the infinite." This is no surprise but only to be expected because there are two radically different notions of infinity, which we often do not keep apart—quantitative infinity and qualitative infinity. Both notions of infinity originate from a denial of finitude. In the case of qualitative infinity, qualitative finitude is denied, whereas quantitative infinity is based on the negation of quantitative finitude.

To understand quantitative infinity, we have to understand quantitative finitude. In this, I assume that mathematics, in particular set theory, is the paradigmatic science of quantity. In set theory, a set is a collection or series of well-defined elements. A set is finite if and only if there is a natural number n that denotes the number of elements in the set. Consequently, a set is infinite if and only if it is not finite, which is to say that a set is infinite if and only if there is no natural number n that is the number of the members of the set.[1]

Since the set of real numbers is such that there is no natural number n, which is the number of the members of the set, it follows that the set of real numbers is infinite. If we suppose that the set of real numbers provides a measurement scale for any kind of quantity, then we can relate the notion of quantitative infinity to any attribute F of which it makes sense to speak of its quantity, because anything that can be of a certain quantity can in principle be measured or modeled on a scale using the real numbers as a measure and thus can in principle be thought of as infinite.[2] For instance, we can ask whether the quantity of objects in the universe is actually or potentially infinite. We can ask whether time is dense or discrete. We can ask whether spatial divisibility infinitesimally approaches zero extension, or we can ask whether there are innumerably many abstract objects.[3]

Assuming that God is omnipotent, omniscient, morally perfect, eternal, simple, and personal,[4] a restriction on the quantitative analysis of the infinity of God is that it only applies to those aspects of the divine being that admit of quantity and therefore can at least in principle be measured on a scale of real numbers. Since, at least prima facie, divine simplicity, eternity, and personality seem to resist analysis in terms of quantity, it follows that the quantitative analysis of the infinity of God is restricted to omniscience, omnipotence, and moral perfection (cf. also Turner 2011, 294).

A Quantitative Explication of Divine Omniscience

There is a straightforward way to analyze the relevant divine attributes in terms of quantitative infinity. I focus on divine omniscience. The analysis of omnipotence and moral perfection *mutatis mutandis* proceeds analogously.[5]

If we treat knowledge as a property, which admits of quantity, then we have to specify the measure of knowledge. Based on the assumption that true propositions are the objects of knowledge and based on the assumption that "every proposition that (one) knows is a proposition of which (one) has explicit representation" (Oppy 2011, 243), it is plausible to suppose that the measure of knowledge is specified by the number of true propositions of which one has explicit representation.

With this measure in mind we can explicate in terms of quantitative infinity the standard analysis of God's omniscience, according to which God knows everything logically possible for him to know. Assuming that there are infinitely many true propositions, we can state the following:

> God is omniscient only if there is no natural number n that is the number of true propositions of which God has explicit representation.
>
> No matter which natural number n of true propositions we pick out, God has explicit representation of more true propositions than n specifies.

In other words:

> God is omniscient only if God has explicit representation of an infinite number of true propositions.

However, it is possible to have explicit representation of infinitely many true propositions without knowing all the true propositions there are. Someone who knows of every natural number that it is a natural number knows infinitely many true propositions without knowing every number. It follows that we have to suppose that in God's case there are not only infinitely many true propositions that he knows, but that there is no true proposition (which it is logically possible for God to know) that He does not know. Consequently, that God is omniscient in terms of quantitative infinity means that

> There is no natural number *n* that denotes the number of true propositions of which God has explicit representation and there is no true proposition (which it is logically possible for God to know) of which God has no explicit representation.[6]

Although this analysis of divine omniscience provides the basic schema of omniscience in terms of infinitely many propositions known to God, there is logical space for further refinement, which I can only briefly mention. For instance, depending on one's view of transfinite sets and power sets of transfinite sets, it could be extended as follows. First, if we assume that there is a one-one mapping between the real numbers and true propositions of the form "*x* is a real number," then we have to assume that God has explicit representation of innumerably many true propositions. Second, if we assume that there is also a one-one mapping between the subsets of a power set of true propositions and true propositions, then we obtain an infinite and never-ending hierarchy of true propositions of which God has explicit representation.

Prima facie there are at least two problems with this kind of analysis of divine omniscience. First, since propositions are abstract entities, this approach presupposes that there is an actually transfinite set of abstract entities. If there is no actually infinite set of abstract objects, then we cannot use the notion of quantitative infinity to analyze divine omniscience. One could at most argue that there are *n* true propositions and that God has explicit representation of *n* true propositions.

Second, to obtain a satisfying account of divine omniscience in terms of infinitely many true propositions known to God, one has to spell out in detail what exactly it is logically possible for God to know. For instance, if, on the one hand, it is not logically possible for God to know propositions involving indexicals, then we obtain the conclusion that, although God knows infinitely many true propositions, and although there is no proposition which he could know that he does not know, there are nevertheless true propositions of which God has no explicit representation (such as the proposition that "I am feeling dizzy right now"). If, on the other, it is logically possible for God to know true propositions involving indexicals, then (assuming that the set of indexical and nonindexical propositions exhausts the set of all true propositions) heavy metaphysical conclusions are unavoidable because, in this case, a panentheistic understanding of God seems to be verified.

A Quasi-Quantitative Approach to Divine Omniscience

Apart from this direct application of the idea of quantitative infinity to those divine attributes that allow of quantitative measure in terms of the extension of the attribute in question, there is another approach to ascribing quantitative infinity to them. According to this second way, we exploit the idea of a mathematical limit. That is, we assume that the divine attribute itself is the limit of a series of degrees, infinitesimally approaching it. This analysis is an attempt to understand the *via eminentiae* in terms of quantitative infinity. As Tapp says, according to this way of considering the divine attributes, we understand them to be "freed from all earthly limits . . . by saying that . . . the degree in which God possesses them exceeds all creaturely degrees infinitely" (Tapp 2011, 104–5).

For this analysis to work in the case of divine omniscience, we have to assume that the degree of God's mode of knowing is 1 and is the limit of an infinite series of possible degrees of knowing that starts with 0 and infinitesimally approaches 1. In contrast to the extensional analysis of God's knowledge, however, we cannot rely on a concept of knowledge that operates by recourse to the number of true propositions of which one has explicit representation, because, according to that measure, either a subject *S* knows a proposition *p* or does not know *p*. Instead, we have to find a measure of degrees of knowing, which allows of infinitely many distinct possible degrees of knowing, and also of a limit of knowing that is realized by the divine mode of knowing.

Since the mode of knowing of God is the limit to an infinite series of possible degrees of knowing, we can, as a first step, define God's mode of knowing as the degree of knowing *simpliciter*. Since knowing *simpliciter* entails that no doubt is possible as to one's epistemic situation, we can introduce the measure of degrees of knowing by way of the doubt one can express as to one's overall epistemic situation: to possess knowledge to degree 1 is to possess knowledge in such a way that any doubt as to one's alleged knowledge is impossible, whereas the closer to 0 the degree of knowing, the more doubt is possible as to one's epistemic situation. Given the fallibility of human knowledge and its openness to skeptical challenges, such a measure of degrees of knowing seems prima facie justified. As Hartshorne says, "Human so-called 'knowledge' of any concrete actuality is mostly ignorance and for the rest mostly guesswork and probability. In the highest sense of 'knowledge,' namely, direct, infallible, concrete, clearly conscious apprehension, we human subjects can scarcely be said to have knowledge" (Hartshorne and Reese 2000, 19).[7]

However, although this way of measuring degrees of knowing seems prima facie attractive, it invites more problems than it solves. First, how are we supposed to specify the measure of doubt one can express as to one's epistemic situation? It seems that apart from the general idea of introducing this measure as a tool to specify degrees of knowing on the *via eminentiae* it cannot be made more precise as a direct quantitative measure on a scale of real numbers but only provides a rough quantitative analogue.[8] Therefore, the suggestion of treating divine omniscience as a limit to a sequence of degrees of knowing, infinitesimally approaching God's mode of knowing, is only a quasi-quantitative analysis that fails to provide a clear measure of degrees of knowing. Although our mode of knowing is somehow "less" than God's mode of knowing (our way of knowing is not the way of knowing *simpliciter*), we cannot specify the difference between them in quantitative terms using the real numbers.

The Failure of Quantitative Analyses of the Infinity of God

According to the quantitative analysis, the extension of the divine attribute in question is infinite. According to the quasi-quantitative way, the divine attribute is considered to be the limit of a series of degrees infinitesimally approaching it. Problems with the first account primarily concern the consistency of an infinite set of corresponding elements upon which the divine attribute is directed. The problem with the second account is to find a reasonable quantitative measure on a scale of real numbers.

However, even if these problems can be solved, there is a serious limit to the use of the quantitative analysis of the infinity of God: quantitative infinity cannot grasp the infinity of God itself because it only applies to properties that admit of quantification, extension, measurement, and degree. Therefore, the idea of quantitative infinity can only help to provide an analysis of God *insofar as* he is infinitely (some property) *F* and not insofar as he *Himself* is infinite. Consequently, if the assertion that God is infinite has a meaning apart from, or in addition to, meaning that some of the divine attributes are quantitatively infinite, then quantitative infinity cannot be the only analysis of God's infinity.[9]

FINITUDE AND INFINITY AS MODES OF ONTOLOGICAL DETERMINACY

The qualitative notion of infinity is not concerned with measurement and number. This notion of infinity is entailed by a denial of qualitative finitude. In order to understand qualitative infinity, we therefore have to be clear about the concept of qualitative finitude. I assume that qualitative finitude is a feature of entities of which it can be said that they are qualitatively finite.

A Minimal Account of Qualitative Finitude

A first specification of qualitative finitude is given by

> An entity is qualitatively finite only if there is a property, which the entity in question does not exemplify.

I take this to be analytic and a priori: we could not say of anything that it is finite if there was nothing which we could exclude as belonging to the entity. For instance, an entity is spatially finite only if there is some spatial region excluded from the space occupied by the entity.

Qualitative finitude, however, not only consists in the fact that there is some property which the entity in question does not exemplify. There also has to be some property that the entity in question exemplifies. If there was no such property, then the entity could not even be said to exist.[10] For instance, in order for an entity to be spatially finite, there not only has to be a region of space that the entity in question does not occupy but also a region of space that the entity in question occupies. Therefore, we can conclude:

> An entity *x* is qualitatively finite only if there is some property that *x* does exemplify and there is some property that *x* does not exemplify.

Qualitative finitude not only consists in the fact that an entity exemplifies at least some property, and does not exemplify some other property. It logically demands that, in respect to any exemplifiable property *F*, the entity in question either exemplifies *F* or does not exemplify *F*, so a qualitatively finite entity is an entity of which the principle of noncontradiction holds true. As Hart says: "The realm of the finite is that realm in which the *principium contradictionis* holds true. . . . The *principium contradictionis* describes that absolute limit by which any finite thing is the thing it is. In a very real sense, finite existence *is* noncontradiction" (Hart 2011, 256). Qualitative finitude, therefore, can be minimally defined:

> *x* is a qualitatively finite entity if and only if there is some property *G* that *x* exemplifies and some property *H* that *x* does not exemplify and such that for every property *F*, *x* either exemplifies *F* or its complement.

Negative Qualitative Infinity and Positive Qualitative Infinity

I suggest understanding qualitative finitude as a *mode of being*, where "mode of being" refers to a particular degree of ontological determinacy. The reason for understanding qualitative finitude as a mode of being of an entity is that qualitative finitude itself is not a property an entity exemplifies *in addition* to other properties it exemplifies. It is, instead, *a way to be*, or a way of being, according to which the entity in question is subject to the law of noncontradiction and is, in this sense, ontologically completely determinate.[11] The mode of being of a qualitatively finite entity is such that it excludes ontological vagueness and implies subordination to the law of noncontradiction.[12]

Keeping this understanding of qualitative finitude in mind, we can turn to qualitative infinity. Although prima facie we are used to thinking that qualitative infinity is the straightforward denial of qualitative finitude, as quantitative infinity is the straightforward denial of quantitative finitude, it turns out that there is no contradictory opposite of finitude because there are two possible ways of denying qualitative finitude. Since the essential feature of a qualitatively finite entity is its subordination to the law of noncontradiction, we can either assume that a nonfinite entity is one which exemplifies *neither* any property *nor* its complement, or we can assume that a nonfinite entity is one which exemplifies *both* every property *and* its complement. That is to say, whereas in reference to finite entities the answer to the question of whether it is *F* is either "Yes, it is *F*" or "No, it is not *F*," there are two distinct answers in respect to nonfinite entities. In the first case, the answer to the question of whether it is *F* is always "No, it is not *F* although it does not exemplify the complement of *F*, either" and in the second case the answer is always "Yes, it is *F*, although it is also not *F*." I refer to the first denial of finitude as *negative-qualitative infinity*, or *negative infinity*, for short, and to the second denial of finitude as *positive-qualitative infinity*, or *positive infinity*, for short.[13]

In the same way in which qualitative finitude is a mode of being on a scale of ontological determinacy, so are negative infinity and positive infinity modes of ontological determinacy. A negatively infinite entity, according to its definition, is not subject to the law of noncontradiction insofar as it is an ontologically absolutely *underdetermined* entity. A positive infinite entity is not subject to the law of noncontradiction insofar as it is an ontologically absolutely *overdetermined* entity. In terms of exclusion, we can specify the difference between qualitatively finite entities and qualitatively infinite entities as follows. Whereas a finite entity is such that *ontological vagueness is excluded* from it, the mode of being of a negative infinite entity is such that *ontological determinacy is excluded* from it. The mode of being of a positive infinite entity is such that *ontological exclusion is excluded* from it.[14] Consequently, a qualitatively negative infinite entity is such that no predication is true of it, whereas a qualitatively positive infinite entity is such that no predication is false of it.[15]

The Epistemological Opaqueness of Infinite Entities

There is an interesting epistemological consequence of the elaborated notions of qualitative finitude and qualitative infinity. We can, at least in principle, discursively grasp finite entities. If we understand what it is to be *F*, then we can understand what it means for an entity to be either *F* or not to be *F*. But qualitatively infinite entities exhaust our understanding, insofar as they transcend the categorical realm of what we can discursively understand.

On the one hand, a negative infinite entity cannot be grasped by the mind because, for any property *F*, the entity neither exemplifies *F* nor its complement. Although we can understand the definition of a negative infinite entity as one possible denial of qualitative finitude, we cannot imagine or conceive of such an entity in the same way in which we conceive or imagine finite entities, in the understanding. Apart from its formal definition, a negative infinite entity is no thing at all.

On the other hand, a positive infinite entity escapes our discursive understanding because, for any property *F*, it unites both the property in question and its negation in its essence. Although we can understand the definition of a positive infinite entity, as one possible denial of qualitative finitude, we cannot imagine or conceive of such an entity in the same way in which we conceive or imagine finite entities, in the understanding. A positively infinite entity is too determined to be a clear and distinct object to the mind. According to its definition, it is everything at once and therefore nothing in particular.

Problems with Qualitatively Infinite Entities

First, one might argue that not only every qualitatively *finite* entity but instead "every *thing*, as regards its possibility, is likewise subject to the principle of *complete* determination, according to which if *all the possible* predicates of *things* be taken together with their contradictory opposites, then one of each pair of contradictory opposites must belong to it" (Kant 1999, A 571–72, B 599–600). However, if one assumes that the notion of a *thing* or *entity* and the notion of a *qualitatively finite entity* are strictly synonymous, then it follows immediately that nonfinite entities are not things and cannot be addressed as entities at all. On this assumption, one's ontology becomes a mere thing-ontology, according to which everything of necessity is subject to the law of noncontradiction. Although this ontological presupposition is widespread in what is called "analytic philosophy," it seems to me to be an unwarranted restriction on philosophical speculation, for two reasons. First, if our concept of an entity is a concept of a thing that is subject to the law of noncontradiction, then concepts like "Being," *actus purus*, and "nothingness" are only prima facie concepts of entities but, secunda facie, turn out not to be concepts of entities at all. For conceptual reasons alone, there could be no such thing as Being or *actus purus*, or nothingness. Though this is exactly what many analytic philosophers take to be the case, the present framework allows an understanding of these expressions as concepts of *qualitatively nonfinite* entities. One might object here that I commit an onto-theological error if I assume that, for instance, Being is some kind of entity. To be clear: Being is some kind of entity for otherwise we could not even talk about Being. However, Being is not a qualitatively finite entity. Confusion arises only if we confuse the notion of an entity with the notion of a finite entity. Second, if there were only things in the sense specified, that is, if there were only finite entities, then God, if he exists, would also have to be a qualitatively finite entity, which is to say that God in principle would be available to the mind as an object of discursive understanding. Although this, again, is what many analytic philosophers seem to suppose, the danger of anthropomorphism is very close because it is only a small step from such an understanding of the divine to the conclusion that God really is little more than a super human being.[16]

A second worry about nonfinite entities concerns the validity of the law of noncontradiction. Prima facie it might seem that by introducing the concept of nonfinite entities I am giving up the validity of the law of noncontradiction. This, however, is not the case. I only *restrict the scope* of the law of noncontradiction: the law of noncontradiction is valid for every qualitatively finite entity, but not valid for nonfinite entities. Since the universe and every thing in the universe is a qualitatively finite entity, it follows that the law of noncontradiction is valid for the universe and everything in it. In other words, *per definitionem* there cannot be any finite entity that violates the law of noncontradiction. This, though, does not contradict the idea of there being entities that are not subject to the law of noncontradiction.[17]

A third and related worry concerns the ontological status of property exemplifications. One might argue that I presuppose that each and every property *F* has a complement and that this complement is *exemplified* by a finite entity if and only if it is not *F*. This, one might object, is problematic because it entails that being *not F* is a *positive* property, which seems inadequate, at least for many properties *F*. If, however, there is no positive property exemplification of entities that are not *F*, then the notion of a qualitatively positive infinite entity becomes problematic also, because, in this case, it is hard to see what the union of being *F* and of being not *F* could be like if there is no property "*not-F-ness*" that could be united with *F-ness*. Now, in order to deal with this objection adequately we would have to elaborate the ontology of properties in depth. Therefore, let me just briefly outline my answer. Anything that is true of a qualitatively finite entity is a property of that entity. Therefore, if it is true that some entity is not *F*, then there is a corresponding property in virtue of which the entity in question in not *F*—*not-F-ness*. For instance, if *F* is the property of being a rational animal, then the property of being not *F* is the property of not being a rational animal, that is, the property of being a nonrational animal.[18]

ARE THERE QUALITATIVELY INFINITE ENTITIES?

We have discussed some objections to the idea of nonfinite entities. Apart from the consistency of the idea of infinite entities, we have to ask whether there actually are qualitatively infinite entities.

Nothingness and Negative Infinite Entities

Let us briefly deal with negative infinite entities. To be clear, if there is an infinite negative entity, then there can only be one infinite negative entity since the assumption that there is more than one negative infinite entity presupposes that we could distinguish between them by a difference of predication, which contradicts the concept of an infinite negative entity. Keeping this in mind, we can show that there is no negative infinite entity because a negative infinite entity, apart from its definition, is nothing at all. For instance, it is neither spatially extended nor not spatially extended, it is neither in time nor not in time, it is neither alive nor not alive, and so forth. No matter which property *F* we pick out, a negative infinite entity is neither *F* nor not *F*. It is an ontologically utterly underdetermined entity, an ontological void and thus is nothing at all, which is to say that the concept of a qualitatively infinite negative entity expresses the metaphysical idea of nothingness. Therefore, there is no entity the mode of being of which is that of negative qualitative infinity. In order to be at all, an entity needs at least a minimal degree of ontological determinacy.[19]

The Source of Everything and Positive Infinite Entities

Is there an entity the mode of being of which is that of positive infinity? If there is a positive infinite entity, then there is at most one positive infinite entity. The reason is analogous to the reason for the uniqueness of an alleged negative infinite entity. If we assumed that there is more than one infinite positive entity, then we would obtain a contradiction because, in order to distinguish between them, we would have to suppose that there is at least one predication that is true of the one but not of the other. This contradicts the notion of a positive infinite entity.

However, we can distinguish between two ways of proving the existence of a positively infinite entity. We could either argue that such an entity has to exist, of necessity, or we could argue that we have reason to assume that such an entity exists based on reflections on the being and structure of the universe. (I assume that the universe is the realm of qualitative finitude.) According to the first way, we could build on the Anselmian argument that a being-than-which-nothing-greater-can-be-conceived has to exist of necessity and qualify the argument by way of assuming that a being is the greater the more contradictory opposites it unites in its essence. The "ability" to unite contradictory opposites in one's essence, according to this approach, would be considered a great-making property. Pursuing this line of thought, an-entity-than-which-nothing-greater-can-be-conceived turns out to be a positive qualitatively infinite entity because, for every property *F*, the entity in question unites both being *F* and not being *F*, in its essence.

I focus only on the second kind of argument, according to which reflections on the nature of the universe lead to the conclusion that there is a positively infinite entity. The key insight concerns the conceptual relation between the concept of an *infinitely positive entity* and the concept of a *single ultimate source of everything.* A single ultimate source of everything is, first, the source of the existence of finite entities and is, second, the source of the *what-it-is* of every finite entity. On the one hand, if it was not the source of the existence of finite entities, but only of the *what-it-is* of each and every finite entity, then we would have to assume that there is another source of the existence of finite entities, which contradicts the idea of there being a single ultimate source of everything. On the other hand, if it was not the source of the *what-it-is* of finite entities, but only of their existence, then we would equally have to assume that there is another source which is responsible for the *what-it-is* of each and every finite entity, which, again, contradicts the idea of a single ultimate source of everything. Therefore, a single ultimate source of everything has to be the single ultimate source not only of the existence of finite entities but also of the properties that make a finite entity the finite entity it is.[20]

Since we find in the realm of finitude entities that are *F* and entities that are not *F*, it follows that the single ultimate source of everything has to be such that, in its essence, it unites both *F-ness* and its negation, in a way beyond contradiction. For instance, if it only was *F* and not also not *F* at the same time, then it could not be the single ultimate source of entities that are *F* and of entities that are not *F* because we would either have to assume that there is another source of the *non-F-ness* of those entities that are not *F* or we would have to assume that there is no source of their being not *F* at all. Therefore, a single ultimate source unites in its essence properties that are contradictory in the realm of finitude, which is to say that a single ultimate source of everything *is* a qualitatively positive infinite entity. If both entities that are *F* and entities that are not *F* have the same source, insofar as they are *F* and not *F*, respectively, then the ultimate source of everything has to be a positively infinite entity because it has to be *F* and also not *F* in a way beyond contradiction.[21]

Therefore, *if* there is a sound argument for the conclusion that there is a single ultimate source of everything then, at the same time, this is an argument that there is a single qualitatively positive infinite entity. Since the question of whether there is such an ultimate source is way beyond the scope of this chapter, I assume that indeed there is sound argument according to which there is an ultimate source of everything.

GOD AS THE SINGLE ULTIMATE SOURCE OF EVERYTHING

We have clarified the relationship between the concept of a single ultimate source of everything and the concept of a qualitatively positive infinite entity. If we look at the tradition of philosophical and theological reflection on the nature of qualitative infinity, then we can draw two further conclusions. First, in the Christian tradition the single ultimate source of everything is referred to as "God." Second, in the ancient Greek tradition, the idea of an infinite entity is used to express the idea of a qualitatively negative infinite entity.

God as the Single Source of Everything

It is not difficult to show that the single ultimate source of everything is what theologians called "God" and that it is what they, at least implicitly, would consider a qualitatively infinite positive entity. For instance, Pseudo-Dionysius asserts this about God, as the ultimate source of everything:

> What has actually to be said about the Cause of everything is this: Since it is the Cause of all beings, we should posit and ascribe to it all the affirmations we make in regards to beings, and more appropriately, we should negate all these affirmations, since it surpasses all being. Now we should not conclude that the negations are simply the opposites of the affirmations, but rather that the cause of all is considerably prior to this, beyond privations, beyond every denial, beyond every assertion. (Pseudo-Dionysius 1987, 1000b)[22]

That the first cause is such that "we should posit and ascribe to it all the affirmations we make in regards to beings" and that "we should negate all these affirmations" because the first cause is "beyond every denial, beyond every assertion" is simply another way of expressing our definition of a qualitatively positively infinite entity as an entity that in the unity of its being is both *F* and not *F*.

Referring to Pseudo-Dionysius, Aquinas states about God, as the ultimate source of everything:

> Cum ergo Deus sit prima causa effectiva rerum, oportet omnium rerum perfectiones praeexistere in Deo secundum eminentiorem modum. Et hanc rationem tangit Dionysius cap. 5 de Div. Nom. dicens de Deo quod "non hoc quidem est, hoc autem non est; sed omnia est, ut omnium causa." (Aquinas, *ST* 1, q. 4, a. 2).[23]

Again: that God "*non hoc quidem est, hoc autem non est*" but instead "*omnia est, ut omnium causa*" fits well with our concept of a qualitatively positive infinite entity. God is not "this" or "that" but instead as the single ultimate source of everything he is "*omnia*."

Finally, Nicholas de Cusa states the following about God:

> Quia igitur maximum absolute est omnia absolute actu quae esse possunt taliter absque quacumque oppositione, ut in maximo minimum coincidat, tunc super omnem affirmationem est pariter et negationem. Et omne id quod concipitur esse non magis est quam non est. Et omne id quod concipitur non esse non magis non est quam est. Sed ita est hoc quod est omnia et ita omnia quod est nullum. (Nicholas de Cusa, *De docta ignorantia* I, c. 4).[24]

That God "*est hoc quod est omnia et ita omnia quod est nullum*" expresses the idea we found in Pseudo-Dionysius and Thomas. Again, the first source of everything satisfies our definition of a qualitative infinite positive entity.

Therefore, there is evidence that in the Christian theological tradition the ultimate source of everything is referred to as "God" and is, at least implicitly, considered a qualitatively positive infinite entity that is beyond discursive understanding and unites, in its being, opposites that are contradictory in the ontology of qualitatively finite entities.

Qualitative Infinity in the Greek Tradition

The scale of ontological determinacy which enables us to grasp that, in the Christian tradition, God is understood as a positively infinite entity also enables us to understand why classical Greek philosophy could not assume that God is an infinite entity but had to assume that God is qualitatively finite. The reason is that the Greek mind did not understand infinity as positive but as negative infinity—infinity as the exclusion of all ontological determination. As Hart (2011, 258) says, "For the entire classical philosophical tradition of Greece, including Stoicism—the infinite was solely a negative concept. Words like απειρον . . . were more or less entirely opprobrious in connotation: they were used to designate that which was 'indefinite' or 'indeterminate' and, hence 'irrational' or 'unthinkable.' The infinite . . . is pure deficiency."

Because negative infinity expresses the metaphysical idea of nothingness, it seemed clear to the Greeks that God could not be infinite in this sense but had to be considered as a finite entity.

The Qualitative Infinity of God

Based on the distinction between negative and positive qualitative infinity and based on the assumption that there is a single ultimate source of everything: It follows, first, that there is no negative qualitatively infinite entity. It follows, second, that the ultimate source of everything is a qualitatively positive infinite entity. It follows, third, that whereas the Christian tradition refers to this entity as "God," and therefore could accept the infinity of God, the Greek tradition understood infinity as qualitatively negative infinity and consequently denied "infinity" of God.

THE RELATION BETWEEN GOD AND FINITUDE

The relation between God and the realm of finitude is difficult to grasp due to the inner dialectic of the concept of distinction.[25] However, we can as a first step show that God is indistinguishable from the realm of finitude. The reasoning is based on the logic of distinction, as applied to finite entities. A finite entity is an entity that is subject to the law of noncontradiction. We distinguish entities on the basis of their distinct property exemplifications. We say that *this* entity is distinct from *that* entity if and only if they exemplify different properties, by means of which we can distinguish between them. That is to say, we can distinguish between different entities x and y if and only if there is some property F that x exemplifies but y does not exemplify.

It follows that a positively infinite entity cannot be distinguished from any single finite entity or a collection of finite entities *because there is no property* F *which a finite entity exemplifies that a positively infinite entity does not exemplify*. For all properties *F*, if a finite entity is *F*, then the positively infinite entity is also *F*, and if a finite entity is not *F*, then a positively infinite entity is also not *F*. There is nothing which is true of any finite entity that is not *eo ipso* true of a positively infinite entity. Therefore, as the ultimate source of everything, God is not to be distinguished from the realm of finitude. This is to say that God is indistinguishable from the world. Based on this premise, we can show, seemingly paradoxically, that God, at the same time, is utterly distinct from the world and the world is utterly distinct from God.

On the one hand, because distinguishability is an essential feature of finite entities, it follows that anything which is not subject to the category of distinguishability is utterly distinct from any finite entity and the whole realm of finitude. In the case of God, because God is indistinguishable from the realm of finitude, it follows from His being God that, at the same time, he is utterly distinct from the realm of finitude. Therefore, God is, at the same time, indistinguishable from the world and *the more distinct insofar as he is indistinct*.[26]

On the other hand, although God is the more distinct from the world the more he is indistinct, we can clearly distinguish any single finite entity or collection of finite entities from God. The reason is this. Although for every property *F*, God is both *F* and not *F*, a finite entity is either *F* or not *F*. As a consequence, seen from the point of view of qualitative finitude, there is always a difference in predication between what is true of a finite entity and what is true of God. Whereas God is the more distinct from the world the more he is indistinct, the world is plainly distinct from God.

SUMMARY

We started by briefly analyzing the concept of quantitative infinity and how it could be applied in the analysis of divine attributes. We then argued that the quantitative analysis cannot be applied in analyzing what it means to say that God himself is infinite. We turned to qualitative notions of finitude and infinity. We distinguished three modes of being on a scale of ontological determinacy: infinite negative quality, finitude, and infinite positive quality. A finite entity is subject to the law of noncontradiction. A negative infinite entity exemplifies neither a property *F* nor its negation, and a positive entity exemplifies both any property *F* and its negation. We then argued that the concept of an ultimate source of everything entails that a single source of everything is a positively infinite entity and that the theological tradition refers to this entity as "God." We ended by arguing that God is both indistinguishable from the realm of finitude and that He is the more distinct insofar as He is indistinct.

NOTES

I am grateful to the participants of the conference "The Infinity of God" held in Bochum in August 2013. In particular, I am grateful to Christian Tapp, Thomas Schärtl, Ruben Schneider, Christina Schneider, Brian Leftow, William Carroll, and Richard Swinburne for valuable criticism of an earlier version of this chapter. I am also grateful to the audience of a presentation of this chapter at Oxford University, Blackfriars Hall, in particular to Grahame Locke and Ralph Weir.

1. Cf. Cocchiarella 2008 (10): "A set is infinite, of course, if, and only if, it is not finite, where by finite we mean that the set has finitely many members—i.e., that for some natural number *n*, the set has *n* many members." Instead of defining the infinity of a set in terms of the denial of quantitative finitude, we could also state that a set is infinite if and only if there is a one-one mapping between its elements and the elements of one of its subsets. Cf. Dotterer 1918 (294).

2. Cf. Oppy 2006 (136): "A measurement scale is a function from the properties of objects and events to a number system. A useful measurement scale captures a homomorphism between structural features of the measured properties . . . and structural features of numbers. . . . In the standard case, the system of numbers is the real numbers." Cf. also Rescher 1955 (109): "How to measure quantity? Measurement is comparison with a unit. First it must be decided what kind of unit is suitable; next such a unit is selected (arbitrarily); then, finally, an appropriate comparison can be made."

3. Cf. Oppy 2006 (7–19) for a more detailed analysis of different ways to specify the idea of an infinite quantity and for some traditional problems concerning infinite quantities.

4. Cf. Göcke 2013b for problems concerning the justification of a particular concept of God. Cf. also Copan and Moser 2003 for a discussion of the rationality of classical theism.

5. As Oppy (2011, 242) says, "In every case, the attribution of one of these properties to God brings with it quantification over a domain of objects, and, in each case, there is then a serious question to address concerning the measure or cardinality of that domain." Cf. also Steinhart 2009 (261).

6. With regard to the relation between God's knowledge and our knowledge, the quantitative analysis of divine omniscience presupposes an unequivocal use of the term and leads to the following: If God is infinite in His knowledge, and if human beings are finite with respect to their knowledge, then the knowledge of human beings is a proper part of the knowledge of God. Every true proposition of which we have explicit representation, and which is logically consistent for God to have explicit representation of, is something that God actually knows. Nevertheless, there are infinitely many propositions, which we do not know, but God knows them. The difference between human and divine knowledge only consists in the different cardinalities of the set of true propositions known to human beings and known to God.

7. There is a direct consequence of this approach to the infinity of divine omniscience. We cannot deploy an unequivocal notion of knowing with respect to God and other minds. If God's mode of knowing is the paradigmatic example of what it means to know something, while any other degree of knowing realized in the realm of finitude falls short of God's mode of knowing, then we can speak of God's mode of knowing only by way of analogy. Here "analogous" means "not unequivocal and not heteronymous."

8. As Tapp (2011, 105) argues, "One problem is the presupposition that all these properties come in objective degrees, that these degrees are objectively comparable, that there exists not only a maximum degree, but also an infinite one etc."

9. It is not always clear whether philosophers and theologians assume that there really is a meaning to the assertion that God is infinite, in addition to the quantitative infinity of some divine attributes. Sometimes the idea of the infinity of God is simply used as an umbrella term for the essence of God. For instance, Gellman (1994, 53) presupposes that the infinity of God can be understood as follows: "Now it is a tenet of these same religions that God is omnipotent, omniscient, perfectly good, and eternal. Let us abbreviate this into saying that these religions teach that God is infinite." However, as I am going to argue, there is a deep conceptual connection between a qualitative notion of infinity and the concept of a first source of everything, which we call "God."

10. I am assuming here that existence is a second-order property and that something exists if and only if it exemplifies some property *F*.

11. We can explicitly show that neither qualitative finitude nor qualitative infinity is a first-order property of entities. For the sake of argument, let us suppose that a property *F* is a first-order property if and only if its complement is a first-order property, which is to say that finitude is a first-order property if and only if qualitative infinity is a first-order property. Let us furthermore suppose that for any *x* and any first-order property *F*, either *x* exemplifies *F* or its complement. With respect to finitude, this means that an entity is infinite if and only if it is not finite. Since *x* is infinite if and only if *x* is not finite, and since *x* is finite if and only if there is some first-order property that it does not exemplify, it follows that *x* is infinite if and only if it is not the case that there is some first-order property that *x* does not exemplify. However, since, by definition, *x* can be infinite if and only if it is not finite, it follows

that there is a first-order property which it does not exemplify, the alleged first-order property of being qualitatively finite, and therefore *x* turns out to be a finite entity after all. Therefore, since the assumption that finitude and infinity are first-order properties of entities leads to a contradiction, it follows that finitude and infinity are not first-order properties.

12. I bracket epistemological concerns according to which we might or might not know which property of any pair of properties a finite entity exemplifies.

13. I am considering only the extreme cases of nonfinite entities according to which a negative infinite entity is such that, for every property *F*, it is neither *F* nor not *F*, and according to which a positive infinite entity is such that, for every property *F*, it is both *F* and not *F*. In principle, one could also demarcate nonfinite entities by taking into account only certain properties *F*, *P*, *G* . . . of which it is true that an entity exemplifies both the property and its complement, or neither the property nor its complement.

14. It is obvious that, according to our definition of qualitative finitude, the set of natural numbers is a qualitatively finite entity since, for instance, it does not exemplify the property of being a physical object. Because infinite sets are the paradigmatic example of quantitative infinites, it follows that the notion of qualitative infinity is of a radically different nature than the notion of quantitative infinity. Cf. Hart 2011 (256): "The set of even numbers, for instance, while endless, is nevertheless definite: it includes and excludes particular members; it is a particular kind of thing, it is bounded by its own nature. In the purely metaphysical sense, in fact, the mathematical infinite remains within the realm of the finite."

15. Cf. Turner 2011 (296–97): "If we are to speak of the divine infinity, it seems necessary to distinguish between infinity understood as the simple negation of the finite such as yields its corresponding contradictory, namely, that of the infinite understood as endlessness—whether of mathematical or of temporal seriality—and the infinity that is yielded by the negation of the negation between the finite and the infinite as so understood in mathematics. Logically, the difference consists in that, in the case of 'ordinary' negation, the notion of infinity amounts to the negation of serial finiteness, an endlessly extruded series each of whose parts is finite, because mutually exclusive, each being a 'this' rather than a 'that.' In the case of the divine being, however, we have to speak of an infinity that transcends that known by force of 'ordinary' negation of the finite, because the negation required is the negation of that 'ordinary' negation itself. As such, this 'transcendent' notion of the infinite is such as to exclude all exclusion."

16. Of course, what I said is not an argument but based on arguments and intuitions I lack the space to present clearly. The thing-ontology of analytic philosophy and philosophical systems that allow more than just entities subject to the law of noncontradiction represent two entirely different philosophical paradigms.

17. A detailed formal exploration of these matters would need to rely on a paraconsistent logic: "The contemporary logical orthodoxy has it that, from contradictory premises, anything can be inferred. Let $\vDash$ be a relation of logical consequence, defined either semantically or proof-theoretically. Call $\vDash$*explosive* if it validates $\{A, \neg A\}$

$\vDash B$ for every *A* and *B* (*ex contradictione quodlibet* (ECQ)). Classical logic, and most standard 'non-classical' logics too such as intuitionist logic, are explosive. Inconsistency, according to received wisdom, cannot be coherently reasoned about. *Paraconsistent logic* challenges this orthodoxy. A logical consequence relation, $\vDash$, is said to be *paraconsistent* if it is not explosive. Thus, if $\vDash$ is paraconsistent, then even if we are in certain circumstances where the available information is inconsistent, the inference relation does not explode into *triviality*" (http://plato.stanford.edu/entries/logic-paraconsistent/). On my account, the existence of nonfinite entities is nonexplosive.

18. Of course, much more ought to be said about this. The account proposed should be thought of as a form or model and not as a completed and fully elaborated theory of what exactly a nonfinite entity is like. Questions I cannot deal with here include the question of just which properties exist. In principle, the account is consistent with different ontologies and with a compositional semantics of properties and predicates. For instance, if it is a property *to be dressed for dinner*, then a positive infinite entity also is dressed for dinner. However, at the same time, it is also not dressed for dinner. It unites both properties in its essence.

19. One might object that we talked about negative infinite entities and that therefore there is at least a minimal sense of being that a negatively infinite entity can be said have, which contradicts the above conclusion that there is no such entity. However, this is not what we did: we defined a concept of a negative infinite entity and its mode of being based on a denial of qualitative finitude and then argued that although this concept is coherent, there is no corresponding reality to the notion of a qualitatively infinite entity.

20. That there is a single ultimate source of everything does not entail that there is *only* a single ultimate source of everything and no other cause or source in addition to it. However, I bracket the issue of secondary causes.

21. As Thomas says about God, "What are diverse and exclusive in themselves pre-exist in God as one, without detriment to his simplicity" (*ST* 1, q. 4, a. 2, ad. 1). Cf. also Turner 2004 (188–89): "For note that the specific difference 'rational' divides the genus 'animal' into exclusive species ('rational' and 'nonrational'), such that, if the one then not the other: if any animal exists, then it is either a rational animal or a non-rational animal. Both belong to the same genus, but, of course, there cannot exist an animal which is, just, generically an animal, being neither rational nor non-rational. But if, *per impossibile*, a generic animal could exist, it could not be *neither* rational *nor* non-rational, for then it would have none of the character of either; it would have to be *both rational and non-rational* in some way which excluded both specifications, in order to exclude the disjunction between them, and thus contain the notions of both in some non-exclusive way: by, to use an expression of Eckhart's (though not of Thomas') 'negating the negation' between them."

22. Cf. also Pseudo-Dionysius, *Divine Names* 5: "For He is not This without being That; nor doth He possess this mode of being without that. On the contrary He *is* all things as being the Cause of them all, and as holding together and anticipating in Himself all the beginnings and all the fulfilments of all things; and He is above

them all in that He, anterior to their existence, super-essentially transcends them all. Hence all attributes may be affirmed at once of Him, and yet He is No Thing. He possesses all shape and form, and yet is formless and shapeless, containing beforehand incomprehensibly and transcendently the beginning, middle, and end of all things, and shedding upon them a pure radiance of that one and undifferenced causality whence all their fairness comes."

23. Cf. Turner on Aquinas: "In God all specification of this and that is excluded—'there is no kind of being that God is,' or, as we might put it, if 'specificness' is excluded from God, then 'exclusion' is excluded from God. . . . That is why we cannot comprehend God: the 'darkness' of God is the simple excess of light. God is not too indeterminate to be known; God is unknowable because too comprehensively determinate, too actual. It is in that excess of actuality that the divine unknowability consists" (Turner 2004, 188–89). That in God "'exclusion' is excluded" and that "God is not too indeterminate to be known" but "too comprehensively determinate, too actual" supports our definition of a positively infinite entity of which ontological exclusion is excluded and which therefore we cannot comprehend. Cf. also Burns 1998a and Burns 1998b for a further analysis of the infinity of God in Thomas Aquinas.

24. Cf. Achtner 2011 (34–35); and also Flasch 1998 (46).

25. The question concerning the relation between God and the world is also discussed in the debate between panentheists, pantheists, and classical theists. Cf. Göcke 2012 and Göcke 2013a for further analysis of panentheism and its relation to classical theism.

26. This is what Meister Eckhart already had in mind when he analyzed the dialectic of the term "unum," which in our context is the term of a positively infinite entity. Colledge sums up this Eckhartian motif: "Since indistinction is the distinguishing mark of unum, what sets it off from everything else, to conceive of God as unum, or Absolute Unity, is to conceive of him as simultaneously distinct and indistinct, indeed, the more distinct insofar as he is indistinct. While this may sound like a mere wordplay, Eckhart's intention is totally serious. If we paraphrase the language of indistinction and distinction into the language of immanence and transcendence, we see better what the Meister was after—a way of speaking about God as simultaneously totally immanent to creatures as their real existence and by that very fact absolutely transcendent to them as esse simpliciter or esse absolutum" (Eckhart 1981, 34).

REFERENCES

Achtner, Wolfgang. 2011. "Infinity as a Transformative Concept in Science and Theology." In *Infinity: New Research Frontiers*, edited by Michael Heller and W. Hugh Woodin, 19–54. Cambridge: Cambridge University Press.

Aristotle. 1984. *Physics*. In *The Complete Works of Aristotle*, 2 vols., edited by Jonathan Barnes. Princeton, NJ: Princeton University Press.

Bombieri, Enrico. 2011. "The Mathematical Infinity." In *Infinity: New Research Frontiers*, edited by Michael Heller and W. Hugh Woodin, 55–75. Cambridge: Cambridge University Press.

Burns, Robert M. 1998a. "Divine Infinity in Thomas Aquinas: I. Philosophico-Theological Background." *The Heythrop Journal* 39 (1): 57–69.

———. 1998b. "Divine Infinity in Thomas Aquinas: II. A Critical Analysis." *The Heythrop Journal* 39 (2): 123–39.

Cocchiarella, Nino B. 2008. "Infinity in Ontology and Mind." *Axiomathes* 18 (1): 1–24.

Copan, Paul, and Paul K. Moser, eds. 2003. *The Rationality of Theism*. Oxford: Routledge.

Dombrowski, Daniel A. 2007. "Oppy, Infinity, and the Neoclassical Concept of God." *International Journal for Philosophy of Religion* 61 (1): 25–37.

Dotterer, Ray H. 1918. "The Definition of Infinity." *Journal of Philosophy, Psychology and Scientific Methods* 15 (11): 294–301.

Eckhart, Meister. 1981. *The Essential Sermons, Commentaries, Treatises, and Defense.* Translation and introduction by Edmund Colledge and Bernard McGinn. New York: Paulist Press.

Flasch, Kurt. 1998. *Nikolaus von Kues: Geschichte einer Entwicklung.* Frankfurt: Klostermann.

Gellman, Jerome I. 1994. "Experiencing God's Infinity." *American Philosophical Quarterly* 31 (1): 53–61.

Göcke, Benedikt Paul. 2012. "Classical Theism and Panentheism." *Sophia: International Journal for Philosophy of Religion, Metaphysical Theology and Ethics* 52 (1): 61–75.

———. 2013a. "On the Importance of Karl Christian Friedrich Krause's Panentheism." *Zygon* 48 (2): 364–79.

———. 2013b. "An Analytic Theologian's Stance on the Existence of God." *European Journal for Philosophy of Religion* 5 (2): 129–46.

Hart, David Bentley. 2011. "Notes on the Concept of the Infinite in the History of Western Metaphysics." In *Infinity: New Research Frontiers*, edited by Michael Heller and W. Hugh Woodin, 255–74. Cambridge: Cambridge University Press.

Hartshorne, Charles, and William L. Reese. 2000. *Philosophers Speak of God.* New York: Humanity.

Kant, Immanuel. 1999. *Critique of Pure Reason.* Edited by Paul Guyer and Allen Wood. Cambridge: Cambridge University Press.

Oppy, Graham. 2006. *Philosophical Perspectives on Infinity.* Cambridge: Cambridge University Press.

———. 2011. "God and Infinity: Directions for Future Research." In *Infinity: New Research Frontiers*, edited by Michael Heller and W. Hugh Woodin, 233–54. Cambridge: Cambridge University Press.

Owens, Joseph. 1951. *The Doctrine of Being in the Aristotelian "Metaphysics."* Toronto: Pontifical Institute of Mediaeval Studies.

Pseudo-Dionysius. 1987. *The Complete Works*. New York: Paulist Press.

Rescher, Nicholas. 1955. "Leibniz' Conception of Quantity, Number, and Infinity." *The Philosophical Review* 64 (1): 108–14.

Russell, Robert John. 2011. "God and Infinity: Theological Insights from Cantor's Mathematics." In *Infinity: New Research Frontiers*, edited by Michael Heller and W. Hugh Woodin, 275–89. Cambridge: Cambridge University Press.

Steinhart, Eric. 2009. "A Mathematical Model of Divine Infinity." *Theology and Science* 7 (3): 261–74.

Tapp, Christian. 2011. "Eternity and Infinity." In *God, Eternity, and Time*, edited by Christian Tapp and Edmund Runggaldier, 99–116. Farnham: Ashgate.

Thomas Aquinas. 1981. *Summa Theologica*. 5 vols. Edited by the Fathers of the English Dominican Province. New York: Benzinger.

Turner, Denys A. 2004. *Faith, Reason and the Existence of God*. Cambridge: Cambridge University Press.

———. 2011. "A (Partially) Skeptical Response to Hart and Russell." In *Infinity: New Research Frontiers*, edited by Michael Heller and W. Hugh Woodin, 290–98. Cambridge: Cambridge University Press.

CONTRIBUTORS

William E. Carroll is research fellow in theology and science at Blackfriars Hall, University of Oxford, and member of the Aquinas Institute. His research concerns the reception of Aristotelian thought in mediaeval Islam, Judaism, and Christianity and the development of the doctrine of creation. He also examines ways in which developments in contemporary cosmology and evolutionary biology are used in discussions about creation and divine agency. He is the author of *Galileo: Science and Faith*; *Creation and Science* (Catholic Truth Society, 2009); and coauthor with Steven Baldner of *Aquinas on Creation* (Pontifical Institute of Mediaeval Studies, 1987).

Philip Clayton is Ingraham Professor at Claremont School of Theology and an affiliated faculty member at Claremont Graduate University. He received a joint doctorate in philosophy and religious studies from Yale University and has been visiting professor at Harvard, Cambridge, and Munich. Clayton has published twenty-four books, largely in theology and philosophy, comparative theologies, and the religion-and-science field. These include *Mind and Emergence*; *The Reemergence of Emergence*; *In Quest of Freedom: The Emergence of Spirit in the Natural World*; *Religion and Science: The Basics* (Vandenhoek & Ruprecht, 2009), and *The Oxford Handbook of Religion and Science* (Oxford University Press, 2008).

Adam Drozdek is an associate professor at Duquesne University. He holds a PhD in philosophy from Jagiellonian University. He has authored several books and numerous articles on the subject of philosophy and theology.

Benedikt Paul Göcke is a research fellow at the Ian Ramsey Centre for Science and Religion and a member of the Faculty of Theology at University of Oxford. He is also a member of Blackfriars Hall, Oxford. Göcke is author of *A Theory of the Absolute* (Macmillan, 2014); *Alles in Gott?* (Friedrich Pustet, 2012); and editor of *After Physicalism* (Notre Dame University Press, 2012). He has published articles in *International Journal for Philosophy of Religion*, *Zygon*, *Sophia*, *European Journal for Philosophy of Religion*, and *Theologie und Philosophie*, among others.

William Hasker (PhD, Edinburgh) is professor emeritus of philosophy at Huntington University, where he taught from 1966 until 2000. He was the editor of *Christian Scholar's Review* from 1985 to 1994, and the editor of *Faith and Philosophy* from 2000 until 2007. He has contributed numerous articles to journals and reference works, and is the author of *Metaphysics* (InterVarsity, 1983); *God, Time, and Knowledge* (Cornell University Press, 1989); *The Emergent Self* (Cornell University Press, 1999); *Providence, Evil, and the Openness of God* (Routledge, 2004); *The Triumph of God over Evil* (InterVarsity, 2008); and *Metaphysics and the Tri-Personal God* (Oxford University Press, 2013).

Bruce Hedman has been on the faculty in the Department of Mathematics at the University of Connecticut since 1982, and pastor of the Abington Congregational Church since 1989. In 1998 he was invited by Thomas F. Torrance to spend a sabbatical term at the University of Edinburgh, during which he edited some unpublished papers of Colin Maclaurin's. He earned his PhD under Harold Kuhn at Princeton University in 1977, and graduated from Princeton Seminary in 1980. Recently, he has become interested in Jungian psychology, and is a member of the International Association for Jungian Studies. This year he received an SFF grant to study Jungian interpretations of Haida myths and totems.

Paul Helm was professor of the history and philosophy of religion, King's College, London, 1993–2000. He is the author of *Eternal God* (Oxford University Press, 1988, 2010), and has written a number of articles on God and time. He has a more general interest in the relation between Christian theology and philosophy.

Franz Krainer has studied philosophy and theology. He is director of the Youth Educational Centre of the Evangelische Gesellschaft für Deutschland.

Bernhard Lang (Dr. theol. habil., emeritus professor) taught Bible and religious studies at the universities of Tübingen, Mainz, and Paderborn in Germany, the Université de Paris IV (Sorbonne), and the University of St. Andrews, Scotland, specializing in the Old Testament, the history of Christian thought, and religion in world literature. He has published more than thirty books, of which *Heaven: A History* (with Colleen McDannell) (Yale University Press, 1998, 2001); *The Hebrew God: Portrait of an Ancient Deity* (Yale University Press, 2002); and *Joseph in Egypt: A Cultural Icon from Grotius to Goethe* (Yale University Press, 2010) are perhaps the best known. He was awarded an honorary degree of Doctor of Theology by the University of Aarhus, Denmark, and is an honorary member of the (British) Society for Old Testament Study. Lang lives in Berlin, Germany.

Brian Leftow has since 2002 been Nolloth Professor of the Philosophy of the Christian Religion at Oxford University. He had previously taught at Fordham University (New York). He is the author of *Anselm's Proofs* (Oxford University Press, forthcoming); *God and Necessity* (Oxford University Press, 2012); *Time and Eternity* (Cornell University Press, 1991); and more than 100 articles in philosophy of religion, medieval philosophy, and metaphysics.

Kenneth Pearce received his PhD from the University of Southern California in 2014 and is currently Ussher Assistant Professor in Berkeley Studies (Early Modern Philosophy) at Trinity College Dublin. His research focuses on issues at the intersection of metaphysics, theology, and philosophy of science in early modern philosophy, and on metaphysical issues in analytic philosophy of religion. He is the author of *Language and the Structure of Berkeley's World* (Oxford University Press, 2017), and has written numerous articles that have appeared in journals such as *Pacific Philosophical Quarterly*, *Philosophers' Imprint*, and *Oxford Studies in Philosophy of Religion*.

Ken Perszyk is associate professor in the Office of the Vice-Chancellor at the University of Waikato in Hamilton, New Zealand, and academic director of their Tauranga campus. He has written a number of papers on Molinism and is the editor of *Molinism: The Contemporary Debate* (Oxford University Press, 2011). He has coauthored (with John Bishop) a series of papers on the euteleological conception of God, with a book in preparation.

Thomas Schärtl studied theology and philosophy in Regensburg and Munich, earned his doctorate in theology at the University of Tübingen in 2001, and his habilitation in philosophy at the Munich School of Philosophy in 2007. From 2006 to 2009 he served as assistant professor of systematic theology at the Catholic University of America in Washington, DC; from 2009 to 2015 he was professor of philosophy at the University of Augsburg (Germany). Since 2015 he has held the chair of philosophical foundations of theology at the University of Regensburg. Areas of research and interest include philosophical theology, philosophy of religion, theology of God, metaphysics, and epistemology.

Christina Schneider teaches mathematics and statistics at the University of Munich and philosophy at the Munich School of Philosophy. Her main areas of interest include analytic metaphysics and philosophy of religion, mathematical logic and mathematical modeling in philosophy, philosophy of science, and history of philosophy (esp. Leibniz). Her publications include *Leibniz' Metaphysik: Ein formaler Zugang* (Philosophia Verlag, 2001); *Untersuchungen zur Ontologie* (ed. with G. Imaguire) (Philosophia Verlag, 2006); *Metaphysische Freiheit: Kohärenz und Theorie* (Philosophia Verlag, 2009).

Ruben Schneider is research assistant in philosophy at the Catholic Faculty of Theology, Ruhr-University Bochum (previously Free University of Berlin). He studied philosophy at the Jesuit School of Philosophy in Munich and received his doctorate under the supervision of Prof. Dr. Godehard Brüntrup, SJ. His areas of interest cover analytic metaphysics, philosophy of language, history of philosophy (classic Scholasticism, Molinism, German idealism), and philosophy of religion. Recent publications include

"Complete Concept Molinism" (with G. Brüntrup), *European Journal for Philosophy of Religion* 5 no. 1 (2013): 93–108, and "Why Consequence-Style Arguments in Philosophy of Religion Are Useless—A Modal Analysis," in M. Szatkowski, ed., *Analytically Oriented Thomism* (Editiones Scholasticae 2015), 321–36.

Richard Swinburne is a fellow of the British Academy. He was Nolloth Professor of the Philosophy of the Christian Religion at the University of Oxford from 1985 until 2002. He continues to lecture in many different countries. He has written many books on the meaning and justification of theism and of particular Christian doctrines, and on other philosophical topics, such as the mind/body problem, probability, and epistemic justification. He is best known for his trilogy on the philosophy of theism: *The Coherence of Theism*, 2nd ed. (Oxford University Press, 2016); *The Existence of God*, 2nd ed. (Oxford University Press, 2004); and *Faith and Reason*, 2nd ed. (Oxford University Press, 2005).

Christian Tapp currently holds an endowed chair for Christian philosophy at Innsbruck University (2015–2019). Having completed his studies in mathematics, philosophy, and theology, he was awarded two doctorates from LMU Munich, one in science (2004, Dr. rer. nat.) and one in philosophy (2007, Dr. phil.). From 2008 to 2013 he was research group leader and junior professor (assistant professor) at Bochum University (Germany). Since 2013 he has been full professor for interdisciplinary questions in philosophy and theology. His main interests are logic and philosophy of science, metaphysics and philosophy of religion. See http://www.christian-tapp.de.

Robert M. Wallace holds a BA (first class) from Oxford University and a PhD in philosophy from Cornell University. He translated and introduced several major volumes by Hans Blumenberg, and is the author of *Hegel's Philosophy of Reality, Freedom, and God* (Cambridge University Press, 2005), and he has written numerous articles on Blumenberg and on Hegel.

INDEX

absolute infinity
  Aquinas on, 54, 66–69, 72n.15, 73n.17
  Cantor on, 11, 165n.12, 167, 170–71, 176, 181, 181n.1
  Way of Absolute Infinity, 15, 378–79, 381, 382
Acts 17:28, 126, 127–28
actual infinity, 2, 334n.15
  Aquinas on, 55, 261
  Aristotle on, 21, 22, 24–25, 55, 260–61, 334n.11
  Cantor on, 167–68, 170–71, 172, 181, 269, 334n.11, 380–81
  Cusanus on, 261
  and Hilbert's Hotel, 269–70, 274n.21
  and *kalam* cosmological argument, 261, 270
  vs. potential infinity, 80, 153, 260–61, 320
  vs. relative infinity, 66–67, 167–68
Advaita Vedanta, 126
Allison, Henry E., 103
Alston, William, 274n.15, 330, 335n.29, 336n.33, 337nn.41–42
Amaury of Bène, on God's immensity, 59
Amos
  5:8, 196
  9:2–3, 197
  9:6, 196
  9:7, 197
analytic philosophy, 80, 395, 405n.16
Anaxagoras, on Mind, 46
Anaximander, on *to apeiron*, 23–24, 35n.2, 334n.9
Anselm, St., 235, 269, 278, 334n.17
  De Concordia, 286
  on God's atemporality, 286, 287, 288, 289, 290–91, 292, 294
  on God's justice and mercy, 301, 303–4, 307–8
  on God's perfection, 261–62, 267, 268, 320–21, 378, 382, 382n.5
  ontological argument in *Proslogion*, 117n.10, 397
aoriston, 23, 27, 28–29, 31, 35n.2, 35n.4
apeiron, 5, 17, 22, 30, 34–35, 259, 265, 333n.1, 401
  Anaximander on *to apeiron*, 23–24, 35n.2, 334n.9
  vs. *aoriston*, 28–29, 31, 35n.2, 35n.4
apophatic theology. *See* negative theology
Apuleius, Isis in *Metamorphoses*, 204
Aquinas, Thomas
  on absolute infinity, 54, 66–69, 72n.15, 73n.17
  on actual infinity, 55, 261
  on analogical predication, 324, 330, 337n.43
  on analogical sense of "infinite," 54, 61, 65–66, 73n.17
  on angels, 66–67

Aquinas, Thomas (*cont.*)
on beatific vision, 58, 72n.7
on Being Itself, 332
on composition, 341
on contradictory tasks, 255n.11
on divine names, 70–71, 108
on extensive vs. intensive infinity, 61, 65
on first principles, 64, 72n.14
on form and matter, 60, 61, 62, 64–65, 66–67, 72n.16, 73n.19
on God as creator, 9, 54–71, 108, 117n.14, 171, 179
on God as subsistent being (*ipsum esse subsistens*), 9, 54, 60, 62, 63, 65–66, 67–69, 73n.18, 114, 382
on God as Trinity, 356
on God as ultimate source of everything, 399–400
on God as unmoved mover/uncaused cause, 55
on God's determination, 63, 67–68, 74n.21
on God's essence, 60, 61, 66, 67–68, 70, 73n.17
on God's eternity, 55, 56
on God's goodness, 61, 63
on God's incomprehensibility, 407n.23
on God's infinity, 9, 13, 51, 54–71, 71n.1, 72n.10, 73n.17, 167, 261, 262, 273n.7, 273n.11, 276, 277, 296–98, 378, 407n.23
on God's omnipotence, 55, 59, 60, 61–62, 71n.1, 165n.13, 212, 234, 255n.11, 296–97
on God's perfection, 61, 62–63, 65, 66, 67, 69, 70, 72n.13, 73nn.17–18, 378, 382
on God's simplicity, 62, 63, 261, 262–63, 266, 273n.7, 273n.9, 337n.43, 350, 406n.21
on God's unreceived being, 9, 54, 65, 67, 69
on God's virtues, 300–301
on human reason, 9, 56
on infinite, perfect whiteness, 61, 65, 72n.15, 73n.19, 297–98
on infinity of succession, 55
on mode of signification, 329
on negative vs. privative sense of infinity, 62
on quantitative infinity, 60–61, 72n.12
on relative infinity, 64–65, 66–67, 167–68
on way of negation (*via negationis*) regarding God, 62, 69–71, 73n.17
Aquinas, Thomas, works of
Commentary on the Book of Causes, 68
Commentary on the "Divine Names" of Pseudo-Dionysius, 68
De ente et essentia, 60, 74n.20
De potentia Dei, 69, 70, 296
De veritate, 61, 67, 72n.12, 114
Scriptum super IV libros Sententiarum, 9, 54–55, 56, 59, 60, 61, 62, 65, 69, 71n.1, 74n.21, 117n.14
Summa contra Gentiles, 9, 60, 61–62, 73n.18, 74n.20, 296, 297–98
Summa theologiae, 9, 55, 60, 61, 62–67, 70, 72n.5, 72n.8, 72n.10, 72n.13, 73n.19, 74n.22, 108, 114, 164n.3, 165n.13, 167, 171, 179, 234, 254n.2, 255n.11, 261, 273n.11, 277, 300, 341, 356, 378, 400, 406n.21
Argument from Evil and God's infinity, 333n.1
Aristotle
on actual infinity, 21, 22, 24–25, 55, 260–61, 334n.11
on conceivability of infinity, 34, 42

definition of infinity, 24–25
De generatione animalium, 32
on eternity of the world, 261
on God as lacking infinity, 21, 24
on incommensurability between the infinite and the finite, 58
on infinite division, 260
on infinite space, 260
on infinity, 13, 21, 22, 24–25, 26, 30, 32, 34, 55, 58, 59, 64, 164n.3, 260–61, 334n.11
on infinity as lack of form and determination, 21, 24, 30, 32, 34
on infinity as lack of perfection, 21, 24, 26, 30, 34
on knowledge, 70
perfection in metaphysics of, 378
Physics, 55, 57, 58, 64
and Platonic tradition, 130, 134, 141n.1, 144n.14
on potential infinity, 55, 260–61
on quantitative infinity, 22, 24, 64, 164n.3, 260
on temporal infinity, 260
translations of works, 57, 58
on virtues as stable, 308
arts, the, as aspect of ultimate reality, 124, 125, 137–38, 141, 146n.28
Athanasius, St., 139
Augsburg Confession, 317
Augustine, St.
on Epicureans, 42
on God as creator, 43, 44, 51, 202, 288
on God as incorporeal, 9, 38, 47, 49, 52n.7, 53n.11
on God as Trinity, 47–48, 52n.8, 359, 360
on God as truth, 38, 44, 50
on God's essence, 9, 45–51
on God's eternity/atemporality, 9, 37, 44–45, 50, 286, 287, 288, 291
on God's immensity, 52n.10
on God's immutability, 9, 44–45, 47, 50
on God's infinity, 9, 37–38, 45–51, 52nn.9–10, 53n.11, 12, 276, 277–78
on God's knowledge of infinity, 9, 42–44, 45, 51, 170
on God's omnipotence, 46–47, 50, 208
on God's power, 46–47, 50
on imagination and infinity, 49, 52n.5
on infinity of numbers, 40–43, 51
influence of Cicero's *Hortensius* on, 37
influence of Platonists on, 38–40, 44, 46, 50–51, 51n.1
on intelligence and infinity, 41–44
on knowledge of God, 40–41
on Last Judgment, 202
on Manichean God, 37–38, 47, 49, 50–51, 52n.10
on Manichean infinite space, 9, 42, 47, 49, 50–51, 277–78
on natural vs. sensory numbers, 40
on power of numbers as power of infinity, 41
on salvation history, 202–3, 209
vs. Scotus, 359
on self-knowledge, 40–41
on time, 9, 78, 293
on unhappiness and problem of infinity, 40–42, 50
Augustine, St., works of
Confessiones, 37, 38, 44, 50, 52n.8, 52n.10, 53n.11, 72n.8, 142n.7, 208
Contra adversarium legis et prophetarum, 46
Contra epistolam Manichaei, 41, 42
Contra Faustum 25.1, 48–50, 52n.10
Contra Iulianum opus imperfectum, 46
Contra Maximum haereticum, 52n.10
De civitate Dei, 42, 43, 44, 46, 170, 202
De diversis quaestionibus, 41

Augustine, St., works of (*cont.*)
- *De Genesi ad litteram*, 43, 44
- *De Genesi contra Manichaeos*, 44
- *De libero arbitrio*, 41, 42
- *De musica*, 41
- *De natura boni*, 50, 51
- *De ordine*, 41
- *De Trinitate*, 44, 47–48, 50, 52n.7, 52n.10
- *De vera religione*, 41, 42
- *Enarrationes in Psalmos*, 43, 45, 46
- *Epistola* 3.1, 3.4, 40
- *Epistola* 7.1.2, 44
- *Epistola* 118.4.24, 46–47
- *Epistola* 119.2, 46
- *Epistola* 135.1, 46
- *In Ioannis Evangelium*, 47, 52n.7, 52n.10
- letter to Dioscorus, 46–47
- letter to Nebridius, 40
- *Sermo* 4.4.5, 52n.10
- *Sermo* 4.5, 46
- *Sermo* 5.3, 46
- *Sermo* 32.6.6, 46
- *Sermo* 215.2, 208
- *Sermo* 270.3, 52n.3
- *Sermo* 361.5.5, 46
- *Soliloquies*, 40–41

Austin, John L., 248, 252

Avicenna, 60

Babylonia
- Akitu festival, 210n.2
- destruction of Jerusalem, 198–99

backward causation, 12, 215–16, 217, 218–19, 222

Barth, Karl
- *Church Dogmatics*, 209
- on God as almighty, 209

Basinger, David
- *The Openness of God*, 266

beatific vision, 57, 58

Belgic Confession, 318

Berkeley, George, 179
- idealism of, 98, 103, 145n.25

Berkhof, Hendrikus, 193

Berkson, William, 178

Biddle, John, 333n.5

Bishop, John, 333n.7

Boethius
- on God's atemporality, 288, 293
- *Liber de Persona et Duabus Naturis*, 322
- on personhood, 322

Bolyai, János, 177–78

Bolzano, Bernard
- vs. Cantor, 151–52, 160, 162, 259
- on definition of infinite quantities, 151–52
- on God as creator, 164n.4
- on God as unconditioned real, 155, 157, 158
- on God's infinity, 10–11, 153–56, 158, 162–64, 164n.4, 165n.5, 259
- on God's omnipotence, 156, 163, 164
- on God's omniscience, 156, 157, 158–60, 163, 165nn.8–9
- on God's perfection, 155, 156, 157
- on God's powers, 155–56
- on Hegel, 152–53
- on infinity as being larger than every possible set, 161
- on infinity as having no end, limit, or bound, 161–63, 164
- on infinity as indeterminateness, 161
- on infinity as nonaugmentability, 161
- on infinity of truths, 158–60
- on natural theology and God's infinity, 154–56
- on one-one mapping, 152, 162
- *Paradoxien des Unendlichen (PU)*, 150–54, 160, 161, 164

on qualitative infinity, 11, 153, 164
on quantitative infinity, 10–11, 150–58, 161, 162, 164
Religionswissenschaft (RW), 154–58, 164n.4, 165n.5
on revealed theology and God's infinity, 154–55, 157–58, 165n.9
Wissenschaftslehre (WL), 158, 161
Bombieri, Enrico, 16n.6
Bonaventure, St., 51
Bosanquet, Bernard, 336n.31
Bracken, Joseph
on God-world relationship, 375–77
The One and the Many, 375–76
on the Trinity, 375, 376
and *via eminentiae*, 377
Bradley, Francis Herbert, 146–47n.36, 336n.31
Brightman, Robert, 28
Brill Dictionary of Gregory of Nyssa, 27
Brower, Jeffrey, 356
Bruno, Giordano, 261
Burns, Elizabeth, 335n.28
Burrell, David, 69
Byrne, Peter, 100–101, 108

Campenhausen, Hans von, 202
Cantor, Georg
on absolute infinity, 11, 165n.12, 167, 170–71, 176, 181, 181n.1, 334n.14, 380–81
on actual infinity, 167–68, 170–71, 172, 181, 269, 334n.11, 380–81
on augmentability, 165n.12
vs. Bolzano, 151–52, 160, 162, 259
on cardinality, 11, 172–76, 269–70
Christian convictions of, 11, 168, 170–71, 181, 259, 380–81
on completed sets, 170, 172
on concrete infinite/*Transfinitum*, 11, 170–71, 181, 380–81
continuum hypothesis of, 175–76, 177
diagonalization argument of, 173–75
on God and transfinite numbers, 169
on God as creator, 169, 170, 171
on God's infinity, 11, 167, 170–71, 176, 181, 181n.1, 259, 334n.14, 380–81, 382
Grundlagen einer allgemeinen Mannigfaltigkeitslehre, 171–72, 176
influence of, 177–78
letter to Cardinal Franzelin, 170–71, 380–81
on levels of existence, 171
on natural numbers as completed set, 170
on one-to-one correspondence, 172–73, 174–76, 269–70
relations with Roman Catholic scholars, 169–71
set theory of, 11, 24, 80, 151–52, 168, 171–72, 176, 177–78, 379–80, 381
on transfinite numbers, 11, 162, 165n.12, 167, 170–71, 172–73, 176, 177, 180, 181, 181n.1, 379–80
"Über die verschiedenen Standpunkte in bezug auf das aktuelle Unendliche," 380
Caputo, John
The Weakness of God, 372–73
cardinality, 43, 84, 85, 94n.18, 320, 403n.5, 404n.6
Cantor on, 11, 172–76, 269–70
and one-to-one correspondence, 172–73, 174–76, 269–70
Carroll, William, 378
categorematic approaches to God's infinity, 8
infinity as boundlessness/ unlimitedness, 4–5
infinity as denial of finitude, 5
infinity as extra feature of God, 2, 3
infinity as total superabundance and fullness of being, 5, 17n.13

Cathars, 59
Catherine of Sienna, 379
ceteris paribus clauses, 248
Christian *theosis*, 126, 141, 142n.7
Chrysostom, John, 58
classical theism, 3, 33, 341, 367, 407n.25
Clayton, Philip, 263, 264, 265, 273n.10, 334n.9, 334n.14
  on minimal personalist theism, 336n.31
Clement of Alexandria, on God's infinity, 317
Cobb, John, 368
Coburn, Robert C., 336n.33
coextensionality, 14, 353–54, 356–57
Cohen, Paul, 177, 178
Cohn, Norman, 203
Collingwood, R. G.
  Speculum Mentis, 146–47n.36
contingency of the universe, 167, 178–80
2 Corinthians 6:18, 206
cosmological argument, 106–10, 261, 270, 274nn.19–20
Côté, Antoine
  on God's determination, 74n.21
  on God's infinity, 57
Craig, William Lane, 274nn.19–21, 277, 336n.33
  on God's infinity, 263–65, 278, 318, 334n.16
  on the infinite and the finite, 263–65
  on infinity as property of properties, 278, 318, 334n.16
Cross, Richard, 273n.7, 273n.9
Cusanus, Nicolaus
  on actual infinity, 261
  on *coincidentia oppositorum*, 351, 379
  De docta ignorantia, 344–46, 349, 350, 361nn.3–4
  dialectical doctrine of God, 345–47
  on God as source of everything, 400
  on God as Trinity, 349–50
  on God's incomprehensibility, 344–45
  on God's infinity, 14, 344–45, 346–51, 357
  on God's maximal equality, 344–45
  on God's perfection, 346–47
  on God's simplicity, 349–51, 357
  on principle of comparison and knowledge, 344–50, 361nn.3–4
  vs. Scotus, 357, 361n.5

Daniel
  4:32, 189
  7:3–7, 201
  7:14, 200, 201
  7:26, 201
  on God, 189, 199, 200, 201, 203
Danto, Arthur C., 245
Dauben, Joseph Warren, 168, 171–72, 181
Davenport, Anne Ashley, 58–59
Davies, Brian, 300, 331, 333n.5
Davis, Stephen, 312, 313
Day of Atonement, 197–98
Dedekind, Richard, 160, 380
  definition of infinity, 152, 164n.2
deism vs. theism, 224
Derrida, Jacques, 370
Descartes, René, 322, 329, 334n.10, 334n.14
  ontological proof of, 117n.10
  on substance, 371
Deuteronomy
  32:8–9, 196
  32:9, 189
Dilley, Frank, 333n.1
Dumézil, Georges, 199
Dunham, Scott, 52n.6
Dupuis, Jacques, 203

Edwards, Denis, 376
Einstein, Albert, 169, 178, 180
Enlightenment, 179
Ennius
  Annales, 206–7

Epicureans, 42
Eriugena, John Scotus, 57
Euclidean geometry
  Fifth Postulate, 177–78
  Kant on, 179
  space in, 17n.15, 161, 162
Eunomius, 29
Exodus
  15:3, 194
  17:11–12, 195
  20:3, 193
  34:14, 194
  covenantal law in, 193

Faber, Roland, 372
faith, 129, 131–32
Faraday, Michael, 178
Feuerbach, Ludwig, 140, 146n.34
Fichte, J. G., 133, 145n.24, 336n.31
Findlay, John Niemeyer, 146–47n.36
finite powers of creatures
  as absolute, 245–47
  as chancy, 247, 252
  as contingent, 251–52
  vs. God's omnipotence, 12, 233, 234, 236, 238, 239, 240, 241, 244, 245, 247–48, 253, 254n.10
  as manipulable, 249–51
  relationship to efficacy of will, 240–41, 243–44, 245–46
  relationship to freedom of will, 236–39, 240–43, 244, 245–47, 249, 250
  as resistible, 247–49
First Vatican Council, 333n.3, 336n.38
Fischer, John Martin, 141
Fishacre, Richard, on God's infinity, 58
Flint, T. P., 231nn.3–4
Foot, Philippa, 300
Fourth Lateran Council, 336n.38
  on creation *ex nihilo*, 59
Fraenkel, Abraham, 177
Franzelin, Johannes, Cardinal, 170–71, 380–81
Freddoso, A. J., 231nn.3–4
Frege, Gottlob
  on intension, 6
  on "p is true" and "p," 160
Frye, Northrop, 192

Gale, Richard
  on God as temporal, 284–85, 290
  on God's personhood, 336n.33
Garrigou-Lagrange, Reginald, 65
Gauss, Friedrich, 168
Geljon, Albert-Kees, 25, 28, 29
Gellman, Jerome I., 334n.16
  on God's infinity, 404n.9
Genesis, on God, 193, 196
German idealism, 131, 359–60, 365, 371
  See also Fichte, J. G.; Hegel, G. W. F.; Schelling, Friedrich Wilhelm Joseph
Gilson, Étienne, 53n.12
Göcke, Benedikt, 272n.3
God
  aseity of, 12, 230, 321–22, 325, 358, 367
  authority of, 187–88, 331
  of the Bible, 3, 4, 5, 7, 8, 11–12, 13, 15n.1, 17n.10, 42, 57, 59, 188–201, 203, 208, 264, 266–67, 318–19, 335n.27, 337n.46, 374
  categorematic approaches to infinity of, 2, 3, 4–5, 8
  as *causa sui*, 107, 158
  as Consciousness, 360, 362nn.12–13
  distinctness and indistinctness of, 15, 385, 402, 403
  essence of, 9, 10, 12, 21–22, 27, 28, 29, 45–51, 52n.10, 57, 58, 60, 61, 62, 67–68, 70, 71n.4, 73n.17, 104, 106–10, 157, 228–29, 230, 235, 239, 263, 267, 277, 278, 319, 321, 327, 333n.1, 333n.4, 334nn.14–15, 342, 343, 359, 397, 404n.9

God (*cont.*)
existence of, 1, 10, 16n.2, 45, 60, 62, 65, 66, 67–69, 73n.17, 85, 87, 101, 104–10, 114, 116n.9, 155, 164n.4, 230, 232n.12, 234, 270, 302–3, 327, 334n.16, 357, 369, 373, 397–98
holiness of, 264
immensity of, 59, 72n.8
as immutable, 9, 44–45, 57, 62, 94n.17, 155, 258, 267, 287, 288, 327, 334n.15, 336n.37, 354
impassibility of, 13–14, 241, 285, 301–2, 325, 336n.32
as incorporeal, 9, 38, 47, 49, 52n.7, 53n.11, 228, 277–78, 285, 326, 329, 330
infinite reason of, 10, 97–98, 113–14
as just, 278, 299, 300–301, 303–4, 306, 307–8
knowledge of infinity possessed by, 171, 176
love for the world, 264, 267, 273n.14, 278, 318–19, 327–28
as merciful, 299, 301–2, 303–4, 307–8
as national god, 189, 190, 193–96, 198, 199, 200–201, 208
as *pantokrátôr*, 203–6, 331
as paradigmatic subject of predication, 14, 354–57
as personal god, 189–93, 198, 199, 208
positive-qualitative infinity of, 15, 385, 398–400, 401, 402, 403, 404n.9, 407n.23, 407n.26
relationality of, 264, 267, 273n.14, 278, 318–19, 327–28
and salvation, 3, 202
self-identity of, 343
as Spirit, 359–60, 362nn.10–11
as subsistent being (*ipsum esse subsistens*), 9, 54, 60, 62, 63, 65–66, 67–69, 73n.18, 114, 382
substantiality of, 155
syncategorematic approaches to infinity of, 2, 3, 5–8
thanks and praise to, 14, 312–14, 315n.10
theological consensus regarding, 3, 4, 5, 7, 8, 17n.10
as truth, 38, 44, 50
as ultimate source/goal of everything, 3–4, 5, 7, 8, 15, 26, 385, 398–400, 401, 402, 403, 404n.9, 406n.20
unity of, 15, 80, 264, 342, 343, 349–51, 358–59, 364
as universal god, 189, 190, 196–98, 199–200, 208
virtues of, 13–14, 298–308, 311–12, 314
wisdom of, 44, 46, 50, 59, 164n.4, 268, 278, 289, 353, 354, 356–57
See also God as creator; God as Trinity; God's eternity/atemporality; God's goodness/benevolence; God's incomprehensibility; God's omnipotence; God's omnipresence; God's omniscience; God's perfection; God's personhood; God's simplicity; God's transcendence; God's uniqueness; negative theology
God as creator, 3, 29, 52n.6, 114, 129, 228–29, 234, 264, 267, 272, 273n.10, 331–32
Aquinas on, 9, 54–71, 108, 117n.14, 171, 179
Augustine as, 43, 44, 51, 202, 288
the Bible on, 189, 196–98, 200, 202, 208, 318–19
Bolzano on, 164n.4
Cantor on, 169, 170, 171
cosmological argument, 106–10, 261, 270, 274nn.19–20

creation *ex nihilo*, 58, 59, 179, 326, 331, 368, 369
God as primary cause, 68, 70, 71, 106–10, 281, 283–84, 355, 399–400
and Job, 192, 193
Kant on, 10, 99–100, 104–5, 106–10
Leibniz on, 81, 86, 87, 88, 91, 93, 93n.7
relationship to God's infinity, 9, 27, 54–71, 164n.4, 276, 277, 279, 294, 374
and Rowe's argument for atheism, 302–3, 305–7
Whitehead on, 365, 366, 367–68
God as Trinity, 259, 273n.4, 333n.5, 375
Augustine on, 47–48, 52n.8, 359, 360
Cusanus on, 349–50
and God as Spirit, 359–60
Gregory of Nyssa on, 27, 342–43
immanent vs. economic Trinity, 375–76
and Rahner's Rule, 376
relationship to God's infinity, 11, 14, 47–48, 52n.8, 349–50, 351, 356–57, 359–60
relationship to God's personhood, 318, 323, 341–43, 360
relationship to God's simplicity, 266, 341–42, 351, 358–60, 361n.2, 367
Scotus on, 356, 358–59
social conception of, 267, 360
Gödel, Kurt
"On Formally Undecidable Propositions," 177, 178
God's eternity/atemporality, 11, 28, 258
Anselm on, 286, 287, 288, 289, 290–91, 292, 294
Aquinas on, 55, 56
Augustine on, 9, 37, 44–45, 50, 286, 287, 288, 291
Boethius on, 288, 293
vs. eternal duration, 286–88, 291, 292, 293
and God's mind, 288–90, 292
vs. God's sempiternity, 229–30, 266, 267, 279–82, 283, 284–86, 290, 326, 334n.15
Leftow on, 283, 285–86, 293
Leibniz on, 78
relationship to God's infinity, 13, 276, 278–88, 290, 293, 294, 334n.16, 374, 404n.9
relationship to God's personhood, 330, 336n.33
relationship to negative theology, 293
and temporal indexical truths, 282–83, 290
Turretin on, 287
God's goodness/benevolence, 59, 225–26, 231n.10, 235, 254nn.2–3, 277, 352
Aquinas on, 61, 63
God's virtues, 13–14, 298–308, 311–12, 314
Leibniz on, 88
as perfect, 1, 11, 13–14, 26, 28, 50, 63, 78, 237, 239, 298–308, 312, 335n.23, 387–88, 404n.9
relationship to God's infinity, 11, 13–14, 28, 29, 30, 61, 81, 87, 262, 296, 298, 318, 320, 321–22, 332, 334n.16, 349, 353–54, 386–87, 404n.9
relationship to God's wisdom, 353–54, 356–57
God's immanence
vs. God's transcendence, 364–65, 370–71, 374–75, 376, 407n.26
Whitehead on, 15, 364, 365, 366, 374
God's incomprehensibility
Aquinas on, 407n.23
Cusanus on, 344–45
relationship to beatific vision, 57

God's incomprehensibility (*cont.*)
relationship to God's infinity, 8, 25, 26, 29, 30, 31, 34, 35, 344–45
relationship to God's perfection, 70
God's omnipotence, 1, 193, 258, 302, 309, 336n.36, 336n.39, 337n.45
as absolute power over every proposition, 12, 239–44, 245, 246
as almightiness, 11–12, 81, 187–88, 199–203, 208–9
as apocalyptic, 199–201, 209
Aquinas on, 55, 59, 60, 61–62, 71n.1, 165n.13, 212, 234, 255n.11, 296–97
Augustine on, 46–47, 50, 208
Bolzano on, 156, 163
definitions of, 12, 212–30, 230n.2, 231n.4, 231n.8, 233–35, 236–37, 239–44, 245, 272, 304, 331
vs. finite powers of creatures, 12, 233, 234, 236, 238, 239, 240, 241, 244, 253, 254n.10
and hard vs. soft events, 220–23, 229, 231n.7
as Infinite Power/power in general, 12, 233, 235–39, 253, 254n.6
and irrational influences, 225, 227, 229, 231n.10
Leibniz on, 78, 87, 88
and moral internalism, 225–26
relationship to causation, 212, 213, 215–20, 222–23
relationship to God as creator, 228–29
relationship to God's aseity, 12, 230
relationship to God's everlastingness, 12, 224, 228, 229–30
relationship to God's existence, 230
relationship to God's goodness, 225–26, 231n.10, 235
relationship to God's infinity, 6, 9, 11–12, 15n.1, 26, 81, 87, 154, 163, 164, 233–34, 235–39, 253, 268–69, 296–97, 318, 320, 321–22, 332, 334n.16, 386–87, 404n.9
relationship to God's omnipresence, 228
relationship to God's omniscience, 226
relationship to God's personhood, 330–32
relationship to metaphysical impossibility, 213, 214, 220, 225, 226, 228, 234, 255n.11
relationship to perfect efficacy of will, 236–37, 240–41, 243–44, 245, 254n.9
relationship to perfect freedom of will, 236–39, 240–43, 244, 245, 255n.14
See also finite powers of creatures
God's omnipresence, 7, 288, 336n.35
vs. God's omniscience, 326–27, 336n.36
relationship to God's infinity, 276–77, 326–27, 334n.16
relationship to God's omnipotence, 228
God's omniscience, 1, 258
Bolzano on, 156, 157, 158–60, 163, 165nn.8–9
as dynamic omniscience, 267, 271–72
vs. God's omnipresence, 326–27, 336n.36
Leibniz on, 10, 78, 88, 90, 91, 92–93
relationship to God's infinity, 6, 9, 10, 11, 13, 15n.1, 81, 87, 157, 158–60, 163, 165n.9, 318, 320, 321–22, 332, 334n.16, 386–90, 404nn.6–7, 404n.9
relationship to God's personhood, 330–31
true propositions known by God, 274n.22, 282–83, 284, 285–86, 290, 334n.13, 387–89, 404n.6
God's perfection, 71n.3, 74n.23
Anselm on, 261–62, 267, 320–21, 378, 382, 382n.5
Aquinas on, 61, 62–63, 65, 66, 67, 69, 70, 72n.13, 73nn.17–18, 378, 382
Bolzano on, 155, 156, 157

Cusanus on, 346–47
as logically possible, 271, 272
moral perfection, 1, 11, 13–14, 26, 28, 50, 78, 237, 239, 298–308, 312, 335n.23, 387–88, 404n.9
no-negation paradigm of pure perfections, 9–10, 81, 83, 85, 92
omni...paradigm of pure perfections, 9–10, 81, 87, 90, 92
perfect benevolence, 11, 13–14, 302–3, 305, 306–7, 312
perfect efficacy of will, 236–37, 240–41, 243–44, 245, 254n.9
perfect freedom of will, 236–39, 240–43, 244, 245, 255n.14
quantitative vs. qualitative perfection, 305–7
relationship to God's infinity, 1, 9–10, 11, 15, 27–28, 29, 30, 61, 62, 63, 65, 66, 67, 73n.17, 81, 85, 153, 155, 162, 261–62, 268, 276, 296, 297–98, 318, 319, 321, 327, 332, 333n.3, 334n.9, 346–47, 350, 352, 358, 377–78, 379
Scotus on, 357, 358
in virtues, 13–14, 298–308
Way of Unlimited Perfection, 377–78, 379, 382
God's personhood, 333n.6, 335n.25, 337n.46
in analytic philosophy of religion, 318–19, 335n.24
definitions of personhood, 322–25, 335n.21, 335n.23
as incompatible with God's attributes, 325–28, 335n.23, 335n.27, 335n.29, 336n.32
as literal, 323, 330, 332
as metaphorical/figurative, 325, 335n.28
and relationships, 318–19, 327–28
relationship to God as Trinity, 318, 323, 341–43, 360
relationship to God's infinity, 14, 31, 34, 35, 317, 318, 319, 322, 323, 324–32, 336n.31, 386–87
relationship to God's omnipotence, 330–32
relationship to God's omniscience, 330–31
relationship to God's simplicity, 335n.29, 336n.38
God's simplicity, 71, 318, 333n.4
Aquinas on, 62, 63, 261, 262–63, 266, 273n.7, 273n.9, 337n.43, 350, 406n.21
and coextensionality, 14, 353–54
Cusanus on, 349–51, 357
and paradigmatic subjects of predication, 14, 354–57
relationship to God as Trinity, 266, 341–42, 351, 358–60, 361n.2, 367
relationship to God's infinity, 11, 14, 28, 58, 327, 336n.38, 350–57, 359, 386–87
relationship to God's personhood, 335n.29, 336n.38
Scotus on, 262–63, 351–54, 358–59, 361n.7
God's transcendence
vs. God's immanence, 364–65, 370–71, 374–75, 376, 407n.26
Hegel on, 126–28, 130, 139–40, 143nn.10–13, 144n.15
Kalton on horizontal transcendence, 370–71
Leibniz on, 81, 87, 89–90, 93n.7
in postmodern theology, 369–73, 374
in process theology, 365–66, 367–69, 374–75
relationship to God's infinity, 365–66, 373
relationship to *via eminentiae*, 377–78
God's uniqueness, 155, 342
relationship to God's infinity, 14, 167, 170, 329, 349, 351, 357–59, 361n.5
Scotus on, 358, 361n.5

Gonçalves, Francolino J., 197
Goodenough, Ursula
  The Sacred Depth of Nature, 371
Gregory Nazianzus, St., 56
Gregory of Nyssa
  Contra Eunomium, 28, 334n.14
  on Eunomius, 29
  on God as Trinity, 27, 342–43
  on God's essence, 21–22, 27, 28, 29
  on God's existence, 29
  on God's incomprehensibility, 25, 29, 30, 34
  on God's infinite goodness, 29
  on God's infinity, 5, 8, 14, 17n.13, 21–22, 25, 26, 27–30, 32–33, 333n.2, 334n.14, 342–43
  and infinity as ontological category, 32–33
  on infinity as total superabundance and fullness of being, 5, 17n.13
  Life of Moses, 29
  on Moses's spiritual ascent to God, 32
  negative theology of, 28, 29–30
  on personhood and relationships, 327
Gregory of Rimini, 274n.18
Griffin, David Ray, 333n.1, 365, 368, 381
Guerric of Saint-Quentin, on God's infinity, 58
Gutberlet, Fr. Constantin, 170

Haase, Mareile, 204
Hadot, Pierre, 52n.5
Handy, Lowell K., 196
Harris, S. H., 146–47n.36
Harrison, Jonathan, 318
Hart, David Bentley
  on God's transcendence, 71n.2
  on infinity, 386, 401, 405n.14
  on principle of noncontradiction and finitude, 392
Hartshorne, Charles, 333n.1, 334n.15, 336n.31, 368, 390
Hasker, William, 231n.7, 336n.33
  "*The Absence of a Timeless God,*" 290–91
  on God and open future, 285
  on God as atemporal, 290–92
  The Openness of God, 266
  on timeless eternity, 282–83, 290
Hegel, G. W. F.
  on absolute thinking, 114
  on art and Spirit, 137–38, 146n.27, 146n.29
  on ascent from Nature to Spirit, 134–35
  on bad infinity, 33, 126, 129, 152–53, 263
  on being finite as being separate, 126, 128, 129, 132–33, 139, 140, 144n.16, 145n.22
  on determinacy, 132–33
  Encyclopedia of the Philosophical Sciences, 122, 132, 135, 136, 143n.10, 145n.20, 145n.22, 146n.27, 146n.29, 146n.31
  on faith, 132
  "Faith and Knowledge," 132
  on the finite as going beyond itself, 123–24, 127, 128, 133, 135, 139–40, 143n.10, 143n.13, 144n.15, 145n.22, 145n.24, 146n.31
  on fundamental principle of philosophy, 133, 145n.22
  on God's transcendence, 126–28, 130, 139–40, 143nn.10–13, 144n.15
  on the Good itself, 145n.24
  on the infinite as ultimate reality, 123–24, 127–30, 132–34
  on integration of natural and supernatural realm of being, 10, 122–23, 134–35

on integration of science and religion, 10, 122–23, 128
vs. Kant, 133–34, 143n.12, 145nn.24–25
on Kant's transcendental idealism, 111
Phenomenology of Spirit, 145n.20
philosopher's God of, 127–30, 144n.18
on philosophy and Spirit, 139–40, 146n.27
and Platonic tradition, 122, 127, 128–29, 134, 135, 140, 141, 141n.1, 144n.14, 144n.17, 145n.24, 146nn.28–29, 146n.36, 147n.37
on qualitative infinity, 127, 133, 143n.8
on quantitative infinity, 152–53
on rational love, 129–30, 139, 140, 144n.17
on rational self-government/freedom, 124, 127–29, 133, 135, 141, 142nn.3–4, 144nn.17–18, 145n.23, 147n.37
on "reality" (*Realität*), 142n.3, 143n.13, 145n.25
on religion and Spirit, 138–39, 146n.27, 146n.29
Science of Logic, 122, 132–33, 134, 142n.3, 143n.8, 143n.13, 145n.20, 145n.24
on self-determining reality, 10, 122, 132–36, 137, 139, 146n.35
on Spirit, 136–41, 145n.25, 146n.27, 146n.31, 273n.10, 360
on sublation (*Aufhebung*), 146n.28
on "the ought" (*das Sollen*), 133–34, 143n.10, 145n.24
on true infinity, 10, 13, 35, 127, 130, 133, 139, 140, 143n.10, 146n.28, 146n.35, 263–65, 273n.10
Heidegger, Martin, 146n.36, 158
Heine, Heinrich, on Hegel, 140
Heller, Michael, 16n.2
Helm, Paul, 336n.33, 336n.35
Helmholtz, Hermann von, 180
Heman, K. F., 169
Henrich, Dieter, 146–47n.36
Hick, John, 318
Hilary of Poitiers
De Trinitate, 45
on God's infinity and eternity, 45
Hilbert, David, 16n.4, 176, 177, 178
Hilbert's Hotel, 269–70, 274n.21
Hodgson, Peter, 146n.36
Homer's *Odyssey* 1.28, 206
Honoré, Anthony M., 254n.6
Houlgate, Stephen, 146n.36
Hughes, Christopher, 337n.43
human reason, 4, 5, 7, 8, 9, 56, 98
Humbrecht, Thierry-Dominique, 69, 71
Hume, David, 134, 179, 231n.6
on Argument from Evil and God's infinity, 333n.1
Hurka, Thomas, 308

indetermination, 320, 401
aoriston as indeterminate, 23, 27, 28–29, 31, 35n.2, 35n.4
Aristotle on infinity as indeterminate, 21, 24, 30, 32, 34
Bolzano on infinity as indeterminate, 161
epistemic vs. ontological, 31
of the future, 280, 281
God's infinity and indeterminacy, 8, 11, 23, 26, 28, 31, 32–33, 34, 181, 259, 407n.23
of one property vs. all properties, 31, 32
Plato on infinity as indeterminate, 5, 17n.12, 21, 22, 24, 25, 26, 30, 34
Plotinus on infinity as indeterminate, 25–26, 39–40

Irenaeus, 202
Isaiah
1–39, 197
40:26, 42
45:7, 189
Isidorus, 204
Isis as almighty/*pantokráteira*, 203–6, 209

Jaeschke, Walter, 146n.36
Jaki, Stanley L., 179
Jean de Ripa, 52n.10
Jesus, 139
as Incarnation, 300, 336n.30, 336n.39
on kingdom of God as within, 125–26, 127–28
prayer to be spared crucifixion, 193
in Revelation, 200
Job, 191–93, 201
John Damascene, St., 56
De fide orthodoxa, 64, 71n.4, 72n.5
on God's essence, 57, 71n.4, 72n.5
on God's infinity, 58, 64
Joshua
7, 195
10:8, 194
10:11, 194
24:19, 196
Judges 1:19, 195–96
Julian of Norwich, 379
Jupiter as almighty/omnipotent, 203–4, 206–8, 209
Justinian's *Book against Origen*, God's infinity in, 333n.1

kalam cosmological argument, 261, 270, 274nn.19–20
Kalton, Michael, on horizontal transcendence, 370–71
Kant, Immanuel
on categories as *a priori* concepts, 98, 99, 101–2, 103, 115n.4, 179–80, 344
on Euclidean geometry, 179
on God as ground of world order, 10, 99–100, 104–5, 106–10
on God as transcendental idea, 97, 98–101, 102, 104, 105–6, 109–10, 115n.1
on God's essence, 10, 104, 106–10
on God's existence, 10, 99–100, 101, 104–10, 117n.10
vs. Hegel, 111, 133–34, 143n.12, 145nn.24–25
on human reason, 98
on immanence, 143n.12
and infinite divine reason/*intellectus archetypus*, 10, 113–14
on intuition (*Anschauung*), 97, 101–2, 110
on knowledge, 97–98, 101–2, 103, 104, 110–11, 113
on knowledge of God, 98, 108
on practical faith, 131–32, 134
on principle of complete determination, 394
on rational autonomy, 133
on regulative ideas, 98, 100, 110, 131, 344
on science, 131–32
on sensation (*Empfindung*), 97
on simultaneous causation, 231n.6
on the soul, 98–99, 116n.9
on things in themselves/noumena, 98–104, 110, 112–13, 115n.2, 115n.4, 116nn.7–8, 179–80
on transcendental apperception, 103, 115n.2
on transcendental objects, 98–104, 110, 112–13, 115nn.2–3
on transcendental schemata, 98, 101, 102, 115n.4
on understanding (*Verstand*), 98
Kant, Immanuel, works of
Critique of Judgment, 131
Critique of Practical Reason, 105

Critique of Pure Reason, 97, 98–100, 101–4, 109, 110–13, 143n.12
Einzig möglichen Beweisgrund zu einer Demonstration des Daseins Gottes, 104, 106–7, 117n.12
Kant Metaphysik L1 (Pölitz), 106
Metaphysik Schön, 118n.19
Metaphysik Volckmann, 107, 108–9, 117nn.15–16, 118n.17
Opus Postumum, 114, 116n.8
Prolegomena zu einerjeden künftigen Metaphysik, 98, 101, 104
Reflexionen, 117n.13
Karfíková, Lenka, 27
Keller, Catherine, 372
Kierkegaard, Søren, on Hegel, 140
1 Kings 20:23, 195
Klemme, Heiner F., 115n.4, 116n.7
Kloppenborg, John S., 205
Kronecker, Leopold, 169, 174
Kuhn, Johannes von, 360, 362nn.10–11
Kuhn, Thomas, 180

La Croix, Richard, 235
LaCugna, Catherine, 376
LaMountain, George F., 61
laws of logic, 237, 254n.9
law of noncontradiction, 15, 392, 393, 395, 398, 400, 401, 402–3, 405nn.16–17
law of the excluded middle, 169, 174
Le Blanc, Jill, 321, 335n.28
Leftow, Brian, 232n.12, 336n.33, 337n.40, 337n.44, 337n.47, 378
"The Eternal Present," 293
on God's eternity, 283, 285–86, 293
on omnipotence, 230n.2, 231n.8
Legenhausen, Gary, 336n.31
Leibniz, Gottfried Wilhelm, 334n.14, 381
determinism of, 94n.16
on God as creator, 81, 86, 87, 88, 91, 93, 93n.7
on God's eternity, 78
on God's existence, 108, 117n.10
on God's goodness, 88
on God's immutability, 94n.17
on God's independence from creatures, 10, 86
on God's infinity, 9–10, 78, 79, 81, 87, 90, 92
on God's omnipotence, 78, 87, 88
on God's omniscience, 78, 88, 90, 91, 92–93
on God's rationalism, 88–89, 90, 92
on God's transcendence, 81, 87, 89–90, 93n.7
on ideas in God's mind, 87–89, 91, 92
on infinitesimals, 78–79
on innate ideas, 87
on labyrinth of freedom, 79
on labyrinth of the continuum, 79
letter to Bourguet, 86
on mathematical infinity, 78–79
Monadology, 78, 87, 90–91
on monads, 10, 78, 86, 89–92, 94nn.18–19, 95nn.22–25, 170, 176
omni...paradigm of pure perfections, 9–10, 81, 87, 90, 92
on possible worlds, 86, 87–88
Principes de la Nature et de la Grace, 93
on principle of sufficient reason, 87
on principle of the identity of indiscernibles, 89, 90, 92, 354
vs. Spinoza, 9–10, 79–80, 81, 85, 86, 87, 88–89, 90, 94n.17
on substances, 88, 94n.17
Theodicée, 86, 89
transcendent of Absolute, 79
Leo XIII
Aeterni patris, 169
Letter of Aristeas, 205
Levinas, Emanuel, 371
Lewis, David, on finkish dispositions, 251
Lewis, Delmas, 289–90, 291

Lewis, D. K., on subjunctives, 254n.8
Lincoln, Bruce, 203
Lobachevsky, Nikolai Ivanovich, 177–78
Locke, John, 179
  on personhood, 322, 335n.23
Lucas, J. R., on valedictory truths, 272
Lucretius, 207
Luke 17:21, 125–26, 127–28
Luther cases, 237

Maimonides, Moses, 70
Malebranche, Nicolas, on God's infinity, 236, 237
Mander, William, 335n.27, 336n.31
Manicheanism
  Augustine on, 9, 37–38, 42, 47, 49, 50–51, 52n.10, 276, 277–78
  and Augustine's dualistic worldview, 202–3
  Davenport on, 58–59
  God in, 37–38, 47, 48–49, 50–51, 52n.10
  space in, 42, 47, 48, 50–51, 277–78
Mann, William E., 336nn.37–38, 351
Marheineke, Philip, 360, 362nn.12–13
Marion, Jean-Luc, 371
Mark 14:36, 193
Martin, Charles B., 251
Marx, Karl, 140, 370
mathematical infinity, 32, 260, 274n.21
  defined, 22, 152, 265, 320
  finitist approaches to, 24
  vs. God's infinity, 2, 80, 258–59, 269, 272n.3, 320–21
  in intuitionist mathematics, 34
  Leibniz on, 78–79
  of natural numbers, 2, 40, 43, 151, 162, 170, 172, 173–74, 269, 386, 387–88, 405n.14
  vs. nonmathematical quantitative infinity, 31
  See also Bolzano, Bernard; Cantor, Georg; quantitative infinity
mathematical intuitionism, 2, 34
mathematical Platonism, 2
mathematical rationality, 167
Matthew
  10:30, 42
  28:18, 200
Maxwell, James Clerk, 178, 180
McDowell, John, 146n.36
McEar objection, 214, 231nn.3–4, 235, 254n.4
Meister Eckhart, 379, 385, 406n.21, 407n.26
Meixner, Uwe, 354
Mele, Alfred, 141
Menzel, C., 232n.12
Messel, Nils, 201
Minucius Felix
  Octavius, 333n.2
Mittag-Leffler, Gösta, 169, 178
modes of being
  negative-qualitative infinity, 385, 393, 394–95, 396–97, 399, 400–401, 402–3, 405n.13, 406n.19
  and ontological determinacy, 385, 392–93, 394–95, 396–97, 400–401, 402
  positive-qualitative infinity, 15, 385, 393, 394–96, 397–400, 401–3, 405n.13, 406n.18, 406n.22, 407n.23, 407n.26
  and principle of noncontradiction, 15, 392, 393, 395, 398, 400, 401, 402–3, 405nn.16–17
  qualitative finitude, 385, 391–93, 398, 401–3
Moore, Adrian W., 143n.10, 260
Morris, Thomas V., 232n.12
  Anselmian Explorations, 382n.5
Morscher, Edgar, 159
Moses, 195, 198
Mühlenberg, Ekkehard, 27, 28, 35n.4
Murdoch, Iris, 146n.36

Nagasawa, Yujin, 334n.15
Nagel, Thomas, 92
Natterer, Paul, 116n.9
negative theology, 57, 104, 293, 344
  of Gregory of Nyssa, 28, 29–30
  relationship to God's infinity, 8–9, 26–27, 28, 29–30, 34, 35, 56, 69–71, 321, 365–66
Neoplatonism, 5, 17n.13, 21, 22, 34, 343, 379
  See also Plotinus
Nestorian heresy, 300
Neuner, Joseph, 203
Newton, Isaac, 93, 178, 179, 180
Nicene Creed, 206
Nietzsche, Friedrich, 146n.36, 370
non-Euclidean geometries, 177–78, 180
Novatianus
  De trinitate, 333n.2

occasionalism, 283
ontological argument, 105, 106, 109, 117n.10
open theism, 3, 17n.10, 267–73, 273n.14, 367, 368, 374
  dynamic omniscience in, 267, 271–72
  God's infinity in, 13, 258, 265–66, 267–68, 270–71
  origin of, 266
Oppy, Graham, 231n.10, 320, 334n.9, 387, 403n.2
  on God's infinity, 16n.2, 318, 334n.15, 334n.17, 336n.39, 403n.5
ordinal numbers, 11, 152, 162, 163, 173–74
Origen, 333n.1
Ott, Ludwig, 333n.3
Ovid, 207
Owen, H. P.
  on God's infinity, 258, 261
  on infinity in Greek philosophy, 259, 260

panentheism, 3, 10, 17n.10, 26, 33, 336n.35, 389, 407n.25
Pannenberg, Wolfhart, 263–65, 273n.10
pantheism, 90, 273n.10, 407n.25
  Davenport on, 58–59
  relationship to God's infinity, 58–59, 79, 81, 84–85, 92, 127, 170, 263–65, 267–68, 277, 336n.35, 380–81
  of Spinoza, 79, 81, 84–85, 92, 170
paraconsistent logic, 15
paradigmatic subjects, 14, 354–57
Pascal, Blaise, 93
Paul, St.
  2 Corinthians 6:18, 206
  on living in God, 126, 127–28
Philo of Alexandria
  on God as *pantokrátôr*, 205
  on God as unlimited, 25
  on God's incomprehensibility, 25
  on God's infinity, 22, 25, 26, 333n.2
philosophy of mathematics
  actual infinity in, 2
  intuitionism in, 2, 34
  mathematical Platonism in, 2
  transfinite sets in, 2
philosophy of physics
  infinitely large in, 2
  infinitely small in, 2
Pieper, Josef, 54
Pike, Nelson, 330, 336n.33
Pinkard, Terry P., 143n.10
Pinnock, Clark
  on God as Most Moved Mover, 267
  The Openness of God, 266
Pippin, Robert, 146n.36
Placher, Bill, 365, 374
Plantinga, Alvin, 231n.3, 232n.12, 285, 351
Plato, 8, 13, 40, 90, 127, 141n.1, 332
  on the arts, 146n.28
  on cognitive ascent, 130, 135, 140, 141, 144n.14, 145n.24, 146nn.30–31

Plato (*cont.*)
on the Demiurge, 24, 38, 144n.14, 146n.31
on God as lacking infinity, 21, 24, 259
on the Good, 17n.12, 134, 145n.24, 146n.29
on infinity as lack of form and determination, 5, 17n.12, 21, 22, 24, 25, 26, 30, 34, 259
on infinity as lack of perfection, 24, 25, 26, 30, 34
on limit and unlimitedness, 170
Parmenides, 144n.14
on participation in Forms, 144n.14
Philebus, 51n.1, 170
Republic, 130, 135, 145n.24, 146n.31
on self-predication, 354–55
on the soul, 130, 145n.24, 146n.29
Timaeus, 24, 38, 144n.18, 146n.31
Plautus's *Poenulus*, 206
Plotinus, 22, 130, 141n.1
vs. Augustine, 44, 46, 50–51
on infinity as evil, 38, 39, 51
on infinity as indetermination, 25–26, 39–40
on infinity of the Soul, 26, 39
influence on Augustine, 38–40
on the Intellect, 38, 39, 260
on matter as evil, 38, 51, 260
on matter as infinite, 26, 38–39, 260
on the One as above the infinite and the finite, 39–40, 51
on the One as immense, 52n.10
on the One as incomprehensible, 26
on the One as indeterminate, 25–26, 39–40
on the One as infinite, 17n.13, 25–26, 39, 40, 46, 343
on the One as source of everything else, 26
on the One's infinite power, 26
on *proodos* and *epistrophe*, 146n.31
on the Soul, 26, 38, 39
Poincaré, Henri, 51n.1
postmodern theology, 369–73, 376
and God's infinity, 378
vs. process theology, 15, 365, 366, 371, 372, 374, 375, 381
poststructuralism, 369
potential infinity, 80
predestination, 308
pre-Socratics, 21, 23, 34
See also Anaxagoras, on Mind; Anaximander, on *to apeiron*
Prior, Arthur
"The Formalities of Omniscience," 282
"Thank Goodness That's Over," 282
process theology, 3, 17n.10, 333n.1, 364–69
and God's infinity, 15, 366, 367, 374–79
vs. postmodern theology, 15, 365, 366, 371, 372, 374, 375, 381
Proverbs 8, on God, 196
Pruss, Alexander, 233, 234, 236–37, 254n.9, 255n.18
Psalms
22:1, 191
23, 191
63:6–8, 191
74, 199
77, 199
115:3, 188
139:17–18a, 165n.9
144:4, 46
145:5, 42–43
147:5, 72n.8
Pseudo-Dionysius, 57, 58, 66, 379
Divine Names, 56, 68, 70, 74n.25, 406n.22
on God as source of everything, 399–400, 406n.22
Puntel, Lorenz B., 111–12, 118nn.21–22
Pythagoreans, 22, 35n.2

qualitative finitude
as mode of being, 385, 391–93, 396, 398, 401–2, 405n.14
and ontological determinacy, 392–93, 397, 404n.11
and principle of noncontradiction, 392, 393, 395, 398, 400, 401, 402–3, 405n.16
qualitative infinity
Bolzano on, 11, 153, 164
Hegel on, 127, 133, 143n.8
vs. Hegel's true infinity, 35n.1, 127
negative-qualitative infinity, 385, 393, 394–95, 396–97, 399, 400–401, 402–3, 405n.13, 406n.19
as ontological category, 32–33
positive-qualitative infinity, 15, 385, 393, 394–96, 397–400, 401–3, 405n.13, 406n.18, 406n.22, 407n.23, 407n.26
vs. quantitative infinity, 2, 8, 11, 15, 22–23, 24, 31–33, 35n.2, 61, 143n.8, 152–53, 164, 262, 296, 297, 321–22, 385, 386, 391, 405n.14
relationship to God's infinity, 2, 8, 11, 15, 164, 262, 296, 297, 321–22, 385, 397–400, 401–3
quantitative infinity
Aquinas on, 60–61, 72n.12
Aristotle on, 22, 24, 64, 164n.3, 260
Augustine on, 9
Bolzano on, 10–11, 150–58, 161, 162, 164
defined, 22–23, 320, 403n.1
as extensional, 6, 7
as infinite continuum, 6, 7, 17n.15
as infinite extension of discrete units, 6, 7, 17n.15
as intensional, 6, 8
vs. qualitative infinity, 2, 8, 11, 15, 22–23, 24, 31–33, 35n.2, 61, 143n.8, 152–53, 164, 262, 296, 297, 321–22, 385, 386, 391, 405n.14
relationship to God's infinity, 5–6, 8, 11, 15, 80–81, 164, 320, 321, 385–91, 402, 405n.15
of space, 9, 42, 47, 49, 50–51, 260, 277–78
of time, 4–5, 9, 22–23, 31, 44, 259, 260, 269, 270, 276
See also Bolzano, Bernard; Cantor, Georg; mathematical infinity

Rabia of Basra, 125, 128, 142n.5
Rahner, Karl, 126, 128
Rahner's Rule, 376
real numbers, 2, 386, 403n.2
Reese, William L., 390
Reichenbach, Bruce, 311, 312, 314
Reid, Thomas, 254n.3
religion, 142n.5
Advaita Vedanta, 126
as aspect of ultimate reality, 125–26, 138–40, 141
Christian *theosis*, 126, 141, 142n.7
God in Abrahamic religions, 125–28
Roman Catholicism, 59, 126, 139, 142n.7, 169, 258, 268, 273n.14, 318, 333n.3, 336n.38, 376
truth in, 128
Zoroastrianism, 202–3
Rescher, Nicholas, 403n.2
Revelation
4:8, 206
11:17, 206
19:13, 200
20:2, 7, 201
on God, 12, 199, 200, 201, 203, 206
on Jesus, 200
Rice, Richard
The Openness of God, 266
Robinson, Howard, 287
Rocca, Gregory, 66, 67, 68, 70
Rogers, Katherin, 276
Rolt, Clarence E., 56

Roman Catholicism
catechism and Mass, 126, 139, 142n.7, 258, 268, 273n.14
First Vatican Council, 333n.3, 336n.38
Fourth Lateran Council, 59, 336n.38
God's infinity in, 170–71, 318, 333n.3, 380–81
Leo XIII's *Aeterni patris*, 169
Rorty, Richard, 370
Rosen, Stanley, 146n.33
Rowe, William, argument for atheism, 302–3, 305–7
Russell, Bertrand, 146n.36, 178
Russell's paradox, 176–77

Sala, Giovanni B., 104, 106
1 Samuel 2:6–7, 189
Sanders, John
on dynamic omniscience, 267
The Openness of God, 266
Sartre, Jean-Paul, 146n.36
Schärtl, Thomas, 273n.4
Schelling, Friedrich Wilhelm Joseph, 336n.31
Schmucker, Josef, 106, 118n.18
scholarship on infinity of God, 1, 15n.1, 16n.2
Scholasticism, 164n.3
intension in, 6
knowledge of God by analogy, 108
Schöndorf, Harald, 111, 113, 115n.9
Schopenhauer, Arthur
World as Will and Representation, 146n.36
science, 141
vs. appetites and opinions, 124
as aspect of ultimate reality, 124, 130–32
Hegel's integration of science and religion, 10, 122–23, 128
infinity in scientific theories, 11
Kant on, 131–32
self-ness in, 135–36
truth sought by, 124
Scotus, John Duns
doctrine of coextensionality, 352
on God as Trinity, 356, 358–59
on God's existence, 357
on God's infinity, 13, 261–63, 265, 268, 333n.2, 351–52, 353, 356–59, 361n.5, 361n.7, 361n.9
on God's knowledge of contingent truths, 273n.9
on God's perfection, 357, 358
on God's simplicity, 262–63, 351–54, 358–59, 361n.7
on God's uniqueness, 358, 361n.5
on God's unity, 358
on God's will and intellect, 358
Ordinatio, 353, 356, 357, 358–59, 361n.5, 361n.7, 361n.9
on paternity and divinity, 356–57, 361n.9
on univocal meaning of language concerning God, 261, 268, 274n.16
Sellars, Wilfrid, 146n.36
Senor, Thomas, 315nn.9–10
Septuagint, 205
Sertillanges, A.-D., 69
Shults, LeRon, 263–64, 265, 273n.10
simultaneous causation, 12, 215–16, 217–18, 220, 222, 231n.6
Sinclair, James D., 274nn.19–21
singularities, 2
Smith, George Adam, 209n.1
space
divisibility of, 2
infinity of, 2, 4–5, 9, 23, 31, 37, 44, 259, 260, 269, 276, 277–78
Spinoza, Benedict de
on affections, 82, 83, 94n.12
on attributes, 83–85, 94n.12
definition of "finite in its own kind," 82–83

definition of God, 84–85, 87
on essences, 83–85
on God's infinity, 9–10, 81, 83, 85, 92
on God's necessary existence, 85
vs. Leibniz, 9–10, 79–80, 81, 85, 86, 87, 88–89, 90, 94n.17
on modes, 83–84, 85
no-negation paradigm of pure perfections, 9–10, 81, 83, 85, 92
pantheism of, 79, 81, 84–85, 92, 170
and Platonic tradition, 130, 141n.1
on substance, 82, 83–85, 87, 94n.12, 170
Yovel on, 370, 382n.2
Spinoza's *Ethics*, 79, 94n.10, 94n.13
Corollary to proposition 6, 82
First Axiom, 82
Sixth Axiom, 82
Stalnaker, R. C., on subjunctives, 254n.8
Stoicism, 17n.12, 401
Stone Paradox, 239
Strawson, Galen, 255n.21
Streete, Gail Corrington, 204
Stump, Eleonore, 350
Sundermann, Werner, 202
Sweeney, Leo, 27, 29, 53n.11, 58, 60, 72n.15, 317
Swinburne, Richard
The Coherence of Theism, 266
on God's infinity, 279
on God's personhood, 335n.25, 336n.33
on Hilbert's Hotel, 269–70
Symbol of Nicaea, 206
syncategorematic approaches to God's infinity
infinity as feature of at least one divine attribute, 2, 3, 5–8
infinity as mode of givenness of a particular attribute, 6, 7
systems theory, 375, 376
Taliaferro, Charles, 333n.7
Taoism, 126
Tapp, Christian, 181n.1, 272n.3, 389, 404n.8
Taylor, Charles, 141, 146n.36
Hegel, 147n.37
"Responsibility for Self," 147n.37
on theism and Hegel, 143n.11
Taylor, Mark C., 370
Ten Commandments, 193
Teresa of Avila, 379
Te Velde, Rudi, 62, 63, 69
Thatcher, Adrian, 335n.25
theism, 31, 35
Tillich, Paul, 126
time
four-dimensionalist theory of, 273n.12
infinity of, 4–5, 9, 22–23, 31, 44, 259, 260, 269, 270, 276
temporal succession, 216–20
truths regarding the future, 271–72, 285, 334n.13, 366
Tomarchio, John, 61, 65–66, 72nn.15–16, 73n.17
Torrance, Thomas F., 167, 179–80
Trendelenburg, Adolf, 140
truth
Augustine on, 38, 44, 50
regarding the future, 271–72, 285, 334n.13, 366
God as, 38, 44, 50
God's infinity as truths about God, 2–3
infinity of truths, 156, 157, 158–60, 320, 387–90, 404n.6
in mathematics, 41, 87
necessary truths, 87, 255n.16
pursuit of, 124, 127
in religion, 128
temporal indexical truths, 279–80, 282–83, 290, 388–89

truth (*cont.*)
truths known by God, 274n.22, 282–83, 284, 285–86, 290, 334n.13, 387–89, 404n.6
Tugwell, Simon, 57
Turner, Denys, 69, 405n.15, 406n.21, 407n.23
Turretin, Francis, 287

understanding of God, relationship to God's infinity, 16n.7

Van den Brink, Gijsbert, 331–32
Van Inwagen, Peter, 255n.12
Versnel, Hendrik S., 204
via eminentiae
relationship to God's transcendence, 377–78
relationship to quantitative infinity, 389–90
Way of Absolute Infinity, 15, 378–79, 381, 382
Way of Unlimited Perfection, 15, 377–78, 379, 382
via negationis, 62, 69–71, 73n.17
Virgil
Dido in *Aeneid*, 208
Jupiter in *Aeneid*, 206, 207–8, 209
virtues
Aristotle on, 308
of God, 13–14, 298–308, 311–12, 314
perfection of, 13–14, 298–308, 315n.6
stability of, 308–14, 315nn.6–7
Von Wright, Georg Henrik, 299–300

Wallis, John, 170
Warmington, Eric H., 207
Watson, Gary, 141, 254n.10, 255n.17
on "Cozy Clara," 249–50
Webb, Clement C. J., 333n.5
Westminster Confession, 318
Westminster Shorter Catechism, 258, 268, 273n.14
Westphal, Merold, 113
Whedbee, William, 192–93
White, Thomas, 69, 70–71
Whitehead, Alfred North, 177, 371, 372
on actual occasions and concrescence, 368–69
on God's immanence, 364, 365, 366, 374
Process and Reality, 146n.36, 364–65, 366
Religion in the Making, 146n.36
on symmetries between God and world, 15, 364–65, 366–69, 374, 375, 376, 377, 381
Wielenberg, Erik, 213–14, 231n.4
on omnipotence, 304
Wierenga, Edward, 231n.4, 235, 254n.4
William of Auvergne, 57
Williams, Thomas, 262
Wippel, John, 69
Wisdom
7:27, 205
11:20, 42
Wittgenstein, Ludwig
Tractatus Logico-Philosophicus, 146n.36
Wolf, Susan, 141
Wolff, Christian, 104, 106–8, 109, 117n.10
Woodin, W. Hugh, 16n.2

Yandell, Keith, 361n.2
Yates, John C., 333n.7
Yovel, Yirmiyahu
Spinoza and Other Heretics, 370, 382n.2

Zeno's paradoxes, 22
Zermelo, Ernst, 177, 178
Žižek, Slavoj, 370
Zoroastrianism, 202–3

www.ingramcontent.com/pod-product-compliance
Ingram Content Group UK Ltd.
Pitfield, Milton Keynes, MK11 3LW, UK
UKHW020222290926
15470UKWH00001B/327

9 780268 104139